ℐNTELLECTUALS

Books by Paul Johnson

Creators

George Washington

Art: A New History

A History of the American People

The Quest for God

The Birth of the Modern

Intellectuals

A History of the English People

A History of the Jews

Modern Times

A History of Christianity

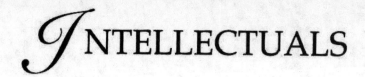

NTELLECTUALS

From Marx and Tolstoy to
Sartre and Chomsky

Paul Johnson

HARPER ● PERENNIAL

NEW YORK ● LONDON ● TORONTO ● SYDNEY

HARPER ● PERENNIAL

A hardcover edition of this book was originally published in 1988 by Harper & Row, Publishers.

P.S.™ is a trademark of HarperCollins Publishers.

HarperCollins books may be purchased for educational, business, or sales promotional use. For information, please e-mail the Special Markets Department at SPsales@harpercollins.com.

First Perennial Library edition published 1990.
First Harper Perennial edition published 2007.

The Library of Congress has catalogued the hardcover edition as follows:

Johnson, Paul.
    Intellectuals.

    Bibliography: p.
    Includes index.
    ISBN 0–06–016050–0
    1.   Intellectuals. I. Title.
HM213.J64   1989   305.5'52   88–45518

ISBN: 978-0-06-125317-1 (pbk.)
ISBN-10: 0-06-125317-0 (pbk.)

24 25 26 27 28 LBC 23 22 21 20 19

To my first grandchild
Samuel Johnson

Contents

1 Jean-Jacques Rousseau: 'An Interesting Madman' 1

2 Shelley, or the Heartlessness of Ideas 28

3 Karl Marx: 'Howling Gigantic Curses' 52

4 Henrik Ibsen: 'On the Contrary!' 82

5 Tolstoy: God's Elder Brother 107

6 The Deep Waters of Ernest Hemingway 138

7 Bertolt Brecht: Heart of Ice 173

8 Bertrand Russell: A Case of Logical Fiddlesticks 197

9 Jean-Paul Sartre: 'A Little Ball of Fur and Ink' 225

10 Edmund Wilson: A Brand from the Burning 252

11 The Troubled Conscience of Victor Gollancz 269

12 Lies, Damned Lies and Lillian Hellman 288

13 The Flight of Reason 306

Notes 343

Index 367

Acknowledgements

$\mathcal{T}$ HIS book is an examination of the moral and judgmental credentials of certain leading intellectuals to give advice to humanity on how to conduct its affairs. I have tried to make it factual and dispassionate and wherever possible I have used the works, letters, diaries, memoirs and reported speech of those under scrutiny. For details of their lives I have made use of a number of biographies, the most important of which are as follows. For Rousseau I found the most useful was Lester G. Crocker, *Jean-Jacques Rousseau: The Quest, 1712–1758* (New York, 1974) and *Jean-Jacques Rousseau: The Prophetic Voice, 1758–1783* (New York, 1973), though I also relished J.H. Huizinga's vigorous polemic, *The Making of a Saint: the Tragi-Comedy of J.J. Rousseau* (London, 1976). For Shelley I have relied most on Richard Holmes's superb book, *Shelley: The Pursuit* (London, 1974), even though I do not agree with him about the illegitimate child. For Marx I turned chiefly to Robert Payne's *Marx* (London, 1968). Ibsen has a model biographer in Michael Meyer, *Henrik Ibsen: i. The Making of a Dramatist, 1828–64* (London, 1967); *ii. The Farewell to Poetry, 1864–82* (London, 1971); *iii. The Top of a Cold Mountain, 1886–1906* (London, 1971), but I also used Hans Heiberg, *Ibsen: Portrait of the Artist* (trans., London, 1969) and Bergliot Ibsen, *The Three Ibsens* (trans., London, 1951). Of the many Tolstoy biographies, I followed most Ernest J. Simmons, *Leo Tolstoy* (London, 1949), but I made use also of Edward Crankshaw's formidably critical account, *Tolstoy: The Making of a Novelist* (London, 1974). For Emerson I used the works of Joel Porte, notably his *Representative Man: Ralph Waldo Emerson in His Time* (New York, 1979). For Hemingway I used the two excellent recent

ix

biographies, Jeffrey Meyers, *Hemingway: A Biography* (London, 1985) and Kenneth S. Lynn, *Hemingway* (London, 1987), as well as the earlier work by Carlos Baker, *Hemingway: A Life Story* (New York, 1969). For Brecht I used Ronald Hayman, *Bertold Brecht: A Biography* (London, 1983) and Martin Esslin's brilliant study, *Bertolt Brecht: A Choice of Evils* (London, 1959). For Russell, the chief source for the biographical facts was Ronald W. Clark, *The Life of Bertrand Russell* (London, 1975). For Sartre I used especially Annie Cohen-Solal: *Sartre: A Life* (trans., London, 1987) and Claude Francis and Fernande Gontier, *Simone de Beauvoir* (trans., London, 1987). The indispensable source for Gollancz is Ruth Dudley Edwards's full and fair-minded account, *Victor Gollancz: A Biography* (London, 1987), and for Lillian Hellman, William Wright's masterly piece of detective work, *Lillian Hellman: The Image, the Woman* (London, 1987), but I also found useful Diane Johnson's book, *The Life of Dashiell Hammett* (London, 1984). For the last chapter I made particular use of David Pryce-Jones, *Cyril Connolly: Diaries and Memoir* (London, 1983), Hilary Mills, *Mailer: A Biography* (New York, 1982), Kathleen Tynan, *The Life of Kenneth Tynan* (London, 1987), Robert Katz and Peter Berling, *Love is Colder than Death: The Life and Times of Rainer Werner Fassbinder* (London, 1987) and Fern Marja Eckman, *The Furious Passage of James Baldwin* (London, 1968). To all these authors I am grateful. References to some of the many other works consulted will be found in the source notes.

1

Jean-Jacques Rousseau: 'An Interesting Madman'

OVER the past two hundred years the influence of intellectuals has grown steadily. Indeed, the rise of the secular intellectual has been a key factor in shaping the modern world. Seen against the long perspective of history it is in many ways a new phenomenon. It is true that in their earlier incarnations as priests, scribes and soothsayers, intellectuals have laid claim to guide society from the very beginning. But as guardians of hieratic cultures, whether primitive or sophisticated, their moral and ideological innovations were limited by the canons of external authority and by the inheritance of tradition. They were not, and could not be, free spirits, adventurers of the mind.

With the decline of clerical power in the eighteenth century, a new kind of mentor emerged to fill the vacuum and capture the ear of society. The secular intellectual might be deist, sceptic or atheist. But he was just as ready as any pontiff or presbyter to tell mankind how to conduct its affairs. He proclaimed, from the start, a special devotion to the interests of humanity and an evangelical duty to advance them by his teaching. He brought to this self-appointed task a far more radical approach than his clerical predecessors. He felt himself bound by no corpus of revealed religion. The collective wisdom of the past, the legacy of tradition, the prescriptive codes of ancestral experience existed to be selectively followed or wholly rejected entirely as his own good sense might decide. For the first time in human history, and with growing confidence and audacity, men arose to assert that they could diagnose the ills of society and cure them with their own unaided intellects: more, that they could devise formulae whereby not merely the structure of society

1

but the fundamental habits of human beings could be transformed for the better. Unlike their sacerdotal predecessors, they were not servants and interpreters of the gods but substitutes. Their hero was Prometheus, who stole the celestial fire and brought it to earth.

One of the most marked characteristics of the new secular intellectuals was the relish with which they subjected religion and its protagonists to critical scrutiny. How far had they benefited or harmed humanity, these great systems of faith? To what extent had these popes and pastors lived up to their precepts, of purity and truthfulness, of charity and benevolence? The verdicts pronounced on both churches and clergy were harsh. Now, after two centuries during which the influence of religion has continued to decline, and secular intellectuals have played an ever-growing role in shaping our attitudes and institutions, it is time to examine *their* record, both public and personal. In particular, I want to focus on the moral and judgmental credentials of intellectuals to tell mankind how to conduct itself. How did they run their own lives? With what degree of rectitude did they behave to family, friends and associates? Were they just in their sexual and financial dealings? Did they tell, and write, the truth? And how have their own systems stood up to the test of time and praxis?

The inquiry begins with Jean-Jacques Rousseau (1712–78), who was the first of the modern intellectuals, their archetype and in many ways the most influential of them all. Older men like Voltaire had started the work of demolishing the altars and enthroning reason. But Rousseau was the first to combine all the salient characteristics of the modern Promethean: the assertion of his right to reject the existing order in its entirety; confidence in his capacity to refashion it from the bottom in accordance with principles of his own devising; belief that this could be achieved by the political process; and, not least, recognition of the huge part instinct, intuition and impulse play in human conduct. He believed he had a unique love for humanity and had been endowed with unprecedented gifts and insights to increase its felicity. An astonishing number of people, in his own day and since, have taken him at his own valuation.

In both the long and the short term his influence was enormous. In the generation after his death, it attained the status of a myth. He died a decade before the French Revolution of 1789 but many contemporaries held him responsible for it, and so for the demolition of the *ancien régime* in Europe. This view was shared by both Louis XVI and Napoleon. Edmund Burke said of the revolutionary elites: 'There is a great dispute among their leaders which of them is the best resemblance of Rousseau . . . He is their standard figure of perfection.' As Robespierre himself put it: 'Rousseau is the one man who, through the loftiness of his soul and the grandeur of his character, showed himself worthy of the role of teacher of mankind.' During the Revolution the National Convention voted to have his ashes transferred to the Panthéon. At the ceremony its president declared: 'It is to Rousseau that is due the health-giving improvement that has transformed our morals, customs, laws, feelings and habits.'[1]

At a much deeper level, however, and over a far longer span of time, Rousseau altered some of the basic assumptions of civilized man and shifted around the furniture of the human mind. The span of his influence is dramatically wide but it can be grouped under five main headings. First, all our modern ideas of education are affected to some degree by Rousseau's doctrine, especially by his treatise *Émile* (1762). He popularized and to some extent invented the cult of nature, the taste for the open air, the quest for freshness, spontaneity, the invigorating and the natural. He introduced the critique of urban sophistication. He identified and branded the artificialities of civilization. He is the father of the cold bath, systematic exercise, sport as character-forming, the weekend cottage.[2]

Second, and linked to his revaluation of nature, Rousseau taught distrust of the progressive, gradual improvements brought about by the slow march of materialist culture; in this sense he rejected the Enlightenment, of which he was part, and looked for a far more radical solution.[3] He insisted that reason itself had severe limitations as the means to cure society. That did not mean, however, that the human mind was inadequate to bring about the necessary changes, because it has hidden, untapped resources of poetic insight and intuition which must be used to overrule the sterilizing dictates of reason.[4] In pursuit of this line of thought, Rousseau wrote his *Confessions*, finished in 1770, though not published until after his death. This third process was the beginning both of the Romantic movement and of modern introspective literature, for in it he took the discovery of the individual, the prime achievement of the Renaissance, a giant stage further, delving into the inner self and producing it for public inspection. For the first time readers were shown the inside of a heart, though – and this too was to be a characteristic of modern literature – the vision was deceptive, the heart thus exhibited misleading, outwardly frank, inwardly full of guile.

The fourth concept Rousseau popularized was in some ways the most pervasive of all. When society evolves from its primitive state of nature to urban sophistication, he argued, man is corrupted: his natural selfishness, which he calls *amour de soi*, is transformed into a far more pernicious instinct, *amour-propre*, which combines vanity and self-esteem, each man rating himself by what others think of him and thus seeking to impress them by his money, strength, brains and moral superiority. His natural selfishness becomes competitive and acquisitive, and so he becomes alienated not only from other men, whom he sees as competitors and not brothers, but from himself.[5] Alienation induces a psychological sickness in man, characterized by a tragic divergence between appearance and reality.

The evil of competition, as he saw it, which destroys man's inborn communal sense and encourages all his most evil traits, including his desire to exploit others, led Rousseau to distrust private property, as the source of social crime. His fifth innovation, then, on the very eve of the Industrial Revolution, was to develop the elements of a critique of capitalism, both in the preface to his play *Narcisse* and in his *Discours sur l'inégalité*, by identifying property and the competition to acquire it as the primary cause of alienation.[6] This was a thought-deposit Marx and others were to mine ruthlessly, together with Rousseau's related idea of cultural evolution. To him, 'natural' meant 'original' or pre-cultural. All culture brings problems since it is man's association with others which brings out his evil propensities: as he puts it in *Émile*, 'Man's breath is fatal to his fellow men.' Thus the culture in which man lived, itself an evolving, artificial construct, dictated man's behaviour, and you could improve, indeed totally transform, his behaviour by changing the culture and the competitive forces which produced it – that is, by social engineering.

These ideas are so wide-ranging as to constitute, almost by themselves, an encyclopaedia of modern thought. It is true that not all of them were original to him. His reading was wide: Descartes, Rabelais, Pascal, Leibnitz, Bayle, Fontenelle, Corneille, Petrarch, Tasso, and in particular he drew on Locke and Montaigne. Germaine de Staël, who believed he possessed 'the most sublime faculties ever bestowed on a man' declared: 'He has invented nothing.' But, she added, 'he has infused all with fire.' It was the simple, direct, powerful, indeed passionate, manner in which Rousseau wrote which made his notions seem so vivid and fresh, so that they came to men and women with the shock of a revelation.

Who, then, was this dispenser of such extraordinary moral and intellectual power, and how did he come to acquire it? Rousseau was a Swiss, born in Geneva in 1712 and brought up a Calvinist. His father Isaac was a watchmaker but did not flourish in his trade, being a troublemaker, often involved in violence and riots. His mother, Suzanne Bernard, came from a wealthy family, but died of puerperal fever shortly after Rousseau's birth. Neither parent came from the tight circle of families which formed the ruling oligarchy of Geneva and composed the Council of

Two Hundred and the Inner Council of Twenty-Five. But they had full voting and legal privileges and Rousseau was always very conscious of his superior status. It made him a natural conservative by interest (though not by intellectual conviction) and gave him a lifelong contempt for the voteless mob. There was also a substantial amount of money in the family.

Rousseau had no sisters but a brother, seven years his senior. He himself strongly resembled his mother, and thus became the favourite of his widowed father. Isaac's treatment of him oscillated between lachrymose affection and frightening violence and even the favoured Jean-Jacques deplored the way his father brought him up, later complaining in *Émile*: 'The ambition, greed, tyranny and misguided foresight of fathers, their negligence and brutal insensitivity, are a hundred times more harmful to children than the unthinking tenderness of mothers.' However, it was the elder brother who became the chief victim of the father's savagery. In 1718 he was sent to a reformatory, at the father's request, on the grounds that he was incorrigibly wicked; in 1723 he ran away and was never seen again. Rousseau was thus, in effect, an only child, a situation he shared with many other modern intellectual leaders. But, though indulged in some ways, he emerged from childhood with a strong sense of deprivation and – perhaps his most marked personal characteristic – self-pity.[7]

Death deprived him quickly of both his father and his foster-mother. He disliked the trade of engraving to which he was apprenticed. So in 1728, aged fifteen, he ran away and became a convert to Catholicism, in order to obtain the protection of a certain Madame Françoise-Louise de Warens, who lived in Annecy. The details of Rousseau's early career, as recorded in his *Confessions*, cannot be trusted. But his own letters, and the vast resources of the immense Rousseau industry, have been used to establish the salient facts.[8] Madame de Warens lived on a French royal pension and seems to have been an agent both of the French government and of the Roman Catholic Church. Rousseau lived with her, at her expense, for the best part of fourteen years, 1728–42. For some of this time he was her lover; there were also periods when he wandered off on his own. Until he was well into his thirties, Rousseau led a life of failure and of dependence, especially on women. He tried at least thirteen jobs, as an engraver, lackey, seminary student, musician, civil servant, farmer, tutor, cashier, music-copier, writer and private secretary. In 1743 he was given what seemed the plum post of secretary to the French Ambassador in Venice, the Comte de Montaigu. This lasted eleven months and ended in his dismissal and flight to avoid arrest by the Venetian Senate. Montaigu stated (and his version is to be preferred to Rousseau's own) that his secretary was doomed to poverty on account of his 'vile disposition' and 'unspeakable insolence', the product of his 'insanity' and 'high opinion of himself'.[9]

For some years Rousseau had come to see himself as a born writer. He had great skill with words. He was particularly effective at putting the case for himself in letters, without too scrupulous a regard for the facts; indeed, he might have made a brilliant lawyer. (One of the reasons Montaigu, a military man, came to dislike him so much was Rousseau's habit, when taking dictation, of yawning ostentatiously or even strolling to the window, while the Ambassador struggled for a word.) In 1745 Rousseau met a young laundress, Thérèse Levasseur, ten years his junior, who agreed to become his mistress on a permanent basis. This gave some kind of stability to his drifting life. In the meantime he had met and been befriended by Denis Diderot, the cardinal figure of the Enlightenment and later to be editor-in-chief of the *Encyclopédie*. Like Rousseau, Diderot was the son of an artisan and became the prototype of the self-made writer. He was a kind man and an assiduous nourisher of talent. Rousseau owed a good deal to him. Through him he met the German literary critic and diplomat Friedrich Melchior Grimm, who was well established in society; and Grimm took him to the famous radical *salon* of Baron d'Holbach, known as *'le Maître d'Hôtel de la philosophie'*.

The power of the French intellectuals was just beginning and was to increase steadily in the second half of the century. But in the 1740s and 1750s their position as critics of society was still precarious. The State, when it felt itself threatened, was still liable to turn on them with sudden ferocity. Rousseau later loudly complained of the persecution he suffered, but in fact he had less to put up with than most of his contemporaries. Voltaire was publicly caned by the servants of an aristocrat he had offended, and served nearly a year in the Bastille. Those who sold forbidden books might get ten years in the galleys. In July 1749 Diderot was arrested and put in solitary confinement in the Vincennes fortress for publishing a book defending atheism. He was there three months. Rousseau visited him there, and while walking on the road to Vincennes he saw in the paper a notice from the Dijon Academy of Letters inviting entries for an essay competition on the theme 'Whether the rebirth of the sciences and the arts has contributed to the improvement of morals'.

This episode, which occurred in 1750, was the turning point in Rousseau's life. He saw in a flash of inspiration what he must do. Other entrants would naturally plead the cause of the arts and sciences. He would argue the superiority of nature. Suddenly, as he says in his *Confessions*, he conceived an overwhelming enthusiasm for 'truth, liberty and virtue'. He says he declared to himself: 'Virtue, truth! I will cry increasingly, truth, virtue!' He added that his waistcoat was 'soaked with tears I had shed without noticing it'. The soaking tears may well be true: tears came easily to him. What is certain is that Rousseau decided there and then to write the essay on lines which became the essence of his creed, won the prize by this paradoxical approach, and became famous almost overnight. Here was a case of a man of thirty-nine, hitherto unsuccessful and embittered, longing for notice and fame, at last hitting the right note. The essay is feeble and today almost unreadable. As always, when one looks back on such a literary event, it seems inexplicable that so paltry a work could have produced such an explosion of celebrity; indeed the famous critic Jules Lemaître called this instant apotheosis of Rousseau 'one of the strongest proofs ever provided of human stupidity'.[10]

Publication of the *Discours* on the arts and sciences did not make Rousseau rich, for though it circulated widely, and evoked nearly three hundred printed replies, the number of copies actually sold was small and it was the booksellers who made money from such works.[11] On the other hand it gave him the run of many aristocratic houses and estates, which were open to fashionable intellectuals. Rousseau could, and sometimes did, support himself by music-copying (he had a beautiful writing-hand) but after 1750 he was always in a position to live off the hospitality of the aristocracy, except (as often happened) when he chose to stage ferocious quarrels with those who dispensed it. For occupation, he became a professional writer. He was always fertile in ideas and, when he got down to it, wrote easily and well. But the impact of his books, at any rate in his own lifetime and for long after, varied greatly.[12] His *Social Contract*, generally supposed to encapsulate his mature political philosophy, which he began in 1752 and finally published ten years later, was scarcely read at all in his lifetime and had only been reprinted once by 1791. Examination of five hundred contemporary libraries showed that only one possessed a copy. The scholar Joan Macdonald, who looked at 1114 political pamphlets published in 1789–91, found only twelve references to it.[13] As she observed: 'It is necessary to distinguish between the cult of Rousseau and the influence of his political thought.' The cult, which began with the prize essay but continued to grow in force, centred around two books. The first was his novel *La Nouvelle Héloïse*, subtitled *Letters of Two Lovers* and modelled on Richardson's *Clarissa*. The story of the pursuit, seduction, repentance and punishment of a young woman, it is written with extraordinary skill to appeal both to the prurient interest of readers, especially women – and especially the burgeoning market of middle-class women – and to their sense of morality. The material is often very outspoken for the time, but the final message is highly proper. The Archbishop of Paris accused it of 'insinuating the poison of lust while seeming to proscribe it', but this merely served to increase its sales, as did Rousseau's own cunningly-worded preface, in which he asserts that a girl who reads a single page of it is a lost soul, adding however that 'chaste girls do not read love stories.' In fact both chaste girls and respectable matrons read it and defended themselves by citing its highly moral conclusions. In short it was a natural best-seller, and became one, though most of the copies bought were pirated.

The Rousseau cult was intensified in 1762 with the publication of *Émile*, in which he launched the myriad of ideas, on nature and man's response to it, which were to become the staple fare of the Romantic Age but were then pristine. This book too was brilliantly engineered to secure the maximum number of readers. But in one respect Rousseau was too clever for his own good. It was part of his growing appeal, as the prophet of truth and virtue, to point out the limits of reason and allow for the place of religion in the hearts of men. He thus included in *Émile* a chapter entitled 'Profession of Faith' in which he accused his fellow intellectuals of the Enlightenment, especially the atheists or mere deists, of being arrogant and dogmatic, 'professing even in their so-called scepticism to know everything' and heedless of the damage they do to decent men and women by undermining faith: 'They destroy and trample underfoot all that men revere, steal from the suffering the consolation they derive from religion and take away the only force that restrains the passions of the rich and the powerful.' It was highly effective stuff, but to balance it Rousseau felt it necessary to criticize the established Church too, especially its cult of miracles and encouragement of superstition. This was highly imprudent, especially since Rousseau, to frustrate the book-pirates, took the risk of signing the work. He was already suspect in French ecclesiastical eyes as a double-renegade: having converted to Catholicism, he later returned to Calvinism in order to regain his Genevan citizenship. So now the Paris *Parlement*, dominated by Jansenists, took the strongest objection to the anti-Catholic sentiments in *Émile*, had the book burnt in front of the Palais de Justice and issued a warrant for Rousseau's arrest. He was saved by a timely warning from high-placed friends. Thereafter he was for some years a fugitive. For the Calvinists objected to *Émile* too and even outside Catholic territory he was forced to move on from one town to another. But he was never without powerful protectors, in Britain (where he spent fifteen months in 1766–67) and in France too, where he lived from 1767 onwards. During his last decade the State lost interest in him, and his chief enemies were fellow intellectuals, notably Voltaire. To answer them Rousseau wrote his *Confessions*, completed in Paris where he finally settled in 1770. He did not venture to publish them but they were widely known from the readings he gave at fashionable houses. By the time of his death in 1778 his reputation was on the eve of a fresh upsurge, consummated when the revolutionaries took over.

Rousseau, then, enjoyed considerable success even in his lifetime. To the unprejudiced modern eye he does not seem to have had much to grumble about. Yet Rousseau was one of the greatest grumblers in the history of literature. He insisted that his life had been one of misery and persecution. He reiterates the complaint so often and in such harrowing terms, that one feels obliged to believe him. On one point he was adamant: he suffered from chronic ill health. He was 'an unfortunate wretch worn out by illness... struggling every day of my life between pain and death'. He had 'not been able to sleep for thirty years'. 'Nature,' he added, 'which has shaped me for suffering, has given me a constitution proof against pain in order that, unable to exhaust my forces, it may always make itself felt with the same intensity.'[14] It is true that he always had trouble with his penis. In a letter to his friend Dr Tronchin, written in 1755, he refers to 'the malformation of an organ, with which I was born'. His biographer Lester Crocker, after a careful diagnosis, writes: 'I am convinced that Jean-Jacques was born a victim of hypospadias, a deformity of the penis in which the urethra opens somewhere on the ventral surface.'[15] In adult life this became a stricture, necessitating painful use of a catheter, which aggravated the problem both psychologically and physically. He constantly felt the need to urinate and this raised difficulties when he was living in high society: 'I still shudder to think of myself,' he wrote, 'in a circle of women, compelled to wait until some fine talk had finished... When at last I find a well-lit staircase there are other ladies who delay me, then a courtyard full of constantly moving carriages ready to crush me, ladies' maids who are looking at me, lackeys who line the walls and laugh at me. I do not find a single wall or wretched little corner that is suitable for my purpose. In short I can urinate only in full view of everybody and on some noble white-stockinged leg.'[16]

The passage is self-pitying and suggests, along with much other evidence, that Rousseau's health was not as bad as he makes out. At times, when it suits his argument, he points to his good health. His insomnia was partly fantasy, since various people testify to his snoring. David Hume, who was with him on the voyage to England, wrote: 'He is one of the most robust men I have ever known. He passed ten hours in the night-time above deck in the most severe weather, where all the seamen were almost frozen to death, and he took no harm.'[17]

9

Incessant concern about his health, justified or not, was the original dynamic of the self-pity which came to envelop him and feed on every episode in his life. At quite an early age he developed a habit of telling what he called his 'story', in order to elicit sympathy, especially from well-born women. He called himself 'the unhappiest of mortals', spoke of 'the grim fate which dogs my footsteps', claimed 'few men have shed so many tears' and insisted: 'My destiny is such that no one would dare to describe it, and no one would believe it.' In fact he described it often and many did believe it, until they learned more about his character. Even then, some sympathy often remained. Madame d'Épinay, a patroness whom he treated abominably, remarked, even after her eyes were opened: 'I still feel moved by the simple and original way in which he recounted his misfortunes.' He was what armies call an Old Soldier, a practised psychological con-man. One is not surprised to find that, as a young man, he wrote begging letters, one of which has survived. It was written to the Governor of Savoy and demands a pension on the grounds that he suffers from a dreadful disfiguring disease and will soon be dead.[18]

Behind the self-pity lay an overpowering egoism, a feeling that he was quite unlike other men, both in his sufferings and his qualities. He wrote: 'What could your miseries have in common with mine? My situation is unique, unheard of since the beginning of time...' Equally, 'The person who can love me as I can love is still to be born.' 'No one ever had more talent for loving.' 'I was born to be the best friend that ever existed.' 'I would leave this life with apprehension if I knew a better man than me.' 'Show me a better man than me, a heart more loving, more tender, more sensitive...' 'Posterity will honour me... because it is my due.' 'I rejoice in myself.' '... my consolation lies in my self-esteem.' '... if there were a single enlightened government in Europe, it would have erected statues to me.'[19] No wonder Burke declared: 'Vanity was the vice he possessed to a degree little short of madness.'

It was part of Rousseau's vanity that he believed himself incapable of base emotions. 'I feel too superior to hate.' 'I love myself too much to hate anybody.' 'Never have I known the hateful passions, never did jealousy, wickedness, vengeance enter my heart... anger occasionally but I am never crafty and never bear a grudge.' In fact he frequently bore grudges and was crafty in pursuing them. Men noticed this. Rousseau was the first intellectual to proclaim himself, repeatedly, the friend of all mankind. But loving as he did humanity in general, he developed a strong propensity for quarrelling with human beings in particular. One of his victims, his former friend Dr Tronchin of Geneva, protested: 'How is it possible that the friend of mankind is no longer the friend of men, or scarcely so?' Replying, Rousseau defended his right to administer rebukes to those who deserved it: 'I am the friend of mankind, and men are everywhere. The friend of truth also finds malevolent men everywhere – and I do not need to go very far.'[20] Being an egoist, Rousseau tended to equate hostility to himself with hostility to truth and virtue as such. Hence nothing was too bad for his enemies; their very existence made sense of the doctrine of eternal punishment: 'I am not ferocious by nature,' he told Madame d'Épinay, 'but when I see there is no justice in this world for these monsters, I like to think there is a hell waiting for them.'[21]

Since Rousseau was vain, egotistical and quarrelsome, how was it that so many people were prepared to befriend him? The answer to this question brings us to the heart of his character and historical significance. Partly by accident, partly by instinct, partly by deliberate contrivance, he was the first intellectual systematically to exploit the guilt of the privileged. And he did it, moreover, in an entirely new way, by the systematic cult of rudeness. He was the prototype of that characteristic figure of the modern age, the Angry Young Man. By nature he was not anti-social. Indeed from an early age he wished to shine in society. In particular he wanted the smiles of society women. 'Seamstresses,' he wrote, 'chambermaids, shopgirls did not tempt me. I needed young ladies.' But he was an obvious and ineradicable provincial, in many ways boorish, ill-bred. His initial attempts to break into society, in the 1740s, by playing society's own game, were complete failures; his first play for the favours of a married society woman was a humiliating disaster.[22]

However, after the success of his essay revealed to him the rich rewards for playing the card of Nature, he reversed his tactics. Instead of trying to conceal his boorishness, he emphasized it. He made a virtue of it. And the strategy worked. It was already customary among the better-educated of the French nobility, who were being made to feel increasingly uneasy by the ancient system of class privilege, to cultivate writers as talismans to ward off evil. The contemporary social critic, C.P. Duclos, wrote: 'Among the grandees, even those who do not really like intellectuals pretend to do so because it is the fashion.'[23] Most writers, thus patronized, sought to ape their betters. By doing the reverse, Rousseau became a much more interesting, and so desirable, visitor to their salons, a brilliant, highly intelligent Brute of Nature or 'Bear', as they liked to call him. He deliberately stressed sentiment as opposed to convention, the impulse of the heart rather than manners. 'My sentiments,' he said, 'are such that they must not be disguised. They dispense me from being polite.' He admitted he was 'uncouth, unpleasant and rude on principle.

11

I do not care twopence for your courtiers. I am a barbarian.' Or again: 'I have things in my heart which absolve me from being good-mannered.'

This approach fitted in very well with his prose, which was far more simple than the polished periods of most contemporary writers. His directness admirably suited his outspoken treatment of sex (*La Nouvelle Héloïse* was one of the first novels to mention such articles as ladies' corsets). Rousseau highlighted his ostentatious rejection of social norms by a studied simplicity and looseness of dress, which in time became the hallmark of all the young Romantics. He later recorded: 'I began my reformation with my dress. I gave up gold lace and white stockings and wore a round wig. I gave up my sword and sold my watch.' Next followed longer hair, what he called 'my usual careless style with a rough beard'. He was the first of the hirsute highbrows. Over the years he developed a variety of sartorial ways of drawing public attention to himself. At Neufchâtel he was painted by Allan Ramsay wearing an Armenian robe, a sort of kaftan. He even wore it to church. The locals objected at first but soon got used to it and in time it became a Rousseau hallmark. During his celebrated visit to England he wore it at the Drury Lane Theatre, and was so anxious to respond to the plaudits of the crowd that Mrs Garrick had to hang onto the robe to prevent him falling out of the box.[24]

Consciously or not, he was a superb self-publicist: his eccentricities, his social brutalities, his personal extremism, even his quarrels, attracted a vast amount of attention and were undoubtedly part of his appeal both to his aristocratic patrons and to his readers and cultists. It is a significant fact, as we shall see, that personal public relations, not least through quirks of dress and appearance, was to become an important element in the success of numerous intellectual leaders. Rousseau led the way in this as in so many other respects. Who can say he was wrong? Most people are resistant to ideas, especially new ones. But they are fascinated by character. Extravagance of personality is one way in which the pill can be sugared and the public induced to look at works dealing with ideas.

As part of his technique for securing publicity, attention and favour, Rousseau, who was no mean psychologist, made a positive virtue of that most repellent of vices, ingratitude. To him it seemed no fault. While professing spontaneity, he was in fact a calculating man; and since he persuaded himself that he was, quite literally, the best of moral human beings, it followed logically that others were even more calculating, and from worse motives, than he was. Hence in any dealings with him, they would seek to take advantage, and he must outwit them. The basis on which he negotiated with others, therefore, was quite simple: they gave, he took. He bolstered this by an audacious argument: because of his uniqueness, anyone who helped him was in fact doing a favour to himself. He set the pattern in his response to the letter of the Dijon Academy awarding him the prize. His essay, he wrote, had taken the unpopular line of truth, 'and by your generosity in honouring my courage, you have honoured yourselves still more. Yes, gentlemen, what you have done for my glory, is a crown of laurels added to your own.' He used the same technique when his fame brought him offers of hospitality; indeed it became second nature to him. First he insisted that such benevolence was no more than his due. 'As a sick man I have a right to the indulgence humanity owes to those who are in pain.' Or: 'I am poor, and ... merit special favour.' Then, he goes on, to accept help, which he will only do under pressure, is very distressing to him: 'When I surrender to prolonged entreaties to accept an offer, repeated over and over again, I do so for the sake of peace and quiet rather than my own advantage. However much it may have cost the giver, he is actually in my debt – for it costs me more.' This being so, he was entitled to lay down conditions for accepting, say, the loan of a *cottage orné* or a small château. He undertook no social duties whatever, since 'my idea of happiness is ... never to have to do anything I don't wish to do.' To a host, therefore, he writes: 'I must insist that you leave me completely free.' 'If you cause me the least annoyance you will never see me again.' His thank-you letters (if that is the right word for them) were liable to be disagreeable documents: 'I thank you,' ran one, 'for the visit you persuaded me to make, and my thanks might have been warmer if you had not made me pay for it so dearly.'[25]

As one of Rousseau's biographers has pointed out, he was always setting little traps for people. He would emphasize his difficulties and poverty, then when they offered help affect hurt surprise, even indignation. Thus: 'Your proposal froze my heart. How you misunderstand your own interests when you try to make a valet out of a friend.' He adds: 'I am not unwilling to listen to what you have to propose, provided you appreciate that I am not for sale.' The would-be host, thus wrong-footed, was then induced to reformulate his invitation on Rousseau's terms.[26] It was one of Rousseau's psychological skills to persuade people, not least his social superiors, that common-or-garden words of thanks were not in his vocabulary. Thus he wrote to the Duc de Montmorency-Luxembourg, who lent him a château: 'I neither praise you nor thank you. But I live in your house. Everyone has his own language – I have said everything in mine.' The ploy worked beautifully, the Duchess replying apologetically: 'It is not for you to thank us – it is the Marshal and I who are in *your* debt.'[27]

But Rousseau was not prepared just to lead an agreeable, Harold Skimpole-like existence. He was too complicated and interesting for that. Alongside his streak of cool, hard-headed calculation there was a genuine element of paranoia, which did not permit him to settle for an easy life of self-centred parasitism. He quarrelled, ferociously and usually permanently, with virtually everyone with whom he had close dealings, and especially those who befriended him; and it is impossible to study the painful and repetitive tale of these rows without reaching the conclusion that he was a mentally sick man. This sickness cohabited with a great and original genius of mind, and the combination was very dangerous both for Rousseau and for others. The conviction of total rectitude was, of course, a primary symptom of his illness, and if Rousseau had possessed no talent it might have cured itself or, at worse, remained a small personal tragedy. But his wonderful gifts as a writer brought him acceptance, celebrity, even popularity. This was proof to him that his conviction that he was always right was not a subjective judgment but that of the world – apart, of course, from his enemies.

These enemies were, in every case, former friends or benefactors, who (Rousseau reasoned after he broke with them) had sought, under the guise of amity, to exploit and destroy him. The notion of disinterested friendship was alien to him; and since he was better than other men, and since he was incapable of feeling such an urge, then *a fortiori* it could not be felt by others. Hence the actions of all his 'friends' were carefully analysed by him from the start, and the moment they made a false move he was onto them. He quarrelled with Diderot, to whom he owed most of all. He quarrelled with Grimm. He had a particularly savage and hurtful break with Madame d'Épinay, his warmest benefactress. He quarrelled with Voltaire – that was not so difficult. He quarrelled with David Hume, who took him at his own valuation as a literary martyr, brought him to England and a hero's welcome and did everything in his power to make the visit a success, and Rousseau happy. There were dozens of minor rows, with his Genevan friend Dr Tronchin, for instance. Rousseau marked most of his major quarrels by composing a gigantic letter of remonstrance. These documents are among his most brilliant works, miracles of forensic skill in which evidence is cunningly fabricated, history rewritten and chronology confused with superb ingenuity in order to prove that the recipient is a monster. The letter he wrote to Hume, 10 July 1766, is eighteen folio pages (twenty-five of printed text) and has been described by Hume's biographer as 'consistent with the complete logical consistency of dementia. It remains one of the most brilliant and fascinating documents ever produced by a disordered mentality.'[28]

Rousseau gradually came to believe that these individual acts of enmity by men and women who had pretended to love him were not isolated but part of a connected pattern. They were all agents in a ramifying, long-term plot to frustrate, annoy and even destroy him and to damage his work. Working backwards through his life, he decided that the conspiracy went back to the days when, aged sixteen, he was a lackey to the Comtesse de Vercellis: 'I believe that from this time I suffered from the malicious play of secret interests which has thwarted me ever since and which has given me an understandable dislike for the apparent order responsible for it.' In sober fact, Rousseau was rather well treated by the French authorities, compared with other authors. There was only one attempt to arrest him, and the chief censor, Malesherbes, usually did his best to help get his work published. But Rousseau's feeling that he was the victim of an international network grew, especially during his visit to England. He became convinced that Hume was currently masterminding the plot, with the help of dozens of assistants. At one point he wrote to Lord Camden, the Lord Chancellor, explaining that his life was in danger and demanding an armed escort to get him out of the country. But Lord Chancellors are not unused to getting letters from madmen and Camden took no action. Rousseau's actions at Dover, just before his final departure, were hysterical, running on board a ship and locking himself in a cabin, and jumping on a post and addressing the crowd with the fantastic claim that Thérèse was now part of the plot and trying to keep him in England by force.[29]

Back on the Continent, he took to pinning posters to his front door listing his complaints against various sections of society arrayed against him: priests, fashionable intellectuals, the common people, women, the Swiss. He became convinced that the Duc de Choiseul, France's Foreign Minister, had taken personal charge of the international conspiracy and spent much of his time organizing the vast network of people whose task it was to make Rousseau's life a misery. Public events, such as the French seizure of Corsica, for which he had written a constitution, were ingeniously woven into the saga. Oddly enough, it was at Choiseul's request that Rousseau produced, for the Polish nationalists, a similar constitution for an independent Poland, and when Choiseul fell from power in 1770, Rousseau was upset: another sinister move! Rousseau declared that he could never discover the original offence (other than his identification with truth and justice) for which 'they' were determined to punish him. But there was no doubt about the details of the plot; it was 'immense, inconceivable': 'They will build around me an impenetrable edifice of darkness. They will bury me alive in a coffin... If I travel, everything will be prearranged to control me wherever I go.

The word will be given to passengers, coachmen, innkeepers . . . Such horror of me will be spread on my road that at each step I take, at everything I see, my heart will be lacerated.' His last works, the *Dialogues avec moi-même* (begun in 1772) and his *Rêvéries du promeneur solitaire* (1776) reflect this persecution-mania. When he finished the *Dialogues* he became convinced that 'they' intended to destroy them, and on 24 February 1776 he went to Notre Dame Cathedral with the intention of claiming sanctuary for his manuscript and placing it on the High Altar. But the gate to the choir was mysteriously locked. Sinister! So he made six copies and deposited them superstitiously in different hands: one went to Dr Johnson's bluestocking friend, Miss Brooke Boothby of Lichfield, and it was she who first published it in 1780. By that time, of course, Rousseau had gone to his grave, still sure that thousands of agents were after him.[30]

The agonies of mind caused by this form of dementia are real enough to the sufferer and it is impossible, from time to time, not to feel pity for Rousseau. Unhappily, he cannot be thus dismissed. He was one of the most influential writers who ever lived. He presented himself as the friend of humanity and, in particular, as the champion of the principles of truth and virtue. He was, and indeed still is, widely accepted as such. It is necessary, therefore, to look more closely at his own conduct as a teller of truth and a man of virtue. What do we find? The issue of truth is particularly significant because Rousseau became, after his death, best known by his *Confessions*. These were a self-proclaimed effort to tell the whole inner truth about a man's life, in a way never before attempted. The book was a new kind of ultra-truthful autobiography, just as James Boswell's life of Dr Johnson, published ten years later (1791), was a new kind of ultra-accurate biography.

Rousseau made absolute claims for the veracity of this book. In the winter of 1770–71 he held readings of it, in packed salons, lasting fifteen to seventeen hours, with breaks for meals. His attacks on his victims were so unsupportable that one of them, Madame d'Épinay, asked the authorities to have them stopped. Rousseau agreed to desist, but at the last reading he added these words: 'I have said the truth. If anyone knows facts contrary to what I have just said, even if they were proved a thousand times, they are lies and impostures . . . [whoever] examines with his own eyes my nature, my character, morals, inclinations, pleasures, habits, and can believe me to be a dishonest man, is himself a man who deserves to be strangled.' This produced an impressive silence.

Rousseau bolstered his title to be a truth-teller by claiming a superb memory. More important, he convinced readers he was sincere by being the first man to disclose details of his sex life, not in a spirit of macho boasting but, on the contrary, with shame and reluctance. As he rightly says, referring to 'the dark and dirty labyrinth' of his sexual experiences, 'It is not what is criminal which is hardest to tell, but what makes us feel ridiculous and ashamed.' But how genuine was the reluctance? In Turin, as a young man, he roamed the dark back streets and exposed his bare bottom to women: 'The foolish pleasure I took in displaying it before their eyes cannot be described.' Rousseau was a natural exhibitionist, in sexual as in other respects, and there is a certain relish in the way he narrates his sex life. He describes his masochism, how he enjoyed being spanked on his bare bottom by the strict pastor's sister, Mademoiselle Lambercier, being deliberately naughty to provoke punishment, and how he encouraged an older girl, Mademoiselle Groton, to spank him too: 'To lie at the feet of an imperious mistress, to obey her commands, to ask her forgiveness – this was for me a sweet enjoyment.'[31] He tells how, as a boy, he took up masturbation. He defends it because it prevents the young from catching venereal disease and because, 'This vice which shame and timidity find so convenient has more than one attraction for live imaginations: it enables them to subject all women to their whims and to make beauty serve the pleasure which tempts them without obtaining its consent.'[32] He gave an account of an attempt to seduce him by a homosexual at the hospice in Turin.[33] He admitted he had shared the favours of Madame de Warens with her gardener. He described how he was unable to make love to one girl when he discovered she had no nipple on one breast, and records her furious dismissal of him: 'Leave women alone and study mathematics.' He confesses to resuming masturbation in later life as more convenient than pursuing an active love life. He gives the impression, part intentionally, part unconsciously, that his attitude to sex remained essentially infantile: his mistress, Madame de Warens, is always 'Maman'.

These damaging admissions build up confidence in Rousseau's regard for truth, and he reinforces it by relating other shameful, non-sexual episodes, involving theft, lies, cowardice and desertion. But there was an element of cunning in this. His accusations against himself make his subsequent accusations against his enemies far more convincing. As Diderot furiously observed, 'he describes himself in odious colours to give his unjust and cruel imputations the semblance of truth.' Moreover, the self-accusations are deceptive since in every critical one he follows up the bare admission by a skilfully presented exculpation so that the reader ends by sympathizing with him and giving him credit for his forthright honesty.[34] Then again, the truths Rousseau presents often turn out to be half-truths: his selective honesty is in some ways the most dishonest aspect both of his *Confessions* and his letters. The 'facts' he so frankly admits often emerge, in the light of modern scholarship, to be inaccurate, distorted or non-existent. This is sometimes clear even from internal evidence. Thus he gives two quite different accounts of the homosexual advance, in *Émile* and in the *Confessions*. His total-recall memory was a myth. He gives the wrong year for his father's death and describes him as 'about sixty', when he was in fact seventy-five. He lies about virtually all the details of his stay at the hospice in Turin, one of the most critical episodes of his early life. It gradually emerges that no statement in the *Confessions* can be trusted if unsupported by external evidence. Indeed it is hard not to agree with one of Rousseau's most comprehensive modern critics, J.H.Huizinga, that the insistent claims of the *Confessions* to truth and honesty make its distortions and falsehoods peculiarly disgraceful: 'The more attentively one reads and re-reads, the deeper one delves into this work, the more layers of ignominy become apparent.'[35] What makes Rousseau's dishonesty so dangerous – what made his inventions so rightly feared by his ex-friends – was the diabolical skill and brilliance with which they were presented. As his fair-minded biographer, Professor Crocker, puts it: 'All his accounts of his quarrels (as in the Venetian episode) have an irresistible persuasiveness, eloquence and air of sincerity; then the facts come as a shock.'[36]

So much for Rousseau's devotion to truth. What of his virtue? Very few of us lead lives which will bear close scrutiny, and there is something mean in subjecting Rousseau's, laid horribly bare by the activities of thousands of scholars, to moral judgment. But granted his claims, and still more his influence on ethics and behaviour, there is no alternative. He was a man, he said, born to love, and he taught the doctrine of love more persistently than most ecclesiastics. How well, then, did he express his love by those nature had placed closest to him? The death of his mother deprived him, from birth, of a normal family life. He could have no feelings for her, one way or another, since he never knew her. But he showed no affection, or indeed interest in, other members of his family. His father meant nothing to him, and his death was merely an opportunity to inherit. At this point Rousseau's concern for his long-lost brother revived to the extent of certifying him dead, so the family money could be his. He saw his family in terms of cash. In the *Confessions* he describes 'one of my apparent inconsistencies – the union of an almost sordid avarice with the greatest contempt for money'.[37] There is not much evidence of this contempt in his life. When his family inheritance was proved in his favour, he described receiving the draft, and, by a supreme effort of will, delaying opening the letter until the next day. Then: 'I opened it with deliberate slowness and found the money-order in it. I felt many pleasures at once but I swear the keenest was of having conquered myself.'[38]

If that was his attitude to his natural family, how did he treat the woman who became, in effect, his foster-mother, Madame de Warens? The answer is: meanly. She had rescued him from destitution no less than four times, but when he later prospered and she became indigent, he did little for her. By his own account he sent her 'a little' money when he inherited the family fortune in the 1740s, but refused to send more as it would simply have been taken by the 'rascals' who surrounded her.[39] This was an excuse. Her later pleas to him for help went unanswered. She spent her last two years bedridden and her death, in 1761, may have been from malnutrition. The Comte de Charmette, who knew both of them, strongly condemned Rousseau's failure 'to return at least a part of what he had cost his generous benefactress'. Rousseau then went on to deal with her, in his *Confessions*, with consummate humbug, hailing her as 'best of women and mothers'. He claimed he had not written to her because he did not want to make her miserable by recounting his troubles. He ended: 'Go, taste the fruits of your charitableness and prepare for your pupil the place he hopes to take next to you some day! Happy in your misfortunes because Heaven, by ending them, has spared you the cruel spectacle of his.' It was characteristic of Rousseau to treat her death in a purely egocentric context.

Was Rousseau capable of loving a woman without strong selfish reservations? According to his own account, 'the first and only love of all my life' was Sophie, Comtesse d'Houdetot, sister-in-law of his benefactress Madame d'Épinay. He may have loved her, but he says he 'took the precaution' of writing his love-letters to her in such a way as to make their publication as damaging to her as to him. Of Thérèse Levasseur, the twenty-three-year-old laundress whom he made his mistress in 1745 and who remained with him thirty-three years until his death, he said he 'never felt the least glimmering of love for her ... the sensual needs I satisfied with her were purely sexual and were nothing to do with her as an individual.' 'I told her,' he wrote, 'I would never leave her and never marry her.' A quarter of a century later he went through a pseudo-wedding with her in front of a few friends but used the occasion to make a vainglorious speech, declaring that posterity would erect statues to him and 'It will then be no empty honour to have been a friend of Jean-Jacques Rousseau.'

In one way he despised Thérèse as a coarse, illiterate servant-girl, and despised himself for consorting with her. He accused her mother of being grasping and her brother of stealing his forty-two fine shirts (there is no evidence her family was as bad as he paints them). He said that Thérèse not only could not read or write but was incapable of telling the time and did not know what day of the month it was. He never took her out and when he invited people to dine she was not allowed to sit down. She brought in the food and he 'made merry at her expense'. To amuse the Duchesse de Montmorency-Luxembourg he compiled a catalogue of her solecisms. Even some of his grand friends were shocked at the contemptuous way he used her. Contemporaries were divided about her, some considering her a malicious gossip; Rousseau's innumerable hagiographers have painted her in the blackest colours to justify his mean-spirited behaviour to her. But she has also had some vigorous defenders.[40]

Indeed, to do Rousseau justice, he paid her compliments too: 'the heart of an angel', 'tender and virtuous', 'an excellent counsellor', 'a simple girl without flirtatiousness'. He found her 'timorous and easily dominated'. In fact it is not at all clear that Rousseau understood her, probably because he was too self-obsessed to study her. The most reliable portrait of her is provided by James Boswell, who visited Rousseau five times in 1764 and later escorted Thérèse to England.[41] He found her 'a little, lively, neat French girl', bribed her to obtain further access to Rousseau and managed to cadge from her two letters Rousseau had written to her (only one other exists).[42] They reveal him as affectionate and their relationship as intimate. She told Boswell: 'I have been twenty-two years with Monsieur Rousseau. I would not give up my place to be Queen of France.' On the other hand, once Boswell became her travelling companion, he seduced her without the slightest difficulty. His blow-by-blow account of the affair was cut from his manuscript diary by his literary executors, who marked the gap 'Reprehensible Passage'. But they left in a sentence in which Boswell, at Dover, had recorded: 'Yesterday morning had gone to her bed very early (on landing) and had done it once: thirteen in all,' and enough remains of his account to reveal her as a far more sophisticated and worldly woman than most people have supposed. The truth seems to be that she was devoted to Rousseau, in most respects, but had been taught, by his own behaviour, to use him, as he used her. Rousseau's warmest affection went to animals. Boswell records a delightful scene of him playing with his cat and his dog Sultan. He gave Sultan (and his predecessor, Turc) a love he could not find for humans, and the howling of this dog, whom he brought with him to London, almost prevented him from attending the special benefit performance Garrick had set up for him at Drury Lane.[43]

Rousseau kept and even cherished Thérèse because she could do for him things animals could not: operate the catheter to relieve his stricture, for instance. He would not tolerate third parties interfering in his relations with her: he became furious, for instance, when a publisher sent her a dress; he promptly vetoed a plan to provide her with a pension, which might have made her independent of him. Most of all, he would not allow children to usurp his claims on her, and this led him to his greatest crime. Since a large part of Rousseau's reputation rests on his theories about the upbringing of children – more education is the main, underlying theme of his *Discours*, *Émile*, the *Social Contract* and even *La Nouvelle Héloïse* – it is curious that, in real life as opposed to writing, he took so little interest in children. There is no evidence whatever that he studied children to verify his theories. He claimed that no one enjoyed playing with children more than himself, but the one anecdote we have of him in this capacity is not reassuring. The painter Delacroix relates in his *Journal* (31 May 1824) that a man told him he had seen Rousseau in the gardens of the Tuileries: 'A child's ball struck the philosopher's leg. He flew into a rage and pursued the child with his cane.'[44] From what we know of his character, it is unlikely that Rousseau could ever have made a good father. Even so, it comes as a sickening shock to discover what Rousseau did to his own children.

The first was born to Thérèse in the winter of 1746–47. We do not know its sex. It was never named. With (he says) 'the greatest difficulty in the world', he persuaded Thérèse that the baby must be abandoned 'to save her honour'. She 'obeyed with a sigh'. He placed a cypher-card in the infant's clothing and told the midwife to drop off the bundle at the Hôpital des Enfants-trouvés. Four other babies he had by Thérèse were disposed of in exactly the same manner, except that he did not trouble to insert a cypher-card after the first. None had names. It is unlikely that any of them survived long. A history of this institution which appeared in 1746 in the *Mercure de France* makes it clear that it was overwhelmed by abandoned infants, over 3000 a year. In 1758 Rousseau himself noted that the total had risen to 5082. By 1772 it averaged nearly 8000. Two-thirds of the babies died in their first year. An average of fourteen out of every hundred survived to the age of seven, and of these five grew to maturity, most of them becoming beggars and vagabonds.[45] Rousseau did not even note the dates of the births of his five children and never took any interest in what happened to them, except once in 1761, when he believed Thérèse was dying and made a perfunctory attempt, soon discontinued, to use the cypher to discover the whereabouts of the first child.

Rousseau could not keep his conduct entirely secret, and on various occasions, in 1751 and again in 1761 for instance, he was obliged to defend himself in private letters. Then in 1764 Voltaire, angered by Rousseau's attacks on his atheism, published an anonymous pamphlet, written in the guise of a Genevan pastor, called *Le Sentiment des Citoyens*. This openly accused him of abandoning his five children; but it also stated he was a syphilitic and murderer, and Rousseau's denials of all these charges were generally accepted. He brooded on the episode, however, and it was one factor determining him to write his *Confessions*, which were essentially designed to rebut or extenuate facts already made public. Twice in this work he defends himself about the babies, and he returns to the subject in his *Reveries* and in various letters. In all, his efforts to justify himself, publicly and privately, spread over twenty-five years and vary considerably. They merely make matters worse, since they compound cruelty and selfishness with hypocrisy.[46] First, he blamed the wicked circle of godless intellectuals among whom he then moved for putting the idea of the orphanage into his innocent head. Then, to have children was 'an inconvenience'. He could not afford it. 'How could I achieve the tranquillity of mind necessary for my work, my garret filled with domestic cares and the noise of children?' He would have been forced to stoop to degrading work, 'to all those infamous acts which fill me with such justified horror'. 'I know full well no father is more tender than I would have been' but he did not want his children to have any contact with Thérèse's mother: 'I trembled at the thought of entrusting mine to that ill-bred family.' As for cruelty, how could anyone of his outstanding moral character be guilty of such a thing? '... my ardent love of the great, the true, the beautiful and the just; my horror of evil of every kind, my utter inability to hate or injure or even to think of it; the sweet and lively emotion which I feel at the sight of all that is virtuous, generous and amiable; is it possible, I ask, that all these can ever agree in the same heart with the depravity which, without the least scruple, tramples underfoot the sweetest of obligations? No! I feel, and loudly assert – it is impossible! Never, for a single moment in his life, could Jean-Jacques have been a man without feeling, without compassion, or an unnatural father.'

Granted his own virtue, Rousseau was obliged to go further and defend his actions on positive grounds. At this point, almost by accident, Rousseau takes us right to the heart both of his own personal problem and of his political philosophy. It is right to dwell on his desertion of his children not only because it is the most striking single example of his inhumanity but because it is organically part of the process which produced his theory of politics and the role of the state. Rousseau regarded himself as an abandoned child. To a great extent he never really grew up but remained a dependent child all his life, turning to Madame de Warens as a mother, to Thérèse as a nanny. There are many passages in his *Confessions* and still more in his letters which stress the child element.

Many of those who had dealings with him – Hume, for instance – saw him as a child. They began by thinking of him as a harmless child, who could be managed, and discovered to their cost they were dealing with a brilliant and savage delinquent. Since Rousseau felt (in some ways) as a child, it followed he could not bring up children of his own. Something had to take his place, and that something was the State, in the form of the orphanage.

Hence, he argued, what he did was 'a good and sensible arrangement'. It was exactly what Plato had advocated. The children would 'be all the better for not being delicately reared since it would make them more robust'. They would be 'happier than their father'. 'I could have wished,' he wrote, 'and still do wish that I had been brought up and nurtured as they have been.' 'If only I could have had the same good fortune.' In short, by transferring his responsibilities to the State, 'I thought I was performing the act of a citizen and a father and I looked on myself as a member of Plato's Republic.'

Rousseau asserts that brooding on his conduct towards his children led him eventually to formulate the theory of education he put forward in *Émile*. It also clearly helped to shape his *Social Contract*, published the same year. What began as a process of personal self-justification in a particular case – a series of hasty, ill thought-out excuses for behaviour he must have known, initially, was unnatural – gradually evolved, as repetition and growing self-esteem hardened them into genuine convictions, into the proposition that education was the key to social and moral improvement and, this being so, it was the concern of the State. The State must form the minds of all, not only as children (as it had done to Rousseau's in the orphanage) but as adult citizens. By a curious chain of infamous moral logic, Rousseau's iniquity as a parent was linked to his ideological offspring, the future totalitarian state.

Confusion has always surrounded Rousseau's political ideas because he was in many respects an inconsistent and contradictory writer – one reason why the Rousseau industry has grown so gigantic: academics thrive on resolving 'problems'. In some passages of his works he appears a conservative, strongly opposed to revolution: 'Think of the dangers of setting the masses in motion.' 'People who make revolutions nearly always end by handing themselves over to tempters who make their chains heavier than before.' 'I would not have anything to do with revolutionary plots which always lead to disorder, violence and bloodshed.' 'The liberty of the entire human race is not worth the life of a single human being.' But his writings also abound with radical bitterness. 'I hate the great, I hate their rank, their harshness, their prejudices, their pettiness, all their vice.' He wrote to one grand lady: 'It is the wealthy class, your class, that steals from mine the bread of my children,' and he admitted to 'a certain resentment against the rich and successful, as if their wealth and happiness had been gained at my expense'. The rich were 'hungry wolves who, once having tasted human flesh, refuse any other nourishment'. His many powerful aphorisms, which make his books so sharply attractive especially to the young, are radical in tone. 'The fruits of the earth belong to us all, the earth itself to none.' 'Man is born free and is everywhere in chains.' His entry in the *Encyclopédie* on 'Political Economy' sums up the attitude of the ruling class: 'You need me for I am rich and you are poor. Let us make an agreement: I will allow you to have the honour of serving me, provided you give me whatever you have left for the trouble I shall take to command you.'

However, once we understand the nature of the state Rousseau wished to create, his views begin to cohere. It was necessary to replace the existing society by something totally different and essentially egalitarian; but, this done, revolutionary disorder could not be permitted. The rich and the privileged, as the ordering force, would be replaced by the State, embodying the General Will, which all contracted to obey. Such obedience would become instinctive and voluntary since the State, by a systematic process of cultural engineering, would inculcate virtue in all. The State was the father, the *patrie*, and all its citizens were the children of the paternal orphanage. (Hence the supposedly puzzling remark of Dr Johnson, who cut clean through Rousseau's sophistries, 'Patriotism is the last refuge of a scoundrel.') It is true that the citizen-children, unlike Rousseau's own babies, originally agree to submit to the State/orphanage by freely contracting into it. They thus constitute, through their collective will, its legitimacy, and thereafter they have no right to feel constrained, since, having wanted the laws, they must love the obligations they impose.[47]

Though Rousseau writes about the General Will in terms of liberty, it is essentially an authoritarian instrument, an early adumbration of Lenin's 'democratic centralism'. Laws made under the General Will must, by definition, have moral authority. 'The people making laws for itself cannot be unjust.' 'The General Will is always righteous.' Moreover, provided the State is 'well-intentioned' (i.e., its long-term objectives are desirable) interpretation of the General Will can safely be left to the leaders since 'they know well that the General Will always favours the decision most conducive to the public interest.' Hence any individual who finds himself in opposition to the General Will is in error: 'When the opinion that is contrary to my own prevails, this simply proves that I was mistaken and that what I thought to be the General Will, was not so.' Indeed, 'if my particular opinion had carried the day I should have achieved the opposite of what was my will and I should not therefore have been free.' We are here almost in the chilly region of Arthur Koestler's *Darkness at Noon* or George Orwell's 'Newspeak'.

Rousseau's state is not merely authoritarian: it is also totalitarian, since it orders every aspect of human activity, thought included. Under the social contract, the individual was obliged to 'alienate himself, with all his rights, to the whole of the community' (i.e., the State). Rousseau held that there was an ineradicable conflict between man's natural selfishness and his social duties, between the Man and the Citizen. And that made him miserable. The function of the social contract, and the State it brought into being, was to make man whole again: 'Make man one, and you will make him as happy as he can be. Give him all to the State, or leave him all to himself. But if you divide his heart, you tear him in two.' You must, therefore, treat citizens as children and control their upbringing and thoughts, planting 'the social law in the bottom of their hearts'. They then become 'social men by their natures and citizens by their inclinations; they will be one, they will be good, they will be happy, and their happiness will be that of the Republic'.

This procedure demanded total submission. The original social contract oath for his projected constitution for Corsica reads: 'I join myself, body, goods, will and all my powers, to the Corsican nation, granting her ownership of me, of myself and all who depend on me.'[48] The State would thus 'possess men and all their powers', and control every aspect of their economic and social life, which would be spartan, anti-luxurious and anti-urban, the people being prevented from entering the towns except by special permission. In a number of ways the State Rousseau planned for Corsica anticipated the one the Pol Pot regime actually tried to create in Cambodia, and this is not entirely surprising since the Paris-educated leaders of the regime had all absorbed Rousseau's ideas. Of course, Rousseau sincerely believed that such a State would be contented since the people would have been trained to like it. He did not use the word 'brainwash', but he wrote: 'Those who control a people's opinions control its actions.' Such control is established by treating citizens, from infancy, as children of the State, trained to 'consider themselves only in their relationship to the Body of the State'. 'For being nothing except by it, they will be nothing except for it. It will have all they have and will be all they are.' Again, this anticipates Mussolini's central Fascist doctrine: 'Everything within the State, nothing outside the State, nothing against the State.' The educational process was thus the key to the success of the cultural engineering needed to make the State acceptable and successful; the axis of Rousseau's ideas was the citizen as child and the State as parent, and he insisted the government should have complete charge of the upbringing of all children. Hence – and this is the true revolution Rousseau's ideas brought about – he moved the political process to the very centre of human existence by making the legislator, who is also a pedagogue, into the new Messiah, capable of solving all human problems by creating New Men. 'Everything,' he wrote, 'is at root dependent on politics.' Virtue is the product of good government. 'Vices belong less to man, than to man badly governed.' The political process, and the new kind of state it brings into being, are the universal remedies for the ills of mankind.[49] Politics will do all. Rousseau thus prepared the blueprint for the principal delusions and follies of the twentieth century.

Rousseau's reputation during his lifetime, and his influence after his death, raise disturbing questions about human gullibility, and indeed about the human propensity to reject evidence it does not wish to admit. The acceptability of what Rousseau wrote depended in great part on his strident claim to be not merely virtuous but the most virtuous man of his time. Why did not this claim collapse in ridicule and ignominy when his weaknesses and vices became not merely public knowledge but the subject of international debate? After all the people who assailed him were not strangers or political opponents but former friends and associates who had gone out of their way to assist him. Their charges were serious and the collective indictment devastating. Hume, who had once thought him 'gentle, modest, affectionate, disinterested and exquisitely sensitive', decided, from more extensive experience, that he was 'a monster who saw himself as the only important being in the universe'. Diderot, after long acquaintance, summed him up as 'deceitful, vain as Satan, ungrateful, cruel, hypocritical and full of malice'. To Grimm he was 'odious, monstrous'. To Voltaire, 'a monster of vanity and vileness'. Saddest of all are the judgments passed on him by kind-hearted women who helped him, like Madame d'Épinay, and her harmless husband, whose last words to Rousseau were 'I have nothing left for you but pity.' These judgments were based not on the man's words but on his deeds, and since that time, over two hundred years, the mass of material unearthed by scholars has tended relentlessly to substantiate them. One modern academic lists Rousseau's shortcomings as follows: he was a 'masochist, exhibitionist, neurasthenic, hypochondriac, onanist, latent homosexual afflicted by the typical urge for repeated displacements, incapable of normal or parental affection, incipient paranoiac, narcissistic introvert rendered unsocial by his illness, filled with guilt feelings, pathologically timid, a kleptomaniac, infantilist, irritable and miserly'.[50]

Such accusations, and extensive display of the evidence on which they are based, made very little difference to the regard in which Rousseau and his works were, and are, held by those for whom he has an intellectual and emotional attraction. During his life, no matter how many friendships he destroyed, he never found any difficulty in forming new ones and recruiting fresh admirers, disciples and grandees to provide him with houses, dinners and the incense he craved. When he died he was buried on the Île des Peupliers on the lake at Ermononville and this rapidly became a place of secular pilgrimage for men and women from all over Europe, like the shrine of a saint in the Middle Ages. Descriptions of the antics of these *dévotés* make hilarious reading: 'I dropped to my knees... pressed my lips to the cold stone of the monument... and kissed it repeatedly.'[51] Relics, such as his tobacco pouch and jar, were carefully preserved at 'the Sanctuary', as it was known. One recalls Erasmus and John Colet visiting the great shrine of St Thomas à Becket at Canterbury in *c.* 1512 and sneering at the excesses of the pilgrims. What would they have found to say of 'Saint Rousseau' (as George Sand was reverently to call him), three hundred years after the Reformation had supposedly ended that sort of thing? The plaudits continued long after the ashes were transferred to the Panthéon. To Kant he had 'a sensibility of soul of unequalled perfection'. To Shelley he was 'a sublime genius'. For Schiller he was 'a Christlike soul for whom only Heaven's angels are fit company'. John Stuart Mill and George Eliot, Hugo and Flaubert, paid deep homage. Tolstoy said that Rousseau and the Gospel had been 'the two great and healthy influences of my life'. One of the most influential intellectuals of our own times, Claude Lévi-Strauss, in his principal work, *Tristes Tropiques*, hails him as 'our master and our brother... every page of this book could have been dedicated to him, had it not been unworthy of his great memory'.[52]

It is all very baffling and suggests that intellectuals are as unreasonable, illogical and superstitious as anyone else. The truth seems to be that Rousseau was a writer of genius but fatally unbalanced both in his life and in his views. He is best summed up by the woman who, he said, was his only love, Sophie d'Houdetot. She lived on until 1813 and, in extreme old age, delivered this verdict: 'He was ugly enough to frighten me and love did not make him more attractive. But he was a pathetic figure and I treated him with gentleness and kindness. He was an interesting madman.'[53]

2

Shelley, or the Heartlessness of Ideas

O N 25 June 1811 the nineteen-year-old heir to an English baronetcy wrote to a young schoolmistress in Sussex: 'I am not an aristocrat, or any *crat* at all but vehemently long for the time when man may *dare* to live in accordance with *Nature* & Reason, in consequence with Virtue.'[1] The doctrine was pure Rousseau but the writer, the poet Percy Bysshe Shelley, was to go much further than Rousseau in staking the claims of intellectuals and writers to guide humanity. Like Rousseau, Shelley believed that society was totally rotten and should be transformed, and that enlightened man, through his own unaided intellect, had the moral right and duty to reconstruct it from first principles. But he also argued that intellectuals, and especially poets – whom he saw as the leaders of the intellectual community – occupied a privileged position in this process. In fact, 'Poets are the unacknowledged legislators of the world.'

Shelley set out this challenge, on behalf of his fellow intellectuals, in 1821 in his 10,000-word essay, *A Defence of Poetry*, which became the most influential statement of the social purpose of literature since Antiquity.[2] Poetry, Shelley argued, is more than a display of verbal ingenuity, a mere amusement. It has the most serious aim of any kind of writing. It is prophecy, law and knowledge. Social progress can be achieved only if it is guided by an ethical sensibility. The churches ought to have supplied this but have manifestly failed. Science cannot supply it. Nor can rationalism alone produce moral purpose. When science and rationalism masquerade as ethics they produce moral disasters like the French Revolutionary Terror and the Napoleonic dictatorship. Only poetry can fill the moral vacuum and give to progress a truly creative force. Poetry 'awakens and enlarges the mind itself by rendering it the receptacle of a thousand unapprehended combinations of thought. Poetry lifts the veil from the hidden beauty of the world.' 'The great secret of morals is love; or a going out of our own nature, and an identification of ourselves with the beautiful which exists in thought, action or person, not our own.' It fights egoism and material calculation. It encourages community spirit. 'A man, to be greatly good, must imagine intensely and comprehensively; he must put himself in the place of another and of many others; the pains and pleasures of his species must become his own. The great instrument of moral good is the imagination; and poetry administers to the effect by acting on the cause.' The achievement of poetry is to push forward the moral progress of civilization: in fact poetry, its hand-maiden imagination, and its natural environment liberty, form the tripod on which all civilization and ethics rest. Imaginative poetry is needed to reconstruct society completely: 'We want the creative faculty to imagine that which we know; we want the generous impulse to act that which we imagine; we want the poetry of life.' Shelley, indeed, was not merely presenting the claims of the poet to rule: he was advancing, for the first time, a fundamental critique of the materialism which was to become the central characteristic of nineteenth-century society: 'Poetry and the principle of Self, of which money is the visible incarnation, are the God and the Mammon of the world.'[3]

In his poetry Shelley certainly practised what he preached. He was a great poet – and his poetry can be understood and enjoyed at many levels. But at the deepest level, at the level Shelley himself intended, it is essentially moral and political. He is the most thoroughly politicized of English poets; all his major and many of his shorter poems have a call for social action of some kind, a public message. His longest, *The Revolt of Islam* (nearly 5000 lines), concerns oppression, uprising and freedom. *A Hymn to Intellectual Beauty*, by which he means the spirit of good, embodying the freedom and equality of all human beings, celebrates its triumph over established evil. *Prometheus Unbound* tells of another successful revolution and the victory of the mythic figure who for Shelley (as for Marx and others) symbolizes the intellectual leading humanity to utopia on earth. *The Cenci* repeats the theme of revolt against tyranny, as does *Swellfoot the Tyrant*, an attack on George IV, and *The Mask of Anarchy*, an attack on his ministers. In 'Ozymandias', a mere sonnet, though a powerful one, he celebrates the nemesis of autocracy. In the lyrical 'Lines from the Eugenean Hills', he notes the cycles of tyranny which encompass the world and invites readers to join him in his righteous utopia.[4] 'Ode to the West Wind' is another plea to readers, to spread his political message, to 'Drive my dead thoughts over the universe' and so 'quicken a new birth', 'scatter ... my words among mankind!' 'To a Skylark' is on similar lines, about the poet's difficulty in getting his voice heard, and his message across. Shelley, in his lifetime, was disappointed at the meagre publicity given to his work and desperate that his political and moral doctrines should penetrate society. It is no accident that two of his most passionate poems are pleas that his words circulate widely and be heeded. As an artist, in short, Shelley was remarkably un-egocentric. Few poets have written less for their own personal satisfaction.

But what of Shelley as a man? Until quite recently, the general view was that assiduously propagated by his second wife and widow, Mary Shelley: that the poet was a singularly pure and innocent, unworldly spirit, without guile or vice, devoted to his art and his fellow men, though in no sense a politician; more a hugely intelligent and inordinately sensitive child. This view was reinforced by some contemporary descriptions of his physical appearance: slender, wan, fragile, retaining an adolescent bloom until well into his twenties. The cult of bohemian clothes Rousseau had inaugurated had persisted into the second and third generation of Romantic intellectuals. Byron sported not merely the Levantine or Oriental mode of dress when it suited him, but even in European clothes affected a certain looseness, dispensing with elaborate cravats, leaving his shirt-neck open, even leaving off his coat altogether and wearing shirt-sleeves. This aristocratic disdain for uncomfortable conventions was copied by more plebeian poets like Keats. Shelley also adopted the fashion, but added his own touch: a fondness for schoolboy jackets and caps, sometimes too small for him but peculiarly suited to the impression he wished to convey, of adolescent spontaneity and freshness, a little gawky but charming. The ladies in particular liked it, as they liked the unlaced and unbuttoned Byron. It helped to build up a powerful and persistent but mythic image of Shelley, which found almost marmoreal form in Matthew Arnold's celebrated description of him as a 'beautiful but ineffectual angel, beating in the void his luminous wings in vain'. This occurs in Arnold's essay on Byron, whose poetry he finds far more serious and weighty than Shelley's, which has 'the incurable fault' of 'insubstantiality'. On the other hand as a person Shelley was a 'beautiful and enchanting spirit' and 'immeasurably Byron's superior'.[5] Hard to imagine a more perverse judgment, wrong on all points, suggesting that Arnold knew little of either man and cannot have read Shelley's work with attention. Oddly enough, however, his judgment of Shelley's character was not unlike Byron's own. Shelley, Byron wrote, was 'without exception, the *best* and least selfish man I ever knew. I never knew one who was not a beast in comparison.' Or again: 'he is to my knowledge the *least* selfish and the mildest of men – a man who has made more sacrifices of his fortune and feelings for others than any I have ever heard of.'[6] These comments were made when Shelley's tragic end was fresh in Byron's mind, and so in a spirit of *nil nisi bunkum*. Moreover, most of Byron's knowledge of Shelley was based on what Shelley himself told him. All the same, Byron was a man of the world, a shrewd judge and a fierce castigator of humbug, and his testimony to the impression Shelley made on his more broad-minded bohemian contemporaries carries weight.

The truth, however, is fundamentally different and to anyone who reveres Shelley as a poet (as I do) it is deeply disturbing. It emerges from a variety of sources, one of the most important of which is Shelley's own letters.[7] It reveals Shelley as astonishingly single-minded in the pursuit of his ideals but ruthless and even brutal in disposing of anyone who got in the way. Like Rousseau, he loved humanity in general but was often cruel to human beings in particular. He burned with a fierce love but it was an abstract flame and the poor mortals who came near it were often scorched. He put ideas before people and his life is a testament to how heartless ideas can be.

Shelley was born on 4 August 1792 at Field Place, a large Georgian house near Horsham in Sussex. He was not, like so many leading intellectuals, an only child. But he occupied what is in many ways an even more corrupting position: the only son and heir to a considerable fortune and a title and the elder brother of four sisters, who were from two to nine years younger than him. It is difficult to convey now what this meant at the end of the eighteenth century: to his parents, still more to his siblings, he was the lord of creation.

The Shelleys were a junior branch of an ancient family and connections of the great local landowner, the Duke of Norfolk. Their money, which was considerable, was new and had been made by Shelley's grandfather Sir Bysshe, the first baronet, who was born in Newark, New Jersey, a New World adventurer, rough, tough and energetic. Shelley clearly inherited from him his drive and ruthlessness. His father, Sir Timothy, who succeeded to the title in 1815, was by contrast a mild and inoffensive man, who for many years led a dutiful and blameless life as MP for Shoreham, gradually shifting from moderate Whiggism to middle-of-the-road Tory.[8]

Shelley had an idyllic childhood on the estate, surrounded by doting parents and adoring sisters. He early showed a passion for nature and natural science, experimenting with chemicals and fire-balloons, which remained a lifelong taste. In 1804, when he was twelve, he was sent to Eton, where he remained for six years. He must have worked hard because he acquired great fluency in Latin and Greek and an extensive knowledge of ancient literature which he retained all his days. He was always an avid and rapid reader of both serious material and fiction and, next to Coleridge, was the best-read poet of his time. He was also a schoolboy prodigy. In 1809 when he was sixteen, Dr James Lind, a former royal doctor who was also a part-time Eton master, amateur scientist and radical, introduced him to *Political Justice* by William Godwin, the key left-wing text of the day.[9] Lind also had an interest in demonology and stimulated in Shelley a passion for the occult and mysterious: not just the Gothic fiction which was then fashionable and is ridiculed so brilliantly in Jane Austen's *Northanger Abbey*, but the real-life activities of the Illuminati and other secret revolutionary societies.

The Illuminati had been institutionalized in 1776 by Adam Weishaupt at the German University of Ingoldstadt, as guardians of the rationalist Enlightenment. Their aim was to illuminate the world until (as he argued) 'Princes and nations will disappear without violence from the earth, the human race will become one family and the world the abode of reasonable men.'[10] In a sense this became Shelley's permanent aim but he absorbed the Illuminist material in conjunction with the aggressive propaganda put out by their enemies, especially the sensational Ultra tract by the Abbé Barruel, *Memoirs Illustrating the History of Jacobitism* (London, 1797–98), which attacked not just the Illuminati but the Masons, Rosicrucians and Jews. Shelley was for many years fascinated by this repellent book, which he often recommended to friends (it was used by his second wife Mary when she was writing *Frankenstein* in 1818). It was mixed up in Shelley's mind with a lot of Gothic novels which he also read, then and later.

Thus, from his teens, Shelley's approach to politics was coloured both by a taste for secret societies and by the conspiracy theory of history preached by the Abbé and his kind. He never could shake it off, and it effectively prevented him from understanding British politics or the motives and policies of men like Liverpool and Castlereagh, whom he saw merely as embodied evil.[11] Almost his earliest political act was to propose to the radical writer Leigh Hunt the formation of a secret society of 'enlightened, unprejudiced members' to resist 'the coalition of the enemies of liberty'.[12] Indeed some of Shelley's acquaintances never saw his politics as anything more than a literary joke, a mere projection into real life of Gothic romance. Thomas Love Peacock, in his novel *Nightmare Abbey* (1818), satirized the secret-society mania and portrays Shelley as Scythrop, who 'now became troubled with the *passion for reforming the world*. He built many castles in the air and peopled them with secret tribunals and bands of Illuminati, who were always the imaginary ingredients of his projected regeneration of the human species'. Shelley was partly to blame for this frivolous view of his utopianism. He not only, according to his friend Thomas Jefferson Hogg, insisted on reading aloud 'with rapturous enthusiasm' a book called *Horrid Mysteries* to anyone who would listen, but he also wrote two Gothic novels of his own, *Zastrozzi*, published in his last term at Eton, and, during his first term at Oxford, *St Irvyne or the Rosicrucian*, rightly dismissed by Elizabeth Barrett Browning as 'boarding-school idiocy'.[13]

Shelley was thus famous or notorious while he was still at school and was known as 'the Eton Atheist'. It is important to note this, in view of his later accusations of intolerance against his family. His grandfather and his father, far from trying to curb his youthful writings, which included of course poetry, encouraged them and financed their publication. According to Shelley's sister Helen, old Sir Bysshe paid for his schoolboy poems to be published. In September 1810, just before he went up to Oxford, Sir Bysshe again paid for 1500 copies to be printed of a volume of Shelley's called *Original Poetry by Victor and Cazire*.[14] When Shelley went up to Oxford in the autumn, his father Timothy took him to the leading bookseller, Slatter's, and said: 'My son here has a literary turn. He is already an author and do pray indulge him in his printing freaks.' Timothy, indeed, encouraged him in writing a prize poem on the Parthenon and sent him material.[15] He obviously hoped to steer Shelley away from what he regarded as adolescent fireworks into serious literature. His financing of his son's writing was on the explicit understanding that, while he might express his anti-religious views among his friends, he was not to publish them and so wreck his university career.

There is no doubt Shelley agreed to this, as a surviving letter shows.[16] He then proceeded to break his word, in the most flagrant and comprehensive way. In March 1811, while a first-year undergraduate at University College, Oxford, he wrote an aggressive pamphlet on his religious views. His argument was neither new nor particularly outrageous – it derived directly from Locke and Hume. Since ideas, wrote Shelley, come from the senses, and 'God' cannot derive from sense-impressions, belief is not a voluntary act and unbelief cannot therefore be criminal. To this dull piece of sophistry he affixed the inflammatory title *The Necessity of Atheism*, printed it, put it in the Oxford bookshops and sent copies to all the bishops and the heads of the colleges. In short his behaviour was deliberately provocative and produced, from the University authorities, exactly the response which might have been expected: he was expelled. Timothy Shelley was dismayed, more particularly since he had received a letter from his son denying he would do any of these things. There was a painful meeting between the two in a London hotel, the father begging the son to give up his ideas, at least until he was older; the son insisting they were more precious to him than the peace of mind of his family, the father 'scolding, crying, swearing, and then weeping again', Shelley laughing aloud, 'with a loud demoniacal burst of laughter'; he 'slipped from his seat and fell on his back at full length on the floor'.[17] There ensued negotiations by Shelley to get from his father a guaranteed allowance of £200 a year, followed by the bombshell (in August 1811) that he had married Harriet Westbrook, a sixteen-year-old schoolfellow of his sister Elizabeth.

Thereafter his relations with his family collapsed. Shelley tried to recruit first his mother, then his sisters to his side of the argument, but failed. In a letter to a friend he denounced his entire family as 'a parcel of cold, selfish and calculating animals who seem to have no other aim or business on earth but to eat, drink and sleep'.[18] His letters to various members of his family make extraordinary reading: at times artful and humbugging, in his attempts to extract money, at other times cruel, violent and threatening. His letters to his father escalate from hypocritical pleading to abuse, mixed with insufferable condescension. Thus, on 30 August 1811, he begs: 'I know of no one to whom I can apply with greater certainty of success when in distress than you . . . you are kind to forgive youthful errors.' By 12 October he is contemptuous: 'The institutions of society have made you, tho liable to be misled by prejudice and passion like others, the *Head of the Family*, and I confess it is almost natural for minds not of the highest order to value even the errors whence they derive their importance.' Three days later he accuses Timothy of 'a cowardly, base, contemptible expedient of persecution . . . You have treated me *ill*, *vilely*. When I was expelled for atheism, you wished I had been killed in Spain. The desire of the consummation is very like the crime; perhaps it is well for me that the laws of England punish murder, & that *cowardice* shrinks from their animadversion. I shall take the first opportunity of seeing you – if *you* will not hear my name, *I* shall pronounce it. Think not I am an insect whom injuries destroy – had I money enough I could meet you in London and hollah in your ears Bysshe, Bysshe, Bysshe – aye, Bysshe till you're deaf.' This was unsigned.[19]

To his mother he was still more ferocious. His sister Elizabeth had become engaged to his friend Edward Fergus Graham. His mother approved of the match but Shelley did not. On 22 October he wrote to his mother accusing her of having an affair with Graham and arranging Elizabeth's marriage to cover it up.[20] There seems to have been no factual basis at all for this terrible letter. But he wrote to Elizabeth the same day telling her of the letter and demanding it be shown to his father. To other correspondents he refers to his mother's 'baseness' and depravity'.[21] As a result the family solicitor, William Whitton, was brought in and instructed to open and deal with all letters Shelley sent to the family. He was a kindly man, anxious to make peace between father and son, but ended by being totally alienated by Shelley's arrogance. When he complained that Shelley's letter to his mother was 'not proper' (a mild term in the circumstance), his own letter was returned, scrawled over: 'William Whitton's letter is conceived in terms which justify Mr P.Shelley's returning it for his cool perusal. Mr S. commends Mr W. when he deals with gentlemen (which opportunity perhaps may not often occur) to refrain from opening private letters or impudence may draw down chastisement on contemptibility.'[22]

The family seems to have feared Shelley's violence. 'Had he stayed in Sussex,' Timothy Shelley wrote to Whitton, 'I would have sworn in Especial Constables around me. He frightened his mother and sisters exceedingly and now if they hear a dog bark they run up the stairs. He has nothing to say but the £200 a year.' There was the further fear that Shelley, who was now leading a wandering, bohemian life, would induce one or more of his younger sisters to join him. In a letter dated 13 December 1811 he tried to induce the huntsman at Field Place to smuggle a letter to Helen ('remember, Allen, that I shall *not* forget you') and the letter itself – Helen was only twelve – is distinctly sinister, enough to chill the hearts of a father and mother.[23] He was also anxious to get at his still younger sister Mary. Shelley was soon a member of Godwin's circle, and associated with his emancipated daughter Mary, whose mother was the feminist leader Mary Wollstonecraft, and her still wilder half-sister, Claire Clairmont. Throughout his adult life Shelley persistently sought to surround himself with young women, living in common and shared by whatever men belonged to the circle – at any rate in theory. His sisters appeared to him natural candidates for such a *ménage*, particularly as he conceived it his moral duty to help them 'escape' from the odious materialism of the parental home. He had a plan to kidnap Elizabeth and Helen from their boarding school in Hackney: Mary and Claire were sent to case the joint.[24] Fortunately nothing came of it. But Shelley would not have drawn the line at incest. He, like Byron, was fascinated by the subject. He did not go as far as Byron, who was in love with his half-sister, Augusta Leigh; but Laon and Cythna, hero and heroine of his long poem *The Revolt of Islam*, were brother and sister until the printers objected and forced Shelley to make changes, as were Selim and Zuleika in Byron's *The Bride of Abydos*.[25] Shelley, like Byron, always considered that he had a perpetual dispensation from the normal rules of sexual behaviour.

This made life difficult for the women he associated with. There is no evidence that any of them, with the possible exception of Claire Clairmont, liked the idea of sharing or had the smallest inclination to promiscuity in any form. To Shelley's displeasure they all (like his own family) wanted a normal life. But the poet was incapable of leading one. He thrived on change, displacement, danger and excitement of all kinds. Instability, anxiety seem to have been necessary for his work. He could curl up with a book or a piece of paper anywhere and pour out his verses. He spent his life in furnished rooms or houses, moving about, often dunned by creditors, or the still centre of the anguished personal dramas which beat around him. But he continued to work and produce. His reading was prodigious. His output was considerable and most of it of high quality. But the *mouvementé* existence he found stimulating was disastrous for others, not least his young wife Harriet.

Harriet was a pretty, neat, highly conventional middle-class girl, the daughter of a successful merchant. She fell for the god-like poet, lost her head and eloped with him. Thereafter her life moved inexorably to disaster.[26] For four years she shared Shelley's insecure existence, moving to London, Edinburgh, York, Keswick, North Wales, Lynmouth, Wales again, Dublin, London and the Thames Valley. In some of these places Shelley engaged in unlawful political activities, attracting the attention of local magistrates and police, or even central government; in all he fell foul of tradesmen, who expected their bills to be paid. He also antagonized the neighbours, who were alarmed by his dangerous chemical experiments and affronted by what they saw as the disgusting improprieties of his ménage, which nearly always contained two or more young women. On two occasions, in the Lake District and in Wales, his house was attacked by the local community, and he was forced to decamp. He also fled before his creditors and police.

Harriet did her best to share his activities. She helped to distribute his illegal political leaflets. She was delighted when his first long poem, *Queen Mab*, was dedicated to her. She bore him a daughter, Eliza Ianthe. She conceived another child, his son Charles. But she lacked the capacity to fascinate him for ever. So did every other woman. Shelley's love was deep, sincere, passionate, indeed everlasting – but it was always changing its object. In July 1814 he broke the news to Harriet that he had fallen in love with Godwin's daughter Mary, and was off with her to the Continent (with Claire Clairmont tagging along). The news came to her as an appalling shock, a reaction which surprised, then affronted Shelley. He was one of those sublime egoists, with a strong moralizing bent, who assume that others have a duty not only to fit in with but to applaud his decisions, and when they fail to do so quickly display a sense of outrage.

Shelley's letters to Harriet after he left her follow the same pattern as those to his father, condescension turning to self-righteous anger when she failed to see things his way. 'It is no reproach to me,' he wrote to her on 14 July 1814, 'that you have never filled my heart with an all-sufficing passion.' He had always behaved generously to her and remained her best friend. Next month he invited her to join himself, Mary and Claire in Troyes, 'where you will at least find one firm and constant friend, to whom your interests will always be dear, by whom your feelings will never wilfully be injured. From none can you expect this but me. All else are either unfeeling or selfish.' A month later, finding this tactic did not work, he became more aggressive: 'I deem myself far worthier and better than any of your nominal friends ... my chief study has been to overwhelm you with benefits. Even now, when a violent and lasting passion for another leads me to prefer her society to yours, I am perpetually employed in devising how I can be permanently & truly useful to you ... in return for this it is not well that I should be wounded with reproach & blame – so unexampled and singular an attachment demands a return far different.' The next day he was at it again: 'Consider how far you would desire your future life to be placed within the influence of my superintending mind, whether you still confide sufficiently in my tried and unalterable integrity to submit to the laws which any friendship would create between us.'[27]

These letters were written partly to extract money from Harriet (at this stage she still had some), partly to put pressure on her to conceal his whereabouts from creditors and enemies, partly to stop her from consulting lawyers. They are dotted with references to 'my personal safety' and 'my safety and comfort'. Shelley was an exceptionally thin-skinned person who seems to have been totally insensitive to the feelings of others (a not uncommon combination). When he discovered that Harriet had finally taken legal advice about her rights, his anger exploded. 'In this proceeding, if it be indeed true that your perversity has reached this excess, you destroy your own designs. The memory of our former kindness, the hope that you might still not be utterly lost to virtue & generosity, would influence me, even now, to concede far more than the law will allow. If after you receive this letter you persist in appealing to the law, it is obvious that I can no longer consider you but as an enemy, as one who ... has acted the part of the basest and blackest treachery.' He adds: 'I was an idiot to expect greatness or generosity from you,' and accuses her of 'mean and despicable selfishness' and of seeking 'to injure an innocent man struggling with distress'.[28] By now his self-deception was complete and he had convinced himself that, from start to finish, he had behaved impeccably and Harriet unforgivably. 'I am deeply persuaded,' he wrote to his friend Hogg, 'that thus enabled I shall become a more constant friend, a more useful lover of mankind, [and] a more ardent asserter of truth and virtue.'[29]

It was one of Shelley's many childish characteristics that he was capable of mingling the most hurtful abuse with requests for favours. Thus he followed the letter to his mother accusing her of adultery with another asking her to send onto him 'my Galvanic Machine & Solar Microscope'; his abuse of Harriet was interspersed with pleas not just for money but for clothes: 'I am in want of stockings, hanks & Mrs Wollstonecraft's *Posthumous Works*.' He told her that, without money, 'I must inevitably be starved to death ... My dear Harriet, send quick supplies.'[30] He did not inquire about her condition, though he knew she was pregnant by him. Then, abruptly, the letters ceased. Harriet wrote to a friend: 'Mr Shelley has become profligate and sensual, owing entirely to Godwin's *Political Justice* ... Next month I shall be confined. He will not be near me. No, he cares not for me now. He never asks after me or sends me word how he is going on. In short, the man I once loved is dead. This is a vampire.'[31]

Shelley's son, whom Harriet called Charles Bysshe, was born 30 November 1814. It is not clear whether the father ever saw him. Harriet's elder sister Eliza, who remained devoted to her – and therefore came to be regarded by Shelley with bitter enmity – was determined that the children should not be brought up by Shelley's bohemian women. He, unlike Rousseau, did not regard his children as 'an inconvenience' and fought hard to get them. But, inevitably, the legal battle went against him and they were made Wards of Chancery; thereafter he lost interest. Harriet's life was wrecked. In September 1816 she left the children with her parents and took lodgings in Chelsea. Her last letter was written to her sister: 'The remembrance of all your kindness, which I have so unworthily repaid has often made my heart ache. I know that you will forgive me, because it is not in your nature to be unkind or severe to any.'[32] On 9 November she disappeared. On 10 December her body was discovered in the Serpentine, Hyde Park. The body was swollen and she was said to have been pregnant, but there is no convincing evidence of this.[33] The reaction to the news by Shelley, who had long since circulated the falsehood that he and Harriet had separated by mutual agreement, was to abuse Harriet's family and produce a tissue of lies: 'It seems,' he wrote to Mary, 'that this poor woman – the most innocent of her abhorred & unnatural family – was driven from her father's house, & descended the steps of prostitution until she lived with a groom of the name of Smith, who deserting her, she killed herself. There can be no question that the beastly viper her sister, unable to gain profit from her connection with me, has secured to herself the fortune of the old man – who is now dying – by the murder of this poor creature! ... everyone does *me* full justice – bears testimony to the uprightness and liberality of my conduct to her.'[34] He followed this, two days later, by a peculiarly heartless letter to the sister.[35]

Shelley's hysterical lies may be partly explained by the fact that he was still unnerved by another suicide for which he was responsible. Fanny Imlay was Godwin's stepdaughter by an earlier marriage of his second wife. She was four years older than Mary and described (by Harriet) as 'very plain and very sensible'. Shelley made a play for her as early as December 1812, writing to her: 'I am one of those formidable and long-clawed animals called a *Man*, and it is not until I have assured you that I am one of the most inoffensive of my species, that I live on vegetable food, & never bit since I was born, that I venture to intrude myself upon your attention.'[36] She may have once featured in Shelley's plan to set up a radical community of friends, with sexual sharing – himself, Mary, Claire, Hogg, Peacock and Claire's brother Charles Clairmont. At all events Shelley dazzled her, and Godwin and his wife believed she had fallen tragically in love with him. Between 10–14 September 1814 Shelley was alone in London, Mary and Claire being in Bath, and Fanny visited him at his lodgings in the evening. The likelihood is that he seduced her there. He then went on to Bath. On 9 October the three of them got a very depressed letter from Fanny, postmarked Bristol. Shelley immediately set off to find her, but failed. She had, in fact, already left for Swansea and the next day took an overdose of opium there, in a room at the Mackworth Arms. Shelley never referred to her in his letters; but in 1815 there is a reference to her in a poem ('Her voice did quiver as we parted') which portrays himself ('A youth with hoary hair and haggard eye') sitting near her grave. But it was just an idea; he never visited her grave, which remained unmarked.[37]

There were other sacrifices on the altar of Shelley's ideas. One was Elizabeth Hitchener, a young working-class Sussex woman, the daughter of a smuggler-turned-innkeeper, who by prodigies of effort and sacrifice had become a schoolmistress in Hurstpierpoint. She was known for her radical ideas and Shelley got into correspondence with her. In 1812 Shelley was in Dublin preaching liberty to the Irish, who were unresponsive. Left with a good deal of his subversive material on his hands, he had the bright idea of sending it to Miss Hitchener for distribution in Sussex. He packed it in a large wooden box but, characteristically, paid for it only as far as Holyhead, assuming it would be forwarded and Miss Hitchener would pay for it on arrival. But of course it was opened at the port of entry, the Home Office informed, and a watch set on the schoolmistress. This effectively destroyed her career. But she still had her honour. However, Shelley now invited her to join his little community and, much against the advice of her father and friends, she agreed. He also persuaded her to loan him £100, presumably her life savings.

At this stage he was loud in her praises: 'though deriving her birth from a very humble source she contracted during youth a very deep and refined habit of thinking; her mind, naturally inquisitive and pene-trating, overleapt the bounds of prejudice.'[38] In letters to her he called her 'my rock' in his storm, 'my better genius, the judge of my reasonings, the guide of my actions, the influencer of my usefulness'. She was 'one of those beings who carry happiness, reform, liberty wherever they go'.[39] She joined the Shelleys in Lynmouth, where it was reported, 'She laughs and talks and writes all day,' and distributed Shelley's leaflets. But Har-riet and her sister soon grew to dislike her. Shelley himself was not averse to a certain competitive tension among his women. But in this case he soon shared their disapproval. He seems to have seduced Miss Hitchener during long walks on the shore, but later felt revulsion. When Harriet and Eliza turned on her, he decided she must go. In any case he had by now made contact with the Godwin household, whose young ladies he found more exciting. So she was sent back to Sussex, to carry on the cause there, with promise of a salary of £2 a week. But she was treated there as a laughing-stock, the discarded mistress of a gent. Shelley wrote sneeringly to Hogg: 'The Brown Demon, as I call our late tormentor and schoolmistress, must receive our stipend. I pay it with a heavy heart; but it must be so. She was deprived by our misjudging haste of a situation where she was going on smoothly; and now she says that her reputation is gone, her health ruined, her peace of mind destroyed by my barbarity: a complete victim of all the woes mental and bodily that heroine ever suffered!' He then could not resist adding: 'She is an artful, superficial, ugly, hermaphroditical beast of a woman.' In fact she received only the first instalment of her wages, and her £100 loan was never repaid. Thus she retreated into the obscurity from which Shelley plucked her, a scorched victim of his flame.

A similar, even humbler, case was that of Dan Healey, a fifteen-year-old lad Shelley brought back from Ireland as a servant. We hear little of the Shelley servants, though there were usually three or four of them. In a letter to Godwin, Shelley defended his leisured existence on the grounds that 'if I was employed at the loom or the plough, & my wife in culinary business and housewifery, we should in the present state of society quickly become very different beings, & I may add, less useful to our species.'[40] So servants there had to be, whether or not Shelley could afford them. He usually employed locals at very low wages, but Dan was different because Shelley had found him useful in Dublin at sticking up unlawful posters. At Lynmouth, in the summer of 1812, he again used him to post his broadsheets on walls and barns. Dan was told that, if questioned by the authorities, he was to tell a story about 'meeting two gentlemen on the road'. On 18 August he was arrested in Barnstaple, and told his story. This did him no good at all. He was convicted under the Act of 39 George III c79 and sentenced to fines totalling £200 or, in default, six months' imprisonment. Instead of paying the fine, as everyone (including the authorities) expected, Shelley ran for it, borrowing 29 shillings from his cleaning woman and £3 from a neighbour for his getaway money.* So Dan went to gaol. On his release he returned to Shelley's service but was sacked six months later, the formal reason being that his conduct had become 'unprincipled' – he may have learned bad habits in gaol – the real one being that the Shelleys had to economize. He was owed £10 in wages, which were never paid.[41] Thus another bruised victim receded into the darkness.

It must be argued on Shelley's behalf that he was very young when all these things happened. In 1812 he was only twenty. He was twenty-two when he deserted Harriet and ran off with Mary. We often forget how young this generation of English poets were when they transformed the literature of the English-speaking world; how young indeed when they died – Keats twenty-five, Shelley twenty-nine, Byron thirty-six. When Byron, having fled England for good, first met Shelley on the shores of Lake Geneva on 10 May 1816, he was still only twenty-eight; Shelley was twenty-four; Mary and Claire a mere eighteen. Mary's novel *Frankenstein*, which she wrote by the Lake during the long early summer nights was, you might say, the work of a schoolgirl. Yet if they were in a sense children, they were also adults rejecting the world's values and presenting alternative systems of their own, rather like the students of the 1960s. They did not think of themselves as too young for responsibilities or demand the indulgence due to youth – quite the contrary. Shelley in particular insisted on the high seriousness of his mission to the world. Intellectually he matured very quickly. The immensely powerful poem *Queen Mab*, though still youthful in some ways, was written when Shelley was twenty and published the following year. From 1815–

---

* Byron's efforts on behalf of *his* servants who got into trouble make a notable contrast to Shelley's indifference. He instantly paid the fine levied on his bearded factotum, G-B ('Tita') Falcieri.

16 onwards, when he was moving towards his mid-twenties, his work was approaching its zenith. By this stage it showed not only remarkable breadth of reading but great profundity of thought. There is no doubt that Shelley had a strong mind, which was also subtle and sensitive. And, young as he was, he had accepted the duties of parentage.

Let us look now at his children. Altogether he had seven, by three different mothers. The first two, Ianthe and Charles, were born to Harriet and made wards of court. Shelley contested this move bitterly, losing it in part because the court was horrified by some of the views he had put forward in *Queen Mab*, and he saw the action as primarily an ideological attempt to get him to recant his revolutionary aims.[42] After the decision went against him he continued to brood on the injustice and to hate Lord Chancellor Eldon, but he showed no further concern for the children. He was obliged by the court's decision to pay £30 a quarter for the children, who were lodged with foster-parents. This was deducted at source from his allowance. He never made any use of such visiting rights as the court allowed him. He never wrote to them, though the elder, Ianthe, was nine at the time of his death. He did not inquire about their welfare, except formally, and the only letter we have to the foster-father, Thomas Hume, dated 17 February 1820, is essentially about his own wrongs and is a heartless document.[43] There are no other references to these children in any other letters or diaries of his which survive. He seems to have exiled them from his mind, though they make a ghostly appearance in his autobiographical poem *Epipsychidion* (which dismisses Harriet as 'the planet of that hour'):

> Marked like twin babes, a sister and a brother,
> The wandering hopes of one abandoned mother

By Mary he had four children, three of whom died, his son Percy Florence, born in 1819, alone surviving to carry on the line. Mary's first child, a girl, died in early infancy. Their son William, aged four, caught gastro-enteritis in Rome; Shelley sat up with him for three consecutive nights, but the child died. Shelley's exertions may have been prompted partly by guilt at his role in the death of his daughter Clara, who was still a baby, the previous year. In August 1818 Mary and the baby were in the comparative cool of the summer resort, Bagni di Lucca. Shelley, who was at Este in the hills above Venice, insisted that Mary and the baby join him immediately, a fearful five-day journey in the hottest season of the year. Shelley did not know that little Clara was unwell even before the journey began; but she was clearly ill on her arrival, and her condition did not improve. Nevertheless, three weeks later, and again entirely to suit his own convenience – he was intoxicated by his exchanges of radical views with Byron – he sent peremptory instructions to Mary to join him with the baby in Venice. Poor Clara, according to her mother, was 'in a frightful state of weakness and fever', and the journey lasted from 3.30 in the morning until five in the afternoon on a broiling day. By the time they reached Padua, Clara was obviously very ill; Shelley insisted they continue to Venice. On the journey, Clara developed 'convulsive motions of the mouth and eyes'; she died an hour after Venice was reached.[44] Shelley admitted that 'this unexpected stroke' (it was surely foreseeable) had reduced Mary 'to a kind of despair'; it was an important stage in the deterioration of their relationship.

A further stage was reached that winter when an illegitimate child, a girl baptized Elena, was born to Shelley in Naples. He registered the child as his own and gave the mother's name as Mary Godwin Shelley. But his wife was certainly not the mother: shortly afterwards, Shelley began to be blackmailed by a former servant, Paolo Foggi, who had married their children's nurse, Elise, and the grounds for his threats were that Shelley had made a criminally false declaration by naming Mary as the mother. It is possible that Elise was the mother. But there are a number of powerful arguments against this. Elise herself had a different story. In 1820 she told Richard Hoppner, the British Consul in Venice, who had hitherto held a high opinion of Shelley despite his reputation, that the poet had deposited in the Naples Foundling Hospital a baby girl he had had by Claire Clairmont. Hoppner was disgusted by Shelley's behaviour and when he confided in Byron the latter replied: 'Of the facts however there can be little doubt – it is just like them.'[45] He knew all about Shelley and Claire Clairmont. She was the mother of his own illegitimate daughter Allegra. She had set her cap at him in the spring of 1816 before he left England. Byron, who had some scruples about seducing a virgin, had slept with her only after she had told him she had already slept with Shelley.[46] He had a very low opinion of Claire's morals since she had not only, in effect, seduced him but offered to procure Mary Shelley for him too;[47] that was one reason he would not allow her to bring up Allegra, though separating her from her mother proved fatal to the child. Byron was satisfied Allegra was his own and not Shelley's since he was sure she was not having sex with Shelley at that time. But he evidently believed they had since resumed their intermittent affair, when Mary was away. Elena was the result. Various other explanations have been produced by Shelley apologists but the Claire–Shelley parentage is by far the most likely.[48] Mary was devastated by the episode – she had never liked Claire and resented her continued presence in their household. If the baby remained with them Claire would become a permanent member of it, and possibly her affair with Shelley would be resumed. In response to Mary's distress, Shelley decided to abandon the baby and, following the example of his hero Rousseau, made use of the orphanage. There, not surprisingly, the child died, aged eighteen months, in 1820. The next year, shrugging off the criticism of Hoppner and others, Shelley in a letter to Mary summed up the business in one hard-boiled and revealing sentence: 'I speedily regained the indifference which the opinions of any thing or any body but our own consciousness amply merits.'[49]

Was Shelley, then, promiscuous? Certainly not in the same sense as Byron, who claimed, in September 1818, that in two and a half years he had spent over £2500 on Venetian women and slept with 'at least two hundred of one sort or another – perhaps more'; and later listed twenty-four of his mistresses by name.[50] On the other hand Byron in some ways had a finer sense of honour than Shelley; he was never sly or humbugging. To the sexual reformer and feminist J.H.Lawrence, Shelley wrote: 'If there is any enormous and desolating crime, of which I should shudder to be accused, it is seduction.'[51] This was his theory; but not his practice. In addition to the cases already mentioned, there was also a love affair with a well-born Italian woman, Emilia Viviani; he told Byron all about her but added: 'Pray do not mention anything of what I told you, as the whole truth is not known and Mary might be very much annoyed by it.'[52] What Shelley seems to have desired was a woman to provide his life with stability and comfort, and a license to pursue side-affairs; in return he would (at any rate in principle) allow his wife the same liberty. Such an arrangement, as we shall see, was to become a recurrent aim of leading male intellectuals. It never worked, certainly not in Shelley's case. The liberty he took himself caused first Harriet, then Mary, anguish; and they simply did not want the reciprocal freedom.

Evidently Shelley often discussed these ideas with his radical friend Leigh Hunt. The painter and diarist Benjamin Robert Haydon records that he had heard Shelley 'hold forth to Mrs Hunt & other women present . . . on the wickedness & absurdity of *Chastity*'. During the discussion, Hunt shocked Haydon by saying 'he would not mind any young man, if he were agreeable, sleeping with his wife.' Haydon added: 'Shelley courageously adopted and acted on his own principles – Hunt defended them without having energy to practise them and was content with a smuggering fondle.'[53] What the women thought is not recorded. When Shelley told Harriet she could sleep with his friend Hogg, she flatly refused. When he offered the same facility to Mary, she appeared to assent but finally decided she did not like the man.[54] The surviving evidence shows that Shelley's own experiments in free love were just as furtive and dishonest as those of most ordinary adulterers and involved him in the usual tangle of concealment and lies.

It was the same story with his money-dealings. They were immensely complicated and harrowing, and I can attempt only the barest summary here. In theory Shelley did not believe in private property at all, let alone inheritance and the primogeniture from which he benefited. In *A Philosophical View of Reform*, he set down his socialist principles: 'Equality of possessions must be the last result of the utmost refinements of civilization; it is one of the conditions of that system of society towards which, with whatever hope of ultimate success, it is our duty to tend.'[55] But in the meantime it was necessary for privileged but enlightened men, like himself, to hang onto their inherited wealth in order to further the cause. This was to become a familiar, indeed almost universal, self-justification among wealthy radical intellectuals, and Shelley employed it to extract as much money from his family as he possibly could. Unfortunately for him, in his very first letter to his mentor Godwin, introducing himself, he proudly announced: 'I am the son of a man of fortune in Sussex. . . . I am heir by entail to an estate of £6000 per annum.'[56] This must have made Godwin prick up his ears. He was not only the leading radical philosopher but a financial muddler of genius and one of the most shameless financial scroungers who ever lived. Truly staggering amounts of money, from a variety of well-meaning friends, disappeared into his labyrinthine system of debts, leaving nothing to show. He seized upon the then young and innocent Shelley and never let go. He not only took Shelley's family money but corrupted him thoroughly into all the seedy devices of an early nineteenth-century debtor: post-dated bonds, discounted paper and, not least, the notorious *post-obit* borrowings whereby young heirs to an entailed estate could raise heavily discounted sums, at enormous rates of interest, in anticipation of their father's death. Shelley adopted all these ruinous procedures and a very large percentage of what he thus raised went straight into Godwin's financial black hole.[57]

Not a penny was ever repaid, or seems to have done Godwin's needy family the slightest good. At long last Shelley turned on the sponger. 'I have given you,' he wrote, 'within a few years the amount of a considerable fortune, & have destituted myself for the purpose of realizing it of nearly four times the amount. Except for the *good will* which this transaction seems to have produced between you and me, this money, for any *advantage* that it ever conferred on you, might as well have been thrown into the sea.' The loss of money was not the only harm Shelley suffered from his contact with Godwin. Harriet was quite right in thinking the great philosopher had coarsened and hardened her husband in a number of ways, especially in his attitude to money. She related that when Shelley, who had already deserted her for Mary, came to see her after the birth of her son William, 'he said he was glad it was a boy, as it would make money cheaper.'[58] By this he meant he could get a lower interest rate on a *post-obit* loan; it was not the remark of a twenty-two-year-old poet-idealist but of a shifty, chronic debtor.

Godwin was not the only bloodsucker in Shelley's life. There was another perpetually cadging intellectual, Leigh Hunt. A quarter-century later, Thomas Babington Macaulay summed up Hunt to the editor of the *Edinburgh Review*, Napier, by saying that he had answered a letter of Hunt's, 'not without fear of becoming one of the numerous persons of whom he asks £20 whenever he wants it'.[59] Eventually he was immortalized as Harold Skimpole in *Bleak House*, Dickens confessing to a friend, 'I suppose he is the most exact portrait that was ever painted in words ... It is an absolute reproduction of the real man.'[60] In Shelley's time Hunt was only beginning his long career of borrowing, using Rousseau's well-tried technique of persuading his victims that he was doing them a favour by taking advantage of their generosity. When Shelley died Hunt moved on to Byron, who eventually gave him his quittance in no uncertain terms; he considered Hunt had plundered Shelley. Alas, he did something worse, persuading Shelley that to men of advanced ideas like themselves, paying what you owed was not a moral necessity: to work for humanity was enough in itself.

Thus Shelley, the man of truth and virtue, became a lifelong absconder and cheat. He borrowed money everywhere and from all kinds of people, most of whom were never repaid. Whenever the Shelleys moved on, usually in some haste, they left behind little groups of once-trusting and now angry people. Young Dan Healey was not the only Irishman Shelley defrauded. He evidently borrowed a substantial sum from John Lawless, the republican editor who had befriended him in Dublin. He could not afford to lose the money and, after Shelley's departure, wrote anxiously to Hogg inquiring his whereabouts. Shortly afterwards he was arrested for debt. Shelley not only made no attempt to get him out of jail by paying the money he owed, but abused him for complaining: 'I am afraid,' he wrote to a mutual friend in Dublin, Catherine Nugent, 'that he has practised on you, as he did upon us.'[61] Worse, in Lynmouth, Shelley signed bills using his name ('the Honourable Mr Lawless'); this was forgery, then a capital offence.[62]

Another group of people Shelley defrauded were the Welsh, during his sojourn there. He arrived in 1812, leasing a farm and engaging servants ('can you hire a trustworthy undermaidservant, as we shall require three in all'), but was soon arrested for debts of £60 to £70 in Caernarvon.

John Williams, who sponsored his Welsh venture, and Dr William Roberts, a country doctor, stood bail for him, and the debt, plus costs, was paid by a London solicitor, John Bedwell. All three came to regret their generosity. More than thirty years later, in 1844, Dr Roberts was still trying to retrieve, from the Shelley estate, the £30 the poet owed him. Bedwell likewise demanded his money in vain, Shelley writing to Williams a year later: 'I have received a very unpleasant and dictatorial letter from Mr Bedwell, which I have answered in an unbending spirit.' Shelley liked to take a high tone. Williams's brother Owen, a farmer, had lent Shelley £100; we find Shelley writing to Williams demanding that Owen produce a further £25, adding, 'I shall know by your compliance with this request whether the absence of friends is a cooler for friendship or not.' Shelley's relations with Williams collapsed the following year in a welter of recriminations over the money the poet owed him. Neither Williams nor Owen was ever repaid anything. Yet Shelley was fierce and moralistic with anyone (Godwin and Hunt excepted) who owed *him* money. Another Welshman, John Evans, got two dunning notes, Shelley reminding him he was owed cash 'which being a debt of honour ought to be of all others the most imperious, & to press the necessity of its immediate repayment, to lament also the apathy and backwardness of defaulters in such a case'.[63]

It was not clear what Shelley meant by debts of honour. He did not scruple to borrow from women, ranging from washerwomen and charladies, and his landlady in Lynmouth – she eventually got back £20 of the £30 he owed her, having wisely hung onto his books – to his Italian friend Emilia, from whom he got 220 crowns. He owed money to tradesmen of every kind. In April 1817, for instance, he and Hunt agreed to pay one Joseph Kirkman for a piano; it was duly delivered but four years later it was unpaid-for. Likewise Shelley got Charter, the famous Bond Street coachmaker, to make him a fine vehicle, costing £532.11.6, which he continued to use until his death. Charter eventually took the poet to court, but he was still trying to collect the money in the 1840s. A particular group Shelley exploited were the small printer-booksellers who published his poems on credit. This began with £20 borrowed from Slatter, the Oxford bookseller, lent when Shelley was sent down. Slatter evidently liked him and was anxious to save him from going to rapacious moneylenders; as a result Shelley involved him in an appallingly expensive mess. In 1831, Slatter's brother, a plumber, wrote to Sir Timothy: 'having suffered very much in consequence of an honest endeavour to save your son from going to Jews for the purpose of obtaining money at an enormous rate of interest . . . we have lost upwards of £1300.' They were eventually arrested for debt, and never seem to have been paid. The Weybridge printer who published *Alastor* was still trying to get Shelley to pay for it four and a half years later; there is no evidence he was ever reimbursed. To a third bookseller Shelley wrote (December 1814): 'If you would furnish me with books, I would grant you a *post-obit* bond in the proportion of £250 for every £100 worth of books provided.' He told him his father's and grandfather's ages were sixty-three and eighty-five, when they were in fact sixty-one and eighty-three. A fourth bookseller-publisher, Thomas Hookham, not only printed *Queen Mab* on credit but advanced Shelley money. He too remained unpaid and, for the crime of sympathizing with Harriet, became an object of hatred, Shelley writing to Mary on 25 October 1814: 'If you see Hookham, do not insult him openly. I have still hopes ... I will make this remorseless villain loathe his own flesh – in good time. He shall be cut down in his season. His pride shall be trampled into atoms. I will wither up his selfish soul by piecemeal.'[64]

What is the common denominator in all this – in Shelley's sexual and financial misdemeanours, in his relations with father and mother, wives and children, friends, business associates and tradesmen? It is, surely, an inability to see any point of view other than his own; in short, lack of imagination. Now this is very curious, for imagination lies at the very heart of his theory of political regeneration. According to Shelley, imagination, or 'Intellectual Beauty', was required to transform the world; and it was because poets possessed this quality in the highest degree, because poetic imagination was the most valuable and creative of all human accomplishments, that he entitled them the world's natural legislators, albeit unacknowledged. Yet here was he, a poet – and one of the greatest of poets – capable, perhaps, of imaginative sympathy with entire classes, downtrodden agricultural labourers, Luddites, Peterloo rioters, factory hands, people he had never set eyes on; capable of feeling for, in the abstract, the whole of suffering humanity, yet finding it manifestly impossible, not once but scores, hundreds of times, to penetrate imaginatively the minds and hearts of all those people with whom he had daily dealings. From booksellers to baronets, from maidservants to mistresses, he simply could not see that they were entitled to a viewpoint which differed from his own; and, confronted with their (to him) intransigence, he fell back on abuse. A letter he wrote to John Williams, on 21 March 1813, perfectly encapsulates Shelley's imaginative limitations. It begins with a verbal assault on the unfortunate Bedwell; it continues with a savage attack on the still more unhappy Miss Hitchener ('a woman of desperate views & dreadful passions but of cool & undeviating revenge ... I laughed heartily at her day of tribulation'); it concludes with a pledge to humanity – 'I am ready to do anything for my country and my friends that will serve them'; and it signs off '& among the rest for you, whose affectionate friend I continue to remain'. This was the very Williams he was in the process of cheating and who would shortly become another embittered debtor.[65]

Shelley devoted his life to political progress, using the marvellous poetic gift bestowed on him, without ever becoming aware of this imaginative disqualification. Nor did he make it good by an attempt to discover the facts about the categories of mankind he wished to aid. He wrote his *An Address to the Irish People* before he even set foot in the country. When he got there he made no systematic effort to investigate conditions or find out what the Irish themselves actually wanted.[66] Indeed he planned secretly to destroy their cherished religion. Shelley likewise remained profoundly ignorant of English politics and public opinion, of the desperate nature of the problems confronting the government in the post-Waterloo period, and of the sincerity of the efforts to solve them. He never tried to inform himself or to do justice to well-meaning and sensitive men like Castlereagh and Sir Robert Peel by precisely that kind of imaginative penetration he said was so essential. Instead, he abused them, in *The Mask of Anarchy*, just as he abused his creditors and discarded women in his letters.

Shelley clearly wanted a total political transformation of society, including the destruction of organized religion. But he was confused about how to get there. At times he preached non-violence, and there are those who see him as the first real evangelist of non-violent resistance, a progenitor of Gandhi.[67] 'Have nothing to do with force or violence,' he wrote in his *Irish Address*; 'Associations for the purposes of violence are entitled to the strongest disapprobation of the real reformist! ... All secret associations are also bad.' But Shelley sometimes sought to organize secret bodies and some of his poetry only makes sense as an incitement to direct action. *The Mask of Anarchy* itself is contradictory: one stanza, lines 340–44, supports non-violence. But the most famous stanza, ending 'You are many, they are few,' which is repeated (ll. 151–54, 369–72) is a plea for insurrection.[68] Byron, who was a rebel like Shelley but more a man of action than an intellectual – he did not believe in transforming society at all, merely in self-determination – was very sceptical of Shelley's utopia. In Shelley's fine poem *Julian and Maddalo*, which records their long conversations in Venice, Maddalo [Byron] says of Shelley's political programme, 'I think you might make such a system refutation-tight/As far as words go,' but in practice thought 'such aspiring theories' were 'vain'.

The fact that in this poem, which dates from 1818–19, Shelley acknowledged Byron's criticism marked a pause in his brash political fundamentalism. Shelley approached Byron with great modesty: 'I despair of rivalling Lord Byron, as well I may, and there is no other with whom it is worth contending ... every word is stamped with immortality.' For a time the power of Byron almost paralysed him: 'the sun has extinguished the glowworm,' as he put it. Certainly, getting to know Byron had a maturing effect on Shelley. But, unlike Byron, who began to see his role as the organizer of oppressed peoples – the Italians, then the Greeks – Shelley began to turn against direct action of any kind. It is very significant that, at the end of his life, he was becoming critical of Rousseau, whom he identified with the horrible excesses of the French Revolution. In his unfinished poem, *The Triumph of Life*, Rousseau is presented as a Virgilian narrative figure, a prisoner of Purgatory because he made the mistake of believing the ideal could be realized in life, and was thus corrupted. But it is not at all clear that Shelley was thereby renouncing actual politics to concentrate on the pure idealism of the imagination.[69]

Certainly, in the months before his death there was no sign of any fundamental change in his character. Claire Clairmont, who lived to be over eighty and to become a sensible woman (the inspiration for Henry James's magnetic story, *The Aspern Papers*), wrote sixty years after these events that 'Harriet's suicide had a beneficial effect on Shelley – he became much less confident in himself and not so wild as he had been before.'[70] This may well be true though Claire, at that great distance in time, was telescoping events. Shelley did indeed become less violently self-centred, but the change was gradual and by no means complete at his death. In 1822 both he and Byron had built themselves boats, the *Don Juan* and the *Bolivar*, and Shelley in particular was obsessed with sailing. For this purpose he insisted on taking a house at Lerici on the Bay of Spezia for the summer. Mary, who was pregnant again, came to hate it, chiefly because it was so hot. The two were drifting apart; she was becoming disillusioned and tired of their unnatural life in exile. Moreover, there was a new menace. Shelley's sailing companion was Edward Williams, a half-pay lieutenant of the East India Company. Shelley was showing increasing interest in Williams's beautiful common-law wife, Jane. Jane was musical, played the guitar and sang well (like Claire), something Shelley liked. There were musical parties under the summer moon. Shelley wrote several poems to and about her. Was Mary to be displaced, just as she had once displaced Harriet?

On 16 June, as Mary had feared, she miscarried, and was again plunged into despair. Two days later Shelley wrote a letter which makes it clear their marriage was virtually at an end: 'I only feel the want of those who can feel, and understand me ... Mary does not. The necessity of concealing from her thoughts that would pain her necessitates this, perhaps. It is the curse of Tantalus that a person possessing such excellent powers and so pure a mind as hers should not excite the sympathy indispensable to their application to domestic life.' Shelley adds: 'I like Jane more and more ... She has a taste for music and an elegance of form and motions that compensate in some degree for the lack of literary refinement.'[71] By the end of the month Mary was finding her position, the heat, the house, insupportable: 'I wish,' she wrote, 'I could break my chains and leave this dungeon.'[72]

She got her manumission in tragic and unexpected manner. Shelley had always been fascinated by speed. In a twentieth-century incarnation he might have become devoted to fast cars or even aircraft. One of his poems, *The Witch of Atlas*, is a hymn to the joys of travelling through space. His boat the *Don Juan* was built for speed and Shelley had it modified to travel even faster. It was only twenty-four feet long but it had twin mainmasts and schooner rigging. He and Williams devised a new topsail rig which dramatically increased the area of canvas; to increase the speed still further, Byron's naval architect, at Shelley's request, created a re-rig and false stern and prow. She was now a very fast and dangerous boat and sailed 'like a witch'.[73] At the time of the disaster she could carry three spinnakers and a storm-sail and floated an extra three inches higher in the water. Shelley and Williams were returning in the refitted boat from Livorno to Lerici. They set off on the afternoon of 8 July 1822 in deteriorating weather, under full sail. The local Italian craft all scurried back into harbour when the storm broke at 6.30. The captain of one of them said he had sighted Shelley's ship in immense waves, still fully-rigged; he invited them to come aboard his own or at least to shorten sail, 'or you are lost'. But one of the two (supposedly Shelley) shouted 'No,' and was seen to stop his companion lowering the sails, seizing him by the arm, 'as if in anger'. The *Don Juan* went down ten miles from shore, still under full sail; both were drowned.[74]

Keats had died in Rome of tuberculosis the year before; Byron was bled to death by his doctors two years later in Greece. So a brief, incandescent epoch in English literature came to an end. Mary took little Percy, the future baronet (Charles had died), back to England and began patiently to erect a mythical monument to Shelley's memory. But the scars remained. She had seen the underside of the intellectual life and had felt the power of ideas to hurt. When a friend, watching Percy learning to read, remarked: 'I am sure he will live to be an extraordinary man,' Mary Shelley blazed up: 'I hope to God,' she said passionately, 'he grows up to be an ordinary one.'

3

Karl Marx:
'Howling Gigantic Curses'

*K*ARL MARX has had more impact on actual events, as well as on the minds of men and women, than any other intellectual in modern times. The reason for this is not primarily the attraction of his concepts and methodology, though both have a strong appeal to unrigorous minds, but the fact that his philosophy has been institutionalized in two of the world's largest countries, Russia and China, and their many satellites. In this sense he resembles St Augustine, whose writings were most widely read among church leaders from the fifth to the thirteenth century and therefore played a predominant role in the shaping of medieval Christendom. But the influence of Marx has been even more direct, since the kind of personal dictatorship he envisaged for himself (as we shall see) was actually carried into effect, with incalculable consequences for mankind, by his three most important followers, Lenin, Stalin and Mao Tse-tung, all of whom, in this respect, were faithful Marxists.

Marx was a child of his time, the mid-nineteenth century, and Marxism was a characteristic nineteenth-century philosophy in that it claimed to be scientific. 'Scientific' was Marx's strongest expression of approval, which he habitually used to distinguish himself from his many enemies. He and his work were 'scientific'; they were not. He felt he had found a scientific explanation of human behaviour in history akin to Darwin's theory of evolution. The notion that Marxism is a science, in a way that no other philosophy ever has been or could be, is implanted in the public doctrine of the states his followers founded, so that it colours the teaching of all subjects in their schools and universities. This has spilled over into the non-Marxist world, for intellectuals, especially academics, are fascinated by power, and the identification of Marxism with massive physical authority has tempted many teachers to admit Marxist 'science' to their own disciplines, especially such inexact or quasi-exact subjects as economics, sociology, history and geography. No doubt if Hitler, rather than Stalin, had won the struggle for Central and Eastern Europe in 1941–45, and so imposed his will on a great part of the world, Nazi doctrines which also claimed to be scientific, such as its race-theory, would have been given an academic gloss and penetrated universities throughout the world. But military victory ensured that Marxist, rather than Nazi, science would prevail.

The first thing we must ask about Marx, therefore, is: in what sense, if any, was he a scientist? That is, to what extent was he engaged in the pursuit of objective knowledge by the careful search for and evaluation of evidence? On the face of it, Marx's biography reveals him as primarily a scholar. He was descended on both sides from lines of scholars. His father Heinrich Marx, a lawyer, whose name originally was Hirschel ha-Levi Marx, was the son of a rabbi and Talmudic scholar, descended from the famous Rabbi Elieser ha-Levi of Mainz, whose son Jehuda Minz was head of the Talmudic School of Padua. Marx's mother Henrietta Pressborck was the daughter of a rabbi likewise descended from famous scholars and sages. Marx was born in Trier (then Prussian territory) on 5 May 1818, one of nine children but the only son to survive into middle age; his sisters married respectively an engineer, a bookseller, a lawyer. The family was quintessentially middle-class and rising in the world. The father was a liberal and described as 'a real eighteenth-century Frenchman, who knew his Voltaire and Rousseau inside out'.[1] Following a Prussian decree of 1816 which banned Jews from the higher ranks of law and medicine, he became a Protestant and on 26 August 1824 he had his six children baptised. Marx was confirmed at fifteen and for a time seems to have been a passionate Christian. He attended a former Jesuit high school, then secularized, and Bonn University. From there he went on to Berlin University, then the finest in the world. He never received any Jewish education or attempted to acquire any, or showed any interest in Jewish causes.[2] But it must be said that he developed traits characteristic of a certain type of scholar, especially Talmudic ones: a tendency to accumulate immense masses of half-assimilated materials and to plan encyclopedic works which were never completed; a withering contempt for all non-scholars; and extreme assertiveness and irascibility in dealing with other scholars. Virtually all his work, indeed, has the hallmark of Talmudic study: it is essentially a commentary on, a critique of the work of others in his field.

Marx became a good classical scholar and later specialized in philosophy, in the prevailing Hegelian mode. He took a doctorate, but from Jena University, which had lower standards than Berlin; he never seems to have been quite good enough to get an academic post. In 1842 he became a journalist with the *Rheinische Zeitung* and edited it for five months until it was banned in 1843; thereafter he wrote for the *Deutsch-Französische Jahrbücher* and other journals in Paris until his expulsion in 1845, and then in Brussels. There he became involved in organizing the Communist League and wrote its manifesto in 1848. After the failure of the revolution he was forced to move (1849) and settled in London, this time for good. For a few years, in the 1860s and 1870s, he was again involved in revolutionary politics, running the International Working Men's Association. But most of his time in London, until his death on 14 March 1883 – that is, thirty-four years – was spent in the British Museum, finding material for a gigantic study of capital, and trying to get it into publishable shape. He saw one volume through the press (1867) but the second and third were compiled from his notes by his colleague Friedrich Engels and published after his death.

Marx, then, led a scholar's life. He once complained: 'I am a machine condemned to devour books.'[3] But in a deeper sense he was not really a scholar and not a scientist at all. He was not interested in finding the truth but in proclaiming it. There were three strands in Marx: the poet, the journalist and the moralist. Each was important. Together, and in combination with his enormous will, they made him a formidable writer and seer. But there was nothing scientific about him; indeed, in all that matters he was anti-scientific.

The poet in Marx was much more important than is generally supposed, even though his poetic imagery soon became absorbed in his political vision. He began writing poetry as a boy, around two main themes: his love for the girl next door, Jenny von Westphalen, of Prussian–Scotch descent, whom he married in 1841; and world destruction. He wrote a great deal of poetry, three manuscript volumes of which were sent to Jenny, were inherited by their daughter Laura and vanished after her death in 1911. But copies of forty poems have survived, including a verse tragedy, *Oulanen*, which Marx hoped would be the *Faust* of his time. Two poems were published in the Berlin *Athenaeum*, 23 January 1841. They were entitled 'Savage Songs', and savagery is a characteristic note of his verse, together with intense pessimism about the human condition, hatred, a fascination with corruption and violence, suicide pacts and pacts with the devil. 'We are chained, shattered, empty, frightened/Eternally chained to this marble block of being,' wrote the young Marx, '. . . We are the apes of a cold God.' He has himself, in the person of God, say: 'I shall howl gigantic curses at mankind,' and below the surface of much of his poetry is the notion of a general world-crisis building up.[4] He was fond of quoting Mephistopheles' line from Goethe's *Faust*, 'Everything that exists deserves to perish'; he used it, for instance, in his tract against Napoleon III, 'The Eighteenth Brumaire', and this apocalyptic vision of an immense, impending catastrophe on the existing system remained with him throughout his life: it is there in the poetry, it is the background to the *Communist Manifesto* of 1848, and it is the climax of *Capital* itself.

Marx, in short, is an eschatological writer from start to finish. It is notable, for instance, that in the original draft of *The German Ideology* (1845–46) he included a passage strongly reminiscent of his poems, dealing with 'the Day of Judgment', 'when the reflections of burning cities are seen in the heavens . . . and when the "celestial harmonies" consist of the melodies of the *Marseillaise* and the *Carmagnole*, to the accompaniment of thundering cannon, while the guillotine beats time and the inflamed masses scream *Ça ira, ça ira*, and self-consciousness is hanged on the lamppost'.[5] Then again, there are echoes of *Oulanen* in the *Communist Manifesto*, with the proletariat taking on the hero's mantle.[6] The apocalyptic note of the poems again erupts in his horror-speech of 14 April 1856: 'History is the judge, its executioner the proletariat' – the terror, the houses marked with the red cross, catastrophic metaphors, earthquakes, lava boiling up as the earth's crust cracks.[7] The point is that Marx's concept of a Doomsday, whether in its lurid poetic version or its eventually economic one, is an artistic not a scientific vision. It was always in Marx's mind, and as a political economist he worked backwards from it, seeking the evidence that made it inevitable, rather than forward to it, from objectively examined data. And of course it is the poetic element which gives Marx's historical projection its drama and its fascination to radical readers, who want to believe that the death and judgment of capitalism is coming. The poetic gift manifests itself intermittently in Marx's pages, producing some memorable passages. In the sense that he intuited rather than reasoned or calculated, Marx remained a poet to the end.

But he was also a journalist, in some ways a good one. Marx found planning, let alone writing, a major book not only difficult but impossible: even *Capital* is a series of essays glued together without any real form. But he was well suited to write short, sharp, opinionated reactions to events as they occurred. He believed, as his poetic imagination told him, that society was on the verge of collapse. So almost every big news story could be related to this general principle, giving his journalism a remarkable consistency. In August 1851, a follower of the early socialist

Robert Owen, Charles Anderson Dana, who had become a senior executive on the *New York Daily Tribune*, asked Marx to become the European political correspondent of the paper, writing two articles a week at £1 each. Over the next ten years Marx contributed nearly five hundred articles, of which about one hundred and twenty-five were ghosted for him by Engels. They were heavily subbed and rewritten in New York, but the sinewy arguments are pure Marx and therein lies their power. In fact his greatest gift was as a polemical journalist. He made brilliant use of epigrams and aphorisms. Many of these were not his invention. Marat produced the phrases 'The workers have no country' and 'The proletarians have nothing to lose but their chains.' The famous joke about the bourgeoisie wearing feudal coats-of-arms on their backsides came from Heine, as did 'Religion is the opium of the people.' Louis Blanc provided 'From each according to his abilities, to each according to his needs.' From Karl Schapper came 'Workers of all countries, unite!' and from Blanqui 'the dictatorship of the proletariat'. But Marx was capable of producing his own: 'In politics the Germans have *thought* what other nations have *done*.' 'Religion is only the illusory sun around which man revolves, until he begins to revolve around himself.' 'Bourgeois marriage is the community of wives.' 'The revolutionary daring which hurls at its adversaries the defiant words: "*I am nothing and I must be everything*".' 'The ruling ideas of each age have been the ideas of its ruling class.' Moreover he had a rare gift for pointing up the sayings of others and using them at exactly the right stage in the argument, and in deadly combination. No political writer has ever excelled the last three sentences of the *Manifesto*: 'The workers have nothing to lose but their chains. They have a world to gain. Workers of the world, unite!' It was Marx's journalistic eye for the short, pithy sentence which, more than anything else, saved his entire philosophy from oblivion in the last quarter of the nineteenth century.

But if poetry supplied the vision, and journalistic aphorism the highlights of Marx's work, its ballast was academic jargon. Marx was an academic; or rather, and worse, he was a failed academic. An embittered, would-be don, he wanted to astonish the world by founding a new philosophical school, which was also a plan of action designed to give him power. Hence his ambivalent attitude to Hegel. Marx says in his preface to the second German edition of *Capital*: 'I frankly proclaimed myself a disciple of that great thinker' and 'toyed with the use of Hegelian terminology when discussing the theory of value' in *Capital*. But, he says, his own 'dialectical method' is in 'direct opposition' to Hegel's. For Hegel, the thought-process is the creator of the real, whereas 'in my view, on the other hand, the ideal is nothing more than the material when it has been transposed and translated inside the human head.' Hence, he argues, 'in Hegel's writings, dialectic stands on its head. You must turn it the right way up again if you want to discover the rational kernel that is hidden away within the wrappings of mystification.'[8]

Marx, then, sought academic fame by what he saw as his sensational discovery of the fatal flaw in Hegel's method, which enabled him to replace the entire Hegelian system with a new philosophy; indeed, a super-philosophy which would make all existing philosophies out-moded. But he continued to accept that Hegel's dialectic was 'the key to human understanding', and he not only used it but remained its prisoner till the end of his life. For the dialectic and its 'contradictions' explained the culminating universal crisis which was his original poetic vision as a teenager. As he wrote towards the end of his life (14 January 1873), business cycles express 'the contradictions inherent in capitalist society' and will produce 'the culminating point of these cycles, a univer-sal crisis'. This will 'drum dialectics' into the heads even of 'the upstarts of the new German empire'.

What did any of this have to do with the politics and economics of the real world? Nothing whatever. Just as the origin of Marx's philosophy lay in a poetic vision, so its elaboration was an exercise in academic jargonizing. What it needed, however, to set Marx's intellectual machi-nery in motion was a moral impulse. He found it in his hatred of usury and moneylenders, a passionate feeling directly related (as we shall see) to his own money difficulties. This found expression in Marx's first ser-ious writings, two essays 'On the Jewish Questions' published in 1844 in the *Deutsch-Französische Jahrbücher*. Hegel's followers were all in vary-ing degrees anti-Semitic, and in 1843 Bruno Bauer, the anti-Semitic leader of the Hegelian left, published an essay demanding that the Jews aban-don Judaism completely. Marx's essays were a reply to this. He did not object to Bauer's anti-Semitism; indeed he shared it, endorsed it and quoted it with approval. But he disagreed with Bauer's solution. Marx rejected Bauer's belief that the anti-social nature of the Jew was religious in origin and could be remedied by tearing the Jew away from his faith. In Marx's opinion, the evil was social and economic. He wrote: 'Let us consider the real Jew. Not the *Sabbath Jew* ... but the *everyday Jew*.' What, he asked, was 'the profane basis of Judaism? *Practical* need, *self-interest*. What is the worldly cult of the Jew? *Huckstering*. What is his worldly god? *Money*.' The Jews had gradually spread this 'practical' religion to all society:

Money is the jealous god of Israel, beside which no other god may exist. Money abases all the gods of mankind and changes them into commodities. Money is the self-sufficient *value* of all things. It has, therefore, deprived the whole world, both the human world and Nature, of their own proper value. Money is the alienated essence of man's work and existence: this essence dominates him and he worships it. The god of the Jews has been secularized and has become the god of the world.

The Jew had corrupted the Christian and convinced him 'he has no other destiny here below than to become richer than his neighbours' and that 'the world is a stock-exchange.' Political power had become the 'bondsman' of money power. Hence the solution was economic. The 'money-Jew' had become 'the universal *anti-social* element of the present time' and to 'make the Jew impossible' it was necessary to abolish the 'preconditions', the 'very possibility' of the kind of money activities which produced him. Abolish the Jewish attitude to money and both the Jew and his religion, and the corrupt version of Christianity he had imposed on the world, would disappear: 'In emancipating itself from *hucksterism* and *money*, and thus from real and practical Judaism, our age would emancipate itself.'[9]

Thus far Marx's explanation of what was wrong with the world was a combination of student-café anti-Semitism and Rousseau. He broadened it into his mature philosophy over the next three years, 1844–46, during which he decided that the evil element in society, the agents of the usurious money-power from which he revolted, were not just the Jews but the bourgeois class as a whole.[10] To do this he made elaborate use of Hegel's dialectic. On the one hand there was the money-power, wealth, capital, the instrument of the bourgeois class. On the other, there was the new redemptive force, the proletariat. The argument is expressed in strict Hegelian terms, using all the considerable resources of German philosophical jargon at its academic worst, though the underlying impulse is clearly moral and the ultimate vision (the apocalyptic crisis) is still poetic. Thus: the revolution is coming, which in Germany will be philosophic: 'A sphere which cannot emancipate itself without emancipating itself from all the other spheres, which is in short a *total loss* of humanity capable of redeeming itself only by a *total redemption* of humanity. This dissolution of society, as a particular class, is the *proletariat*.' What Marx seems to be saying is that the proletariat, the class which is not a class, the dissolvent of class and classes, is a redemptive force which has no history, is not subject to historical laws and ultimately ends history – in itself, curiously enough, a very Jewish concept, the proletariat being the Messiah or redeemer. The revolution consists of two elements: 'the head of the emancipation is *philosophy*, its heart is the *proletariat.*' Thus the intellectuals would form the elite, the generals, the workers the foot-soldiers.

Having defined wealth as Jewish money-power expanded into the bourgeois class as a whole, and having defined the proletariat in his new philosophical sense, Marx then proceeds, using Hegelian dialectic, to the heart of his philosophy, the events leading up to the great crisis. The key passage ends:

> The proletariat executes the sentence that private property
> pronounced on itself by begetting the proletariat, just as it carries
> out the sentence which wage-labour pronounced for itself by bringing
> forth wealth for others and misery for itself. If the proletariat is
> victorious it does not at all mean that it becomes the absolute side
> of society, for it is victorious only by abolishing itself and its opposite.
> Then the proletariat and its determining opposite, private property,
> disappear.

Marx had thus succeeded in defining the cataclysmic event he had first seen as a poetic vision. But the definition is in German academic terms. It does not actually mean anything in terms of the real world beyond the university lecture room.

Even when Marx goes on to politicize the events, he still uses philosophical jargon: '*Socialism* cannot be brought into existence without *revolution*. When the *organizing activity* begins, when the *soul*, the thing-in-itself appears, then socialism can toss aside all the political veils.' Marx was a true Victorian; he underlined words as often as Queen Victoria herself in her letters. But his underlining does not actually help much to convey his meaning, which remains sunk in the obscurity of the concepts of German academic philosophy. To ram his points home, Marx likewise resorts to a habitual gigantism, stressing the global nature of the process he is describing, but this too is cumbered with jargon. Thus: 'the proletariat can only exist *world-historically*, just as communism, its actions, can only have world-historical existence.' Or: 'Communism is empirically only possible as the act of the ruling people all at once and simultaneously, which presupposes the universal development of productive power and the world commerce which depends on it.' However, even when Marx's meaning is clear, his statements do not necessarily have any validity; they are no more than the *obiter dicta* of a moral philosopher.[11] Some of the sentences I have quoted above would sound equally plausible or implausible if they were altered to say the opposite. Where, then, were the facts, the evidence from the real world, to turn these prophetic utterances of a moral philosopher, these revelations, into a science?

Marx had an ambivalent attitude to facts, as he had to Hegel's philosophy. On the one hand he spent entire decades of his life amassing facts, which accumulated in over a hundred enormous notebooks. But these were the facts to be found in libraries, Blue Book facts. The kind of facts which did not interest Marx were the facts to be discovered by examining the world and the people who live in it with his own eyes and ears. He was totally and incorrigibly deskbound. Nothing on earth would get him out of the library and the study. His interest in poverty and exploitation went back to the autumn of 1842, when he was twenty-four and wrote a series of articles on the laws governing the right of local peasants to gather wood. According to Engels, Marx told him 'it was his study of the law concerning the theft of wood, and his investigation of the Moselle peasantry, which turned his attention from mere politics to economic conditions and thus to socialism.'[12] But there is no evidence that Marx actually talked to the peasants and the landowners and looked at the conditions on the spot. Again, in 1844 he wrote for the financial weekly *Vorwärts* (*Forward*) an article on the plight of the Silesian weavers. But he never went to Silesia or, so far as we know, ever talked to a weaver of any description: it would have been very uncharacteristic of him if he had. Marx wrote about finance and industry all his life but he only knew two people connected with financial and industrial processes. One was his uncle in Holland, Lion Philips, a successful businessman who created what eventually became the vast Philips Electric Company. Uncle Philips' views on the whole capitalist process would have been well-informed and interesting, had Marx troubled to explore them. But he only once consulted him, on a technical matter of high finance, and though he visited Philips four times, these concerned purely personal matters of family money. The other knowledgeable man was Engels himself. But Marx declined Engels's invitation to accompany him on a visit to a cotton mill, and so far as we know Marx never set foot in a mill, factory, mine or other industrial workplace in the whole of his life.

What is even more striking is Marx's hostility to fellow revolutionaries who had such experience – that is, working men who had become politically conscious. He met such people for the first time only in 1845, when he paid a brief visit to London, and attended a meeting of the German Workers' Education Society. He did not like what he saw. These men were mostly skilled workers, watchmakers, printers, shoemakers; their leader was a forester. They were self-educated, disciplined, solemn, well-mannered, very anti-bohemian, anxious to transform society but moderate about the practical steps to this end. They did not share Marx's apocalyptic visions and, above all, they did not talk his academic jargon. He viewed them with contempt: revolutionary cannon-fodder, no more. Marx always preferred to associate with middle-class intellectuals like himself. When he and Engels created the Communist League, and again when they formed the International, Marx made sure that working-class socialists were eliminated from any positions of influence and sat on committees merely as statutory proles. His motive was partly intellectual snobbery, partly that men with actual experience of factory conditions tended to be anti-violence and in favour of modest, progressive improvements: they were knowledgeably sceptical about the apocalyptic revolution he claimed was not only necessary but inevitable. Some of Marx's most venomous assaults were directed against men of this type. Thus in March 1846 he subjected William Weitling to a kind of trial before a meeting of the Communist League in Brussels. Weitling was the poor, illegitimate son of a laundress who never knew his father's name, a tailor's apprentice who by sheer hard work and self-education had won himself a large following among German workers. The object of the trial was to insist on 'correctness' of doctrine and to put down any uppity working-class type who lacked the philosophical training Marx thought essential. Marx's attack on Weitling was extraordinarily aggressive. He was guilty, said Marx, of conducting an agitation without doctrine. This was all very well in barbarous Russia where 'you can build up successful unions with stupid young men and apostles. But in a civilized country like Germany you must realize that nothing can be achieved without our doctrine.' Again: 'If you attempt to influence the workers, especially the German workers, without a body of doctrine and clear scientific ideas, then you are merely playing an empty and unscrupulous game of propaganda, leading inevitably to the setting-up on the one hand of an inspired apostle and, on the other, of open-mouthed donkeys listening to him.' Weitling replied he had not become a socialist to learn about doctrines manufactured in a study; he spoke for actual working men and would not submit to the views of mere theoreticians who were remote from the suffering world of real labour. This, said an eyewitness, 'so enraged Marx that he struck his fist on the table so violently that the lamp shook. Jumping to his feet he shouted, "Ignorance has never helped anybody yet."' The meeting ended with Marx 'still striding up and down the room in violent rage'.[13]

This was the pattern for further assaults, both on socialists of working-class origin and on any leaders who had secured a large following of working men by preaching practical solutions to actual problems of work and wages, rather than doctrinaire revolution. Thus Marx went for the former composer Pierre-Joseph Proudhon, the agricultural reformer Hermann Kriege and the first really important German social democrat and labour organizer, Ferdinand Lassalle. In his *Manifesto Against Kriege*, Marx, who knew nothing about agriculture, especially in the United States where Kriege had settled, denounced his proposal to give 160 acres of public land to each peasant; he said that peasants should be recruited by promises of land, but once a communist society was set up, land had to be collectively held. Proudhon was an anti-dogmatist: 'For Gods sake,' he wrote, 'after we have demolished all the [religious] dogmatism a priori, let us not of all things attempt to instil another kind of dogma into the people . . . let us not make ourselves the leaders of a new intolerance.' Marx hated this line. In his violent diatribe against Proudhon, the *Misère de la Philosophie*, written in June 1846, he accused him of 'infantilism', gross 'ignorance' of economics and philosophy and, above all, misuse of Hegel's ideas and techniques – 'Monsieur Proudhon knows no more of the Hegelian dialectic than its idiom.' As for Lassalle, he became the victim of Marx's most brutal anti-Semitic and racial sneers: he was 'Baron Itzig', 'the Jewish Nigger', 'a greasy Jew disguised under brilliantine and cheap jewels'. 'It is now perfectly clear to me,' Marx wrote to Engels on 30 July 1862, 'that, as the shape of his head and the growth of his hair indicates, he is descended from the Negroes who joined in Moses' flight from Egypt (unless his mother or grandmother on the father's side was crossed with a nigger). This union of Jew and German on a Negro base was bound to produce an extraordinary hybrid.'[14]

Marx, then, was unwilling either to investigate working conditions in industry himself or to learn from intelligent working men who had experienced them. Why should he? In all essentials, using the Hegelian dialectic, he had reached his conclusions about the fate of humanity by the late 1840s. All that remained was to find the facts to substantiate them, and these could be garnered from newspaper reports, government blue books and evidence collected by earlier writers; and all this material could be found in libraries. Why look further? The problem, as it appeared to Marx, was to find the right kind of facts: the facts that fitted. His method has been well summarized by the philosopher Karl Jaspers:

> The style of Marx's writings is not that of the investigator . . . he does not quote examples or adduce facts which run counter to his own theory but only those which clearly support or confirm that which he considers the ultimate truth. The whole approach is one of vindication, not investigation, but it is a vindication of something proclaimed as the perfect truth with the conviction not of the scientist but of the believer.[15]

In this sense, then, the 'facts' are not central to Marx's work; they are ancillary, buttressing conclusions already reached independently of them. *Capital*, the monument around which his life as a scholar revolved, should be seen, then, not as a scientific investigation of the nature of the economic process it purported to describe but as an exercise in moral philosophy, a tract comparable to those of Carlyle or Ruskin. It is a huge and often incoherent sermon, an attack on the industrial process and the principle of ownership by a man who had conceived a powerful but essential irrational hatred for them. Curiously enough, it does not have a central argument which acts as an organizing principle. Marx originally, in 1857, conceived the work as consisting of six volumes: capital, land, wages and labour, the state, trade and a final volume on the world market and crises.[16] But the methodical self-discipline needed to carry through such a plan proved beyond his power. The only volume he actually produced (which, confusingly, is two volumes) really has no logical pattern; it is a series of individual expositions arranged in arbitrary order. The French Marxist philosopher Louis Althusser found its structure so confusing that he thought it 'imperative' that readers ignore Part One and begin with Part Two, Chapter Four.[17] But other Marxist exegetes have hotly repudiated this interpretation. In fact, Althusser's approach does not help much. Engels's own synopsis of *Capital* Volume One merely serves to underline the weakness or rather absence of structure.[18] After Marx died, Engels produced Volume Two from 1500 folio pages of Marx's notes, a quarter of which he rewrote. The result is 600 dull, messy pages on the circulation of capital, chiefly on the economic theories of the 1860s. Volume Three, on which Engels worked from 1885–93, surveys all aspects of capital not already covered but is no more than a series of notes, including 1000 pages on usury, most of them Marx's memoranda. The material nearly all dates from the early 1860s, accumulated at the same time as Marx was working on the first volume. There was, in fact, nothing to have prevented Marx from completing the book himself, other than lack of energy and the knowledge that it simply did not cohere.

The second and third volumes are not our concern, as it is most unlikely that Marx would have produced them in this form, or indeed at all, since he had in effect stopped work on them for a decade and a half. Of Volume One, which was his work, only two chapters really matter, Chapter Eight, 'The Working Day', and Chapter Twenty-Four, towards the end of the second volume, 'Primary Accumulation', which includes the famous Section 7, 'Historical Tendency of Capitalist Accumulation'. This is not a scientific analysis in any sense but a simple prophecy. There will be, Marx says, (1) 'a progressive diminution in the number of the capitalist magnates'; (2) 'a corresponding increase in the mass of poverty, oppression, enslavement, degeneration and exploitation'; (3) 'a steady intensification of the wrath of the working class'. These three forces, working together, produce the Hegelian crisis, or the politico-economic version of the poetic catastrophe he had imagined as a tee-nager: 'The centralization of the means of production and the socialization of labour reach a point where they prove incompatible with their capitalist husk. This bursts asunder. The knell of capitalist private property sounds. The expropriators are expropriated.'[19] This is very exciting and has delighted generations of socialist zealots. But it has no more claim to be a scientific projection than an astrologer's almanac.

Chapter Eight, 'The Working Day', does, by contrast, present itself as a factual analysis of the impact of capitalism on the lives of the British proletariat; indeed, it is the only part of Marx's work which actually deals with the workers, the ostensible subjects of his entire philosophy. It is therefore worth examining for its 'scientific' value.[20] Since, as we have already noted, Marx only really looked for facts which fitted his preconceptions, and since this militates against all the principles of scientific method, the chapter has a radical weakness from the start. But did Marx, in addition to a tendentious selection of facts, also misrepresent or falsify them? That we must now consider.

What the chapter seeks to argue, and it is the core of Marx's moral case, is that capitalism, by its very nature, involves the progressive and increasing exploitation of the workers; thus the more capital employed, the more the workers will be exploited, and it is this great moral evil which produces the final crisis. In order to justify his thesis scientifically, he has to prove that, (1) bad as conditions in pre-capitalist workshops were, they have become far worse under industrial capitalism; (2) granted the impersonal, implacable nature of capital, exploitation of workers rises to a crescendo in the most highly capitalized industries. Marx does not even attempt to do (1). He writes: 'As far as concerns the period from the beginning of large-scale industry in England down to the year 1845, I shall only touch on this here and there, referring the reader for fuller details to Friedrich Engels's *Die Lage der arbeitenden Klasse in England* (Leipzig, 1845).' Marx adds that subsequent government publications, especially factory inspectors' reports, have confirmed 'Engels's insight into the nature of the capitalist method' and showed 'with what an admirable fidelity to detail he depicted the circumstances'.[21]

In short, all the first part of Marx's scientific examination of working conditions under capitalism in the mid-1860s is based upon a single work, Engels's *Condition of the Working Class in England*, published twenty years before. And what scientific value, in turn, can be attached to this single source? Engels was born in 1820, the son of a prosperous cotton manufacturer at Barmen in the Rhineland, and entered the family business in 1837. In 1842 he was sent to the Manchester office of the firm, spending twenty months in England. During that time he visited London, Oldham, Rochdale, Ashton, Leeds, Bradford and Huddersfield as well as Manchester. He thus had direct experience of the textile trades but otherwise knew nothing first-hand about English conditions. For instance, he knew nothing about mining and never went down a mine; he knew nothing of the country districts or rural labour. Yet he devotes two entire chapters to 'The Miners' and 'The Proletariat on the Land'. In 1958 two exact scholars, W.O. Henderson and W.H. Challoner, retranslated and edited Engels's book and examined his sources and the original text of all his quotations.[22] The effect of their analysis was to destroy the objective historical value of the book almost entirely, and reduce it to what it undoubtedly was: a work of political polemic, a tract, a tirade. Engels wrote to Marx, as he was working on the book: 'At the bar of world opinion, I charge the English middle classes with mass murder, wholesale robbery and all the other crimes in the calendar.'[23]

That just about sums up the book: it was the case for the prosecution. A great deal of the book, including all the examination of the pre-capitalist era and the early stages of industrialization, was based not on primary sources but on a few secondary sources of dubious value, especially Peter Gaskell's *The Manufacturing Population of England* (1833), a work of Romantic mythology which attempted to show that the eighteenth century had been a golden age for English yeomen and craftsmen. In fact, as the Royal Commission on Children's Employment of 1842 conclusively demonstrated, working conditions in the small, pre-capitalist workshops and cottages were far worse than in the big new Lancashire cotton mills. Printed primary sources used by Engels were five, ten, twenty, twenty-five or even forty years out of date, though he usually presents them as contemporary. Giving figures for the births of illegitimate babies attributed to night-shifts, he omitted to state that these dated from 1801. He quoted a paper on sanitation in Edinburgh without letting his readers know it was written in 1818. On various occasions he omitted facts and events which invalidated his out-of-date evidence completely.

It is not always clear whether Engels's misrepresentations are deliberate deception of the reader or self-deception. But sometimes the deceit is clearly intentional. He used evidence of bad conditions unearthed by the Factories Enquiry Commission of 1833 without telling readers that Lord Althorp's Factory Act of 1833 had been passed, and had long been in operation, precisely to eliminate the conditions the report described. He used the same deception in handling one of his main sources, Dr J.P.Kay's *Physical and Moral Conditions of the Working Classes Employed in the Cotton Manufacture in Manchester* (1832), which had helped to produce fundamental reforms in local government sanitation; Engels does not mention them. He misinterpreted the criminal statistics, or ignored them when they did not support his thesis. Indeed he constantly and knowingly suppresses facts that contradict his argument or explain away a particular 'iniquity' he is seeking to expose. Careful checking of Engels's extracts from his secondary sources show these are often truncated, condensed, garbled or twisted, but invariably put in quotation marks as though given verbatim. Throughout the Henderson and Challoner edition of the book, footnotes catalogue Engels's distortions and dishonesties. In one section alone, Chapter Seven, 'The Proletariat', falsehoods, including errors of fact and transcription, occur on pages 152, 155, 157, 159, 160, 163, 165, 167, 168, 170, 172, 174, 178, 179, 182, 185, 186, 188, 189, 190, 191, 194 and 203.[24]

Marx cannot have been unaware of the weaknesses, indeed dishonesties, of Engels's book since many of them were exposed in detail as early as 1848 by the German economist Bruno Hildebrand, in a publication with which Marx was familiar.[25] Moreover Marx himself compounds Engels's misrepresentations knowingly by omitting to tell the reader of the enormous improvements brought about by enforcement of the Factory Acts and other remedial legislation since the book was published and which affected precisely the type of conditions he had highlighted. In any case, Marx brought to the use of primary and secondary written sources the same spirit of gross carelessness, tendentious distortion and downright dishonesty which marked Engels's work.[26] Indeed they were often collaborators in deception, though Marx was the more audacious forger. In one particularly flagrant case he outreached himself. This was the so-called 'Inaugural Address' to the International Working Men's Association, founded in September 1864. With the object of stirring the English working class from its apathy, and anxious therefore to prove that living standards were falling, he deliberately falsified a sentence from W.E.Gladstone's Budget speech of 1863. What Gladstone said, commenting on the increase in national wealth, was: 'I should look almost with apprehension and with pain upon this intoxicating augmentation of wealth and power if it were my belief that it was confined to the class who are in easy circumstances.' But, he added, 'the average condition of the British labourer, we have the happiness to know, has improved during the last twenty years in a degree which we know to be extraordinary, and which we may almost pronounce to be unexampled in the history of any country and of any age.'[27] Marx, in his address, has Gladstone say: 'This intoxicating augmentation of wealth and power is entirely confined to classes of property.' Since what Gladstone actually said was true, and confirmed by a mass of statistical evidence, and since in any case he was known to be obsessed with the need to ensure that wealth was distributed as widely as possible, it would be hard to conceive of a more outrageous reversal of his meaning. Marx gave as his sources the *Morning Star* newspaper; but the *Star*, along with the other news-papers and *Hansard*, gives Gladstone's words correctly. Marx's misquo-tation was pointed out. Nonetheless, he reproduced it in *Capital*, along with other discrepancies, and when the falsification was again noticed and denounced, he let out a huge discharge of obfuscating ink; he, Engels and later his daughter Eleanor were involved in the row, attempt-ing to defend the indefensible, for twenty years. None of them would ever admit the original, clear falsification and the result of the debate is that some readers are left with the impression, as Marx intended, that there are two sides to the controversy. There are not. Marx knew Gladstone never said any such thing and the cheat was deliberate.[28] It was not unique. Marx similarly falsified quotations from Adam Smith.[29]

Marx's systematic misuse of sources attracted the attention of two Cambridge scholars in the 1880s. Using the revised French edition of *Capital* (1872–75), they produced a paper for the Cambridge Economic Club, 'Comments on the use of the Blue Books by Karl Marx in Chapter XV of *Le Capital*' (1885).[30] They say they first checked Marx's references 'to derive fuller information on some points', but being struck by the 'accumulating discrepancies' they decided to examine 'the scope and importance of the errors so plainly existing'. They discovered that the differences between the Blue Book texts and Marx's quotations from them were not the result solely of inaccuracy but 'showed signs of a distorting influence'. In one class of cases they found that quotations had often been 'conveniently shortened by the omission of passages which would be likely to weigh against the conclusions which Marx was trying to establish'. Another category 'consists in piecing together fictitious quotations out of isolated statements contained in different parts of a Report. These are then foisted upon the reader in inverted commas with all the authority of direct quotations from the Blue Books them-selves.' On one topic, the sewing machine, 'he uses the Blue Books with a recklessness which is appalling ... to prove just the contrary of what they really establish.' They concluded that their evidence might not be 'sufficient to sustain a charge of deliberate falsification' but cer-tainly showed 'an almost criminal recklessness in the use of authorities' and warranted treating any 'other parts of Marx's work with suspicion'.[31]

The truth is, even the most superficial inquiry into Marx's use of evidence forces one to treat with scepticism everything he wrote which relies on factual data. He can never be trusted. The whole of the key Chapter Eight of *Capital* is a deliberate and systematic falsification to prove a thesis which an objective examination of the facts showed was untenable. His crimes against the truth fall under four heads. First, he uses out-of-date material because up-to-date material does not support his case. Second he selects certain industries, where conditions were particularly bad, as typical of capitalism. This cheat was particularly important to Marx because without it he would not really have had Chapter Eight at all. His thesis was that capitalism produces ever-worsening conditions; the more capital employed, the more badly the workers had to be treated to secure adequate returns. The evidence he quotes at length to justify it comes almost entirely from small, inefficient, under-capitalized firms in archaic industries which in most cases were precapitalist – pottery, dressmaking, blacksmiths, baking, matches, wallpaper, lace, for instance. In many of the specific cases he cites (e.g., baking) conditions were bad precisely because the firm had not been able to afford to introduce machinery, since it lacked capital. In effect, Marx is dealing with pre-capitalist conditions, and ignoring the truth which stared him in the face: the more capital, the less suffering. Where he does treat a modern, highly-capitalized industry, he finds a dearth of evidence; thus, dealing with steel, he has to fall back on interpolated comments ('What cynical frankness!' 'What mealy-mouthed phraseology!'), and with railways he is driven to use yellowing clippings of old accidents ('fresh railway catastrophes'): it was necessary to his thesis that the accident rate per passenger mile travelled should be rising, whereas it was falling dramatically and by the time *Capital* was published railways were already becoming the safest mode of mass travel in world history.

Thirdly, using reports of the factory inspectorate, Marx quotes examples of bad conditions and ill-treatment of workers as though they were the inevitable norm of the system; in fact these were the responsibility of what the inspectors themselves call 'the fraudulent mill-owner', whom they were appointed to detect and prosecute and who was thus in the process of being eliminated. Fourthly the fact that Marx's main evidence came from this source, the inspectorate, betrays his biggest cheat of all. It was his thesis that capitalism was, by its nature, incorrigible and, further, that in the miseries it inflicted on the workers, the bourgeois State was its associate since the State, he wrote, 'is an executive committee for managing the affairs of the governing class a whole'. But if that were true Parliament would never have passed the Factory Acts, nor the State enforced them. Virtually all Marx's facts, selectively deployed (and sometimes falsified) as they were, came from the efforts of the State (inspectors, courts, Justices of the Peace) to improve conditions, which necessarily involved exposing and punishing those responsible for bad ones. If the system had not been in the process of reforming itself, which by Marx's reasoning was impossible, *Capital* could not have been written. As he was unwilling to do any on-the-spot investigating himself, he was forced to rely precisely on the evidence of those, whom he designated 'the governing class', who were trying to put things right and to an increasing extent succeeding. Thus Marx had to distort his main source of evidence, or abandon his thesis. The book was, and is, structurally dishonest.

What Marx could not or would not grasp, because he made no effort to understand how industry worked, was that from the very dawn of the Industrial Revolution, 1760–90, the most efficient manufacturers, who had ample access to capital, habitually favoured better conditions for their workforce; they therefore tended to support factory legislation and, what was equally important, its effective enforcement, because it eliminated what they regarded as unfair competition. So conditions improved, and because conditions improved, the workers failed to rise, as Marx predicted they would. The prophet was thus confounded. What emerges from a reading of *Capital* is Marx's fundamental failure to understand capitalism. He failed precisely because he was unscientific: he would not investigate the facts himself, or use objectively the facts investigated by others. From start to finish, not just *Capital* but all his work reflects a disregard for truth which at times amounts to contempt. That is the primary reason why Marxism, as a system, cannot produce the results claimed for it; and to call it 'scientific' is preposterous.

If Marx, then, though in appearance a scholar, was not motivated by a love of truth, what was the energizing force in his life? To discover this we have to look much more closely at his personal character. It is a fact, and in some ways a melancholy fact, that massive works of the intellect do not spring from the abstract workings of the brain and the imagination; they are deeply rooted in the personality. Marx is an outstanding example of this principle. We have already considered the presentation of his philosophy as the amalgam of his poetic vision, his journalistic skill and his academicism. But it can also be shown that its actual content can be related to four aspects of his character: his taste for violence, his appetite for power, his inability to handle money and, above all, his tendency to exploit those around him.

The undertone of violence always present in Marxism and constantly exhibited by the actual behaviour of Marxist regimes was a projection of the man himself. Marx lived his life in an atmosphere of extreme verbal violence, periodically exploding into violent rows and sometimes physical assault. Marx's family quarrels were almost the first thing his future wife, Jenny von Westphalen, noticed about him. At Bonn University the police arrested him for possessing a pistol and he was very nearly sent down; the university archives show he engaged in student warfare, fought a duel and got a gash on his left eye. His rows within the family darkened his father's last years and led eventually to a total breach with his mother. One of Jenny's earliest surviving letters reads: 'Please do not write with so much rancour and irritation,' and it is clear that many of his incessant rows arose from the violent expressions he was prone to use in writing and still more in speech, the latter often aggravated by alcohol. Marx was not an alcoholic but he drank regularly, often heavily and sometimes engaged in serious drinking bouts. Part of his trouble was that, from his mid-twenties, Marx was always an exile living almost exclusively in expatriate, mainly German, communities in foreign cities. He rarely sought acquaintances outside them and never tried to integrate himself. Moreover, the expatriates with whom he always associated were themselves a very narrow group interested wholly in revolutionary politics. This in itself helps to explain Marx's tunnel-vision of life, and it would be difficult to imagine a social background more likely to encourage his quarrelsome nature, for such circles are notorious for their ferocious disputes. According to Jenny, the rows were perpetual except in Brussels. In Paris his editorial meetings in the Rue des Moulins had to be held behind closed windows so that people outside could not hear the endless shouting.

These rows were not aimless, however. Marx quarrelled with everyone with whom he associated, from Bruno Bauer onwards, unless he succeeded in dominating them completely. As a result there are many descriptions, mainly hostile, of the furious Marx in action. Bauer's brother even wrote a poem about him: 'Dark fellow from Trier in fury raging, / His evil fist is clenched, he roars interminably, / As though ten thousand devils had him by the hair.'[32] Marx was short, broad, black-haired and bearded, with a sallow skin (his children called him 'Moor') and Prussian-style monocle. Pavel Annenkov, who saw him at the 'trial' of Weitling, described his 'thick black mane of hair, his hairy hands and crookedly buttoned frock coat'; he had no manners, was 'proud and faintly contemptuous'; his 'sharp, metallic voice was well suited to the radical judgments he was continually delivering on men and things'; everything he said had a 'jarring tone'.[33] His favourite Shakespeare was *Troilus and Cressida*, which he relished for the violent abuse of Ajax and Thersites. He enjoyed quoting it, and the victim of one passage ('Thou sodden-witted lord: thou hast no more brain than I have in mine elbow') was his fellow revolutionary Karl Heinzen, who retaliated with a memorable portrait of the angry little man. He found Marx 'intolerably dirty', a 'cross between a cat and an ape', with 'dishevelled coal-black hair and dirty yellow complexion'. It was, he said, impossible to say whether his clothes and skin were naturally mud-coloured or just filthy. He had small, fierce, malicious eyes, 'spitting out spurts of wicked fire'; he had a habit of saying: 'I will annihilate you.'[34]

Much of Marx's time, in fact, was spent in collecting elaborate dossiers about his political rivals and enemies, which he did not scruple to feed to the police if he thought it would serve his turn. The big public rows, as for instance at the meeting of the International at the Hague in 1872, adumbrated the *réglements des comptes* of Soviet Russia: there is nothing in the Stalinist epoch which is not distantly prefigured in Marx's behaviour. Occasionally blood was indeed spilt. Marx was so abusive during his row with August von Willich in 1850 that the latter challenged him to a duel. Marx, though a former duellist, said he 'would not engage in the frolics of Prussian officers' but he made no attempt to stop his young assistant, Konrad Schramm, from taking his place, though Schramm had never used a pistol in his life and Willich was an excellent shot. Schramm was wounded. Willich's second on this occasion was a particularly sinister associate of Marx, Gustav Techow, rightly detested by Jenny, who killed at least one fellow revolutionary and was eventually hanged for murdering a police officer. Marx himself did not reject violence or even terrorism when it suited his tactics. Addressing the Prussian government in 1849, he threatened: 'We are ruthless and ask no quarter from you. When our turn comes we shall not disguise our terrorism.'[35] The following year, the 'Plan of Action' he had distributed in Germany specifically encouraged mob violence: 'Far from opposing the so-called excesses, those examples of popular vengeance against hated individuals or public buildings which have acquired hateful memories, we must not only condone these examples but lend them a helping hand.'[36] On occasions he was willing to support assassination, provided it was effective. A fellow revolutionary, Maxim Kovalevsky, who was present when Marx got the news of a failed attempt to murder the Emperor Wilhelm I in the Unter den Linden in 1878, records his fury, 'heaping curses on this terrorist who had failed to carry out his act of terror'.[37] That Marx, once established in power, would have been capable of great violence and cruelty seems certain. But of course he was never in a position to carry out large-scale revolution, violent or otherwise, and his pent-up rage therefore passed into his books, which always have a tone of intransigence and extremism. Many passages give the impression that they have actually been written in a state of fury. In due course Lenin, Stalin and Mao Tse-tung practiced, on an enormous scale, the violence which Marx felt in his heart and which his works exude.

How Marx actually saw the morality of his actions, whether distorting truth or encouraging violence, it is impossible to say. In one sense he was a strongly moral being. He was filled with a burning desire to create a better world. Yet he ridiculed morality in *The German Ideology*; he argued it was 'unscientific' and could be an obstacle to the revolution. He seems to have thought that it would be dispensed with as a result of the quasi-metaphysical change in human behaviour that the advent of communism would bring about.[38] Like many self-centred individuals, he tended to think that moral laws did not apply to himself, or rather to identify his interests with morality as such. Certainly he came to see the interests of the proletariat and the fulfilment of his own views as co-extensive. The anarchist Michael Bakunin noted that he had 'an earnest devotion to the cause of the proletariat though it always had in it an admixture of personal vanity'.[39] He was always self-obsessed; a huge, youthful letter survives, ostensibly written to his father, in reality written to, as well as about, himself.[40] The feelings and views of others were never of much interest or concern to him. He had to run, single-handed, any enterprise in which he was engaged. Of his editorship of the *Neue Rheinische Zeitung*, Engels observed: 'The organization of the editorial staff was a simple dictatorship by Marx.'[41] He had no time or interest in democracy, except in the special and perverse sense he attached to the word; elections of any kind were abhorrent to him – in his journalism he dismissed British general elections as mere drunken orgies.[42]

In the testimony about Marx's political aims and behaviour, from a variety of sources, it is notable how often the word 'dictator' crops up. Annenkov called him 'the personification of a democratic dictator' (1846). An unusually intelligent Prussian police agent who reported on him in London noted: 'The dominating trait of his character is an unlimited ambition and love of power ... he is the absolute ruler of his party ... he does everything on his own and he gives orders on his own responsibility and will endure no contradiction.' Techow (Willich's sinister second), who once managed to get Marx drunk and to pour forth his soul, gives a brilliant pen-portrait of him. He was 'a man of outstanding personality' with 'a rare intellectual superiority' and 'if his heart had matched his intellect and he had possessed as much love as hate, I would have gone through fire for him.' But 'he is lacking in nobility of soul. I am convinced that a most dangerous personal ambition has eaten away all the good in him ... the acquisition of personal power [is] the aim of all his endeavours.' Bakunin's final judgment on Marx struck the same note: 'Marx does not believe in God but he believes much in himself and makes everyone serve himself. His heart is not full of love but of bitterness and he has very little sympathy for the human race.'[43]

Marx's habitual anger, his dictatorial habits and his bitterness reflected no doubt his justified consciousness of great powers and his intense frustration at his inability to exercise them more effectively. As a young man he led a bohemian, often idle and dissolute life; in early middle age he still found it difficult to work sensibly and systematically, often sitting up all night talking, then lying half-asleep on the sofa for most of the day. In late middle age he kept more regular hours but he never became self-disciplined about work. Yet he resented the smallest criticism. It was one of the characteristics he shared with Rousseau that he tended to quarrel with friends and benefactors, especially if they gave him good advice. When his devoted colleague Dr Ludwig Kugelmann suggested in 1874 that he would find no difficulty in finishing *Capital* if only he would organize his life a little better, Marx broke with him for good and subjected him to relentless abuse.[44]

His angry egoism had physical as well as psychological roots. He led a peculiarly unhealthy life, took very little exercise, ate highly spiced food, often in large quantities, smoked heavily, drank a lot, especially strong ale, and as a result had constant trouble with his liver. He rarely took baths or washed much at all. This, plus his unsuitable diet, may explain the veritable plague of boils from which he suffered for a quarter of a century. They increased his natural irritability and seem to have been at their worst while he was writing *Capital*. 'Whatever happens,' he wrote grimly to Engels, 'I hope the bourgeoisie as long as they exist will have cause to remember my carbuncles.'[45] The boils varied in numbers, size and intensity but at one time or another they appeared on all parts of his body, including his cheeks, the bridge of his nose, his bottom, which meant he could not write, and his penis. In 1873 they brought on a nervous collapse marked by trembling and huge bursts of rage.

Still more central to his anger and frustration, and lying perhaps at the very roots of his hatred for the capitalist system, was his grotesque incompetence in handling money. As a young man it drove him into the hands of moneylenders at high rates of interest, and a passionate hatred of usury was the real emotional dynamic of his whole moral philosophy. It explains why he devoted so much time and space to the subject, why his entire theory of class is rooted in anti-Semitism, and why he included in *Capital* a long and violent passage denouncing usury which he culled from one of Luther's anti-Semitic diatribes.[46]

Marx's money troubles began at university and lasted his entire life.

They arose from an essentially childish attitude. Marx borrowed money heedlessly, spent it, then was invariably astounded and angry when the heavily discounted bills, plus interest, became due. He saw the charging of interest, essential as it is to any system based on capital, as a crime against humanity, and at the root of the exploitation of man by man which his entire system was designed to eliminate. That was in general terms. But in the particular context of his own case he responded to his difficulties by himself exploiting anyone within reach, and in the first place his own family. Money dominates his family correspondence. The last letter from his father, written in February 1838 when he was already dying, reiterates his complaint that Marx was indifferent to his family except for the purpose of getting their help and complains: 'You are now in the fourth month of your law course and you have already spent 280 thalers. I have not earned so much throughout the entire winter.'[47] Three months later he was dead. Marx did not trouble to attend his funeral. Instead he started putting pressure on his mother. He had already adopted a pattern of living off loans from friends and gouging periodic sums from the family. He argued that the family was 'quite rich' and had a duty to support him in his important work. Apart from his intermittent journalism, the purpose of which was political rather than to earn money, Marx never seriously attempted to get a job, though he once in London (September 1862) applied for a post as a railway clerk, being turned down on the grounds that his handwriting was too poor. Marx's unwillingness to pursue a career seems to have been the main reason why his family was unsympathetic to his pleas for handouts. His mother not only refused to pay his debts, believing he would then simply contract more, but eventually cut him off completely. Thereafter their relations were minimal. She is credited with the bitter wish that 'Karl would accumulate capital instead of just writing about it'.

All the same, one way or another Marx got considerable sums of money by inheritance. His father's death brought him 6000 gold francs, some of which he spent on arming Belgian workmen. His mother's death in 1856 brought him less than he expected, but this was because he had anticipated the legacy by borrowing from his Uncle Philips. He also received a substantial sum from the estate of Wilhelm Wolf in 1864. Other sums came in through his wife and her family (she also brought with her as part of her wedding portion a silver dinner service with the coat of arms of her Argyll ancestors, crested cutlery and bedlinen). Between them they received enough money, sensibly invested, to provide a competence, and at no point did their actual income fall below £200 a year, three times the average wage of a skilled workman. But neither Marx himself nor Jenny had any interest in money except to spend it. Legacies and loans alike went in dribs and drabs and they were never a penny better off permanently. Indeed they were always in debt, often seriously, and the silver dinner service regularly went to the pawnbrokers along with much else, including the family's clothing. At one point Marx alone was in a position to leave the house, retaining one pair of trousers. Jenny's family, like Marx's own, refused further help to a son-in-law they regarded as incorrigibly idle and improvident. In March 1851, writing to Engels to announce the birth of a daughter, Marx complained: 'I have literally not a farthing in the house.'[48]

By this time, of course, Engels was the new subject of exploitation. From the mid-1840s, when they first came together, until Marx's death, Engels was the main source of income for the Marx family. He probably handed over more than half of what he received himself. But the total is impossible to compute because for a quarter of a century he provided it in irregular sums, believing Marx's repeated assurances that, provided the next donation was forthcoming, he would soon put himself to rights. The relationship was exploitative on Marx's side and unequal altogether since he was always the dominant and sometimes the domineering partner. Yet in a curious way each needed the other, like a pair of stage-comedians in a double act, unable to perform separately, frequently grumbling but always in the end sticking together. The partnership almost broke down in 1863 when Engels felt Marx's insensitive cadging had gone too far. Engels kept two houses in Manchester, one for business entertaining, one for his mistress, Mary Burns. When she died Engels was deeply distressed. He was furious to receive from Marx an unfeeling letter (dated 6 January 1863), which briefly acknowledged his loss and then instantly got down to the more important business of asking for money.[49] Nothing illustrates better Marx's adamantine egocentricity. Engels replied coldly, and the incident almost ended their relationship. In some ways it was never the same again, for it brought home to Engels the limitations of Marx's character. He seems to have decided, about this time, that Marx would never be able to get a job or support his family or indeed get his affairs into any kind of order. The only thing to do was to pay him a regular dole. So in 1869 Engels sold out of the business, securing for himself an income of rather more than £800 a year. Of this £350 went to Marx. For the last fifteen years of his life, therefore, Marx was the pensioner of a *rentier*, and enjoyed a certain security. Nevertheless, he seems to have lived at the rate of about £500 a year, or even more, justifying himself to Engels: 'even looked at commercially, a purely proletarian set-up would be unsuitable here.'[50] Hence the letters requesting additional handouts from Engels continued.[51]

But of course the principal victims of Marx's improvidence and unwillingness to work were his own household, his wife above all. Jenny Marx is one of the tragic, pitiful figures of socialist history. She had the clear Scottish colouring, pale skin, green eyes and auburn hair of her paternal grandmother, descended from the second Earl of Argyll, killed at Flodden. She was a beauty and Marx loved her – his poems prove it – and she loved him passionately, fighting his battles both with her family and his own; it took many years of bitterness for her love to die. How could an egoist like Marx inspire such affection? The answer, I think, is that he was strong, masterful, in youth and early manhood handsome, though always dirty; not least, he was funny. Historians pay too little attention to this quality; it often helps to explain an appeal otherwise mysterious (it was one of Hitler's assets, both in private and as a public speaker). Marx's humour was often biting and savage. Nonetheless his excellent jokes made people laugh. Had he been humourless, his many unpleasant characteristics would have denied him a following at all, and his womenfolk would have turned their backs on him. But jokes are the surest way to the hearts of much-tried women, whose lives are even harder than men's. Marx and Jenny were often heard laughing together, and later it was Marx's jokes, more than anything else, which bound his daughters to him.

Marx was proud of his wife's noble Scottish descent (he exaggerated it) and her position as the daughter of a baron and senior official in the Prussian government. Printed invitations to a ball which he issued in London in the 1860s refer to her as '*née* von Westphalen'. He often asserted that he got on better with genuine aristocrats than with the grasping bourgeoisie (a word, say witnesses, he pronounced with a peculiar rasping contempt). But Jenny, once the horrific reality of marriage to a stateless, workless revolutionary had dawned on her, would willingly have settled for a bourgeois existence, however petty. From the beginning of 1848 and for at least the next ten years, her life was a nightmare. On 3 March 1848 a Belgian expulsion order was issued against Marx and he was taken to prison; Jenny spent the night in a cell too, with a crowd of prostitutes; the next day the family was taken under police escort to the frontier. Much of the next year Marx was on the run or on trial. By June 1849 he was destitute. Next month he confessed to a friend: 'already the last piece of jewellery belonging to my wife has found its way to the pawnshop.'[52] He kept up his own spirits by an absurd, perennial revolutionary optimism, writing to Engels: 'In spite of everything a colossal outbreak of the revolutionary volcano was never more imminent. Details later.' But for her there was no such consolation, and she was pregnant. They found safety in England, but degradation too. She now had three children, Jenny, Laura and Edgar, and gave birth to a fourth, Guy or Guido, in November 1849. Five months later they were evicted from their rooms in Chelsea for nonpayment of rent, being turned out onto the pavement before (wrote Jenny) 'the entire mob of Chelsea'. Their beds were sold to pay the butcher, milkman, chemist and baker. They found refuge in a squalid German boarding-house in Leicester Square and there, that winter, the baby Guido died. Jenny left a despairing account of these days, from which her spirits, and her affection for Marx, never really recovered.[53]

On 24 May 1850 the British Ambassador in Berlin, the Earl of Westmore-land, was given a copy of a report by a clever Prussian police spy describ-ing in great detail the activities of the German revolutionaries centred around Marx. Nothing more clearly conveys what Jenny had to put up with:

[Marx] leads the existence of a Bohemian intellectual. Washing, grooming and changing his linen are things he does rarely, and he is often drunk. Though he is frequently idle for days on end, he will work day and night with tireless endurance when he has much work to do. He has no fixed time for going to sleep or waking up. He often stays up all night and then lies down fully clothed on the sofa at midday, and sleeps till evening, untroubled by the whole world coming and going through their room [there were only two altogether]. . . . There is not one clean and solid piece of furniture. Everything is broken, tattered and torn, with half an inch of dust over everything and the greatest disorder everywhere. In the middle of the [living room] there is a large, old-fashioned table covered with oilcloth and on it lie manuscripts, books and newspapers, as well as the children's toys, rags and tatters of his wife's sewing basket, several cups with chipped rims, knives, forks, lamps, an inkpot, tumblers, Dutch clay pipes, tobacco, ash . . . a junk-shop owner would be ashamed to give away such a remarkable collection of odds and ends. When you enter Marx's room smoke and tobacco fumes make your eyes water . . . Everything is dirty and covered with dust, so that to sit down becomes a hazardous business. Here is a chair with three legs. On another chair the children are playing at cooking. This chair happens to have four legs. This is the one that is offered to the visitor, but the children's cooking has not been wiped away and if you sit down you risk a pair of trousers.[54]

This report, dating from 1850, probably described the lowest point of the family fortunes. But other blows fell in the next few years. A daughter Franziska, born in 1851, died the following year. Edgar, the much-loved son, Marx's favourite whom he called *Musch* (Little Fly), got gastro-enteritis in the squalid conditions and died in 1855, a fearful blow to both of them. Jenny never got over it. 'Every day,' wrote Marx, 'my wife tells me she wishes she were lying in her grave . . .' Another girl, Eleanor, had been born three months before, but for Marx it was not the same thing. He had wanted sons and now he had none; girls were unimportant to him, except as clerical assistants.

In 1860 Jenny caught smallpox and lost what remained of her looks; from that point, until her death in 1881, she faded slowly into the background of Marx's life, a tired, disillusioned woman, grateful for small mercies: her silver back from the pawnshop, a house of her own. In 1856, thanks to Engels, the family was able to move out of Soho to a rented house, 9 Grafton Terrace, Haverstock Hill; nine years later, again thanks to Engels, they took a much better one, I Maitland Park Road. From now on they never had less than two servants. Marx took to reading *The Times* every morning. He was elected to the local vestry. On fine Sundays he led a solemn family walk onto Hampstead Heath, himself striding at the head, wife, daughters and friends behind.

But the *embourgeoisement* of Marx led to another form of exploitation, this time of his daughters. All three were clever. One might have thought that, to compensate for the disturbed and impoverished childhood they endured as children of a revolutionary, he would at least have pursued the logic of his radicalism and encouraged them to have careers. In fact he denied them a satisfactory education, refused to allow them to get any training, and vetoed careers absolutely. As Eleanor, who loved him best, said to Olive Schreiner: 'for long, miserable years there was a shadow between us.' Instead the girls were kept at home, learning to play the piano and paint watercolours, like the daughters of merchants. As they grew older, Marx still went on occasional pub-crawls with his revolutionary friends; but according to Wilhelm Liebknecht, he refused to allow them to sing bawdy songs in his house, as the girls might hear.[55]

Later he disapproved of the girls' suitors, who came from his own revolutionary milieu. He could not, or did not, stop them marrying, but he made things difficult and his opposition left scars. He called Paul Lafargue, Laura's husband, who came from Cuba and had some Negro blood, 'Negrillo' or 'The Gorilla'. He did not like Charles Longuet, who married Jenny, either. In his view both his sons-in-law were idiots: 'Longuet is the last of the Proudhonists and Lafargue is the last of the Bakunists – to hell with both of them!'[56] Eleanor, the youngest, suffered most from his refusal to allow the girls to pursue careers and his hostility to suitors. She had been brought up to regard man – that is, her father – as the centre of the universe. Perhaps not surprisingly, she eventually fell in love with a man who was even more egocentric than her father.

Edward Aveling, writer and would-be left-wing politician, was a philanderer and sponger who specialized in seducing actresses. Eleanor wanted to be an actress, and was a natural victim. By one of history's sharp little ironies, he, Eleanor and George Bernard Shaw took part in the first private reading, in London, of Ibsen's brilliant plea for women's freedom, *A Doll's House*, Eleanor playing Nora. Shortly before Marx died, she became Aveling's mistress, and from then on his suffering slave, as her mother Jenny had once been her father's.[57]

Marx, however, may have needed his wife more than he cared to admit. After her death in 1881 he faded rapidly himself, doing no work, taking the cure at various European spas or travelling to Algiers, Monte Carlo and Switzerland in search of sun or pure air. In December 1882 he exulted at his growing influence in Russia: 'Nowhere is my success more delightful.' Destructive to the end, he boasted that 'it gives me the satisfaction that I damage a power which, next to England, is the true bulwark of the old society.' Three months later he died in his dressing-gown, sitting near the fire. One of his daughters, Jenny, had died a few weeks before. The ends of the other two were also tragic. Eleanor, heartbroken by her husband's conduct, took an overdose of opium in 1898, possibly in a suicide pact from which he wriggled out. Thirteen years later Laura and Lafargue also agreed a suicide pact, and both carried it through.

There was, however, one curious, obscure survivor of this tragic family, the product of Marx's most bizarre act of personal exploitation. In all his researches into the iniquities of British capitalists, he came across many instances of low-paid workers but he never succeeded in unearthing one who was paid literally no wages at all. Yet such a worker did exist, in his own household. When Marx took his family on their formal Sunday walks, bringing up the rear, carrying the picnic basket and other impedimenta, was a stumpy female figure. This was Helen Demuth, known in the family as 'Lenchen'. Born in 1823, of peasant stock, she had joined the von Westphalen family at the age of eight as a nursery-maid. She got her keep but was paid nothing. In 1845 the Baroness, who felt sorrow and anxiety for her married daughter, gave Lenchen, then twenty-two, to Jenny Marx to ease her lot. She remained in the Marx family until her death in 1890. Eleanor called her 'the most tender of beings to others, while throughout her life a stoic to herself'.[58] She was a ferociously hard worker, not only cooking and scrubbing but managing the family budget, which Jenny was incapable of handling. Marx never paid her a penny. In 1849–50, during the darkest period of the family's existence, Lenchen became Marx's mistress and conceived a child. The little boy Guido had recently died, but Jenny, too, was pregnant again. The entire household was living in two rooms, and Marx had to conceal Lenchen's state not only from his wife but from his endless revolutionary visitors. Eventually Jenny found out or had to be told and, on top of her other miseries at this time, it probably marked the end of her love for Marx. She called it 'an event which I shall not dwell upon further, though it brought about a great increase in our private and public sorrows'. This occurs in an autobiographical sketch she wrote in 1865, of which twenty-nine out of thirty-seven pages survive: the remainder, describing her quarrels with Marx, were destroyed, probably by Eleanor.[59]

Lenchen's child was born at the Soho address, 28 Dean Street, on 23 June 1851.[60] It was a son, registered as Henry Frederick Demuth. Marx refused to acknowledge his responsibility, then or ever, and flatly denied the rumours that he was the father. He may well have wished to do a Rousseau and put the child in an orphanage, or have him permanently adopted. But Lenchen was a stronger character than Rousseau's mistress. She insisted on acknowledging the boy herself. He was put out to be fostered by a working-class family called Lewis but allowed to visit the Marx household. He was, however, forbidden to use the front door and obliged to see his mother only in the kitchen. Marx was terrified that Freddy's paternity would be discovered and that this would do him fatal damage as a revolutionary leader and seer. One obscure reference to the event survives in his letters; others have been suppressed by various hands. He eventually persuaded Engels to acknowledge Freddy privately, as a cover-story for family consumption. That, for instance, was what Eleanor believed. But Engels, though prepared as usual to submit himself to Marx's demands for the sake of their joint work, was not willing to take the secret to the grave. Engels died, of cancer of the throat, on 5 August 1895; unable to speak but unwilling that Eleanor (Tussy as she was called) should continue to think her father unsullied, he wrote on a slate: 'Freddy is Marx's son. Tussy wants to make an idol of her father.' Engels's secretary-housekeeper, Louise Freyberger, in a letter to August Bebel, of 2 September 1898, said Engels himself told her the truth, adding: 'Freddy looks ridiculously like Marx and, with that typically Jewish face and blue-black hair, it was really only blind prejudice that could see in him any resemblance to General' (her name for Engels). Eleanor herself accepted that Freddy was her half-brother, and became attached to him; nine of her letters to him have survived.[61] She did not bring him any luck, since her lover Aveling succeeded in borrowing Freddy's life savings, which were never repaid.

Lenchen was the only member of the working class that Marx ever knew at all well, his one real contact with the proletariat. Freddy might have been another, since he was brought up as a working-class lad and in 1888, when he was thirty-six, he got his coveted certificate as a qualified engineer-fitter. He spent virtually all his life in King's Cross and Hackney and was a regular member of the engineers' union. But Marx never knew him. They met only once, presumably when Freddy was coming up the outside steps from the kitchen, and he had no idea then that the revolutionary philosopher was his father. He died in January 1929, by which time Marx's vision of the dictatorship of the proletariat had taken concrete and terrifying shape, and Stalin – the ruler who achieved the absolute power for which Marx had yearned – was just beginning his catastrophic assault on the Russian peasantry.

4

Henrik Ibsen:
'On the Contrary!'

*A*LL writing is hard. Creative writing is intellectual drudgery of the hardest kind. Creative innovation, particularly on a fundamental scale, requires a still more exceptional degree of concentration and energy. To spend one's entire working life continually advancing the creative frontiers in one's art implies a level of self-discipline and intellectual industry which few writers have ever possessed. Yet this was the consistent pattern of Henrik Ibsen's work. It is hard to think of any writer, in any field or age, who was more successfully devoted to it. He not only invented modern drama but wrote a succession of plays which still form a substantial part of its entire repertoire. He found the Western stage empty and impotent and transformed it into a rich and immensely powerful art form, not only in his own country but throughout the world. Moreover, he not only revolutionized his art but changed the social thinking of his generation and the one that came after. What Rousseau had done for the late eighteenth century, he did for the late nineteenth century. Whereas Rousseau persuaded men and women to go back to nature and in so doing precipitated a collective revolution, Ibsen preached the revolt of the individual against the *ancien régime* of inhibitions and prejudices which held sway in every small town, indeed in every family. He taught men, and especially women, that their individual conscience and their personal notions of freedom have moral precedence over the requirements of society. In doing so he precipitated a revolution in attitudes and behaviour which began even in his own lifetime and has been proceeding, in sudden jumps and spasms, ever since. Long before Freud, he laid the foundations of the permissive society. Perhaps not even Rousseau, and certainly not Marx, has had more influence over the way people, as opposed to governments, actually behave. He and his work form one of the keystones of the arch of modernity.

Ibsen's achievement is all the more remarkable if we take into account the double obscurity of his own background. Double because he was not only poor himself but came from a small, poor country with no formal cultural tradition at all. Norway had been powerful and enterprising in the early Middle Ages, 900–1100 A D; then a decline set in, especially after the death of her last wholly Norwegian king, Olaf IV, in 1387; by 1536 it was a province of Denmark and remained so for nearly three centuries. The name of the capital, Oslo, was changed to Christiania, to commemorate a Danish ruler, and all the higher culture was Danish – poetry, novels and plays. From the Congress of Vienna in 1814–15, Norway got what was known as the Eidsvoll Constitution,[1] which guaranteed self-government under the Swedish Crown; but not until 1905 did the country have a separate monarchy. Until the nineteenth century, Norwegian was more a rustic, provincial dialect than a written national language. The first university dated only from 1813 and it was 1850 before the first Norwegian theatre was built in Bergen.[2] In Ibsen's youth and early manhood, the culture was still overwhelmingly Danish. To write in Norwegian was to isolate oneself even from the rest of Scandinavia, let alone the world. Danish remained the language of literature.

The country itself was miserable and dejected. The capital was a small provincial town by European standards, with only 20,000 inhabitants, a muddy, graceless place. Skien, where Ibsen was born on 20 March 1828, was on the coast, a hundred miles to the south, a barbarous area where wolves and leprosy were still common. A few years before, the place had been burnt down by the carelessness of a servant-girl, who was executed for it. As Ibsen described it in an autobiographical fragment, it was superstitious, eerie and brutal, sounding to the roar of its weirs and the screaming and moaning of its saws: 'When later I read of the guillotine, I always thought of those sawblades.' By the town hall was the pillory, 'a reddish-brown post about the height of a man. On the top was a big round knob that had originally been painted black ... From the front of the post hung an iron chain, and from this an open shackle which looked to me like two small arms ready and eager to reach out and grab me by the neck ... Underneath [the town hall] were dungeons with barred windows looking out on the marketplace. Through those bars I saw many bleak and gloomy faces.'[3]

Ibsen was the eldest of five children (four sons, one daughter) of a merchant, Knud Ibsen, whose ancestors were sea-captains. His mother came from a shipping family. But when Ibsen was six his father went bankrupt, and thereafter was a broken man, cadging, bad-tempered and litigious – Old Ekdal in *The Wild Duck*. His mother, once beautiful, a frustrated actress, turned inwards, hid herself away and played with dolls. The family was always in debt and lived mainly on potatoes. Ibsen himself was small and ugly and grew up under the additional shadow of rumoured illegitimacy, said to be the son of a local philanderer. Ibsent intermittently believed this, and would blurt it out himself when drunk; but there is no evidence it is true. After a humiliating childhood, he was sent to the gloomy seaport of Grimstad, as an apothecary's assistant, and there too his luck was poor. His master's business, long failing, toppled over into bankruptcy.[4]

Ibsen's slow ascent from this abyss was an epic of lonely self-education. From 1850 he worked his way through university. His privations, then and for many years after, were extreme. He wrote poetry, blank-verse plays, drama criticism, political commentary. His earliest play, the satire *Norma*, was not produced. The first to get on the stage, the tragedy *Cataline*, also in verse, was a failure. He had no luck with the second to be staged, *St John's Night*. His third play, *The Warrior's Barrow*, failed in Bergen. His fourth, *Lady Ingar of Ostraat*, in prose, was put on anonymously, and that too failed. The first work of his to attract favourable notice, *The Feast at Solhaug*, was a trivial, conventional thing in his view. If he followed his natural inclinations, as in the verse-drama *Love's Comedy*, it was classified as 'immoral' and not put on at all. Yet he gradually acquired immense stage experience. The musician Ole Bull, founder of the first Norwegian-language theatre in Bergen, took him on as house author at £5 a month, and for six years he was a theatrical dogsbody, working on sets, costumes, box office, even directing (though never acting; it was his weakness that he lacked confidence in directing actors). The conditions were primitive: gas-lighting, available in London and Paris from about 1810, did not arrive until the year he left, 1856. He then had another five years at the new Christiania theatre. By prodigies of hard work he inched his way to proficiency at the craft, then began to experiment. But in 1862 the new theatre went bankrupt and he was sacked. He was now married, deep in debt, harassed by creditors, depressed, drinking heavily. He was seen by students lying senseless in the gutter, and a fund was set up to send 'the drunken poet Henrik Ibsen' abroad.[5] He himself was constantly writing petitions, which make pathetic reading today, to the Crown and Parliament for a grant to travel in the south. At last he got one, and for the next quarter-century, 1864–92, he led the life of an exile, in Rome, Dresden and Munich.

The first hint of success came in 1864, when his verse-drama *The*

Pretenders got into the repertory of the revived Christiania theatre. It was Ibsen's custom to publish all his plays first in book form, as indeed did most nineteenth-century poets, from Byron and Shelley onwards. Actual productions did not take place, as a rule, until years later, sometimes many years. But slowly the number of copies of each play printed and sold rose: to 5000, 8000, then 10,000, even 15,000. Stage presentations followed. Ibsen's celebrity came in three great waves. First came his big verse dramas, *Brand* and *Peer Gynt*, in 1866–67 – at the time Marx was publishing *Capital*. *Brand* was an attack on conventional materialism and a plea to follow the private conscience against the rules of society, perhaps the central theme of his life's work. It aroused immense controversy when it was published (1866) and for the first time Ibsen was seen as the leader of the revolt against orthodoxy, not just in Norway but in all Scandinavia; he had broken out of the narrow Norwegian enclave.

The second wave came in the 1870s. With *Brand* he became committed to the play of revolutionary ideas but he reached the systematic conclusion that such plays would have infinitely more impact if presented on stage than if read in the study. That led him to renounce poetry and embrace prose, and with it a new kind of theatrical realism. As he put it, 'verse is for visions, prose for ideas.'[6] The transition, like all Ibsen's advances, took years to accomplish, and at times Ibsen appeared to be inactive, brooding rather than working. A playwright, compared to a novelist, does not actually spend much time in writing. The number of words even in a long play is surprisingly small. A play is conceived not so much logically and thematically as in spasms, individual theatrical incidents, which become the source of the plot, rather than developments from it. In Ibsen's case, the pre-writing phase was particularly arduous because he was doing something entirely new. Like all the greatest artists, he could not bear to repeat himself and each work is fundamentally different, usually a new step into the unknown. But once he had decided what he wanted to happen on stage, he wrote quickly and well. The first important fruits of his new policy, *Pillars of Society* (1877), *A Doll's House* (1879) and *Ghosts* (1881), coincided with the breakdown of the long mid-Victorian boom and a new mood of anxiety and disquiet in society. Ibsen asked disturbing questions about the power of money, the oppression of women, even the taboo subject of sexual disease. He placed fundamental political and social issues literally on the centre of the stage, in simple, everyday language and in settings all could recognize. The passion, anger, disgust but above all interest he aroused were immense and spread in widening circles from Scandinavia. *Pillars* marked his breakthrough to the audiences of central Europe, *A Doll's House* into the Anglo-Saxon world. They were the first modern plays and they began the process of turning Ibsen into a world figure.

But, being Ibsen, he found it hard to settle down to the role of social-purpose playwright, even one with an international following. The third great phase in his progress, which again occurred with accumulating speed after years of slow gestation, saw him turning away from political issues as such and towards the problem of personal liberation, which probably occupied his mind more than any other aspect of human existence. 'Liberation,' he wrote in his notebook, 'consists in securing for individuals the right to free themselves, each according to his particular need.' He constantly argued that formal political freedoms were meaningless unless this personal right was guaranteed by the actual behaviour of people in society. So in this third phase he produced, among others, *The Wild Duck* (1884), *Rosmersholm* (1886), *Hedda Gabler* (1890), *The Master Builder* (1892) and *John Gabriel Borkman* (1896), plays which many found puzzling, even incomprehensible at the time, but which have become the most valued of his works: plays which explore the human psyche and its quest for freedom, the unconscious mind and the fearful subject of how one human being gets control over another. It was Ibsen's merit not merely always to be doing something fresh and original in his art but to be sensitive to notions only half-formulated or even still unexplored. As the Danish critic and his onetime friend Georg Brandes put it, Ibsen stood 'in a sort of mysterious correspondence with the fermenting, germinating ideas of the day ... he had the ear for the low rumbling that tells of ideas undermining the ground.'[7]

Moreover, these ideas had an international currency. Theatregoers all over the world were able to identify themselves or their neighbours with the suffering victims and tortured exploiters of his plays. His assaults on conventional values, his programme of personal liberation, his plea that all human beings should have the chance to fulfil themselves, were welcomed everywhere. From the early 1890s, when he returned home to Christiania in triumph, his plays were increasingly performed all over the world. For the last decade of his life (he died in 1906) the former chemist's assistant was the most famous man in Scandinavia. Indeed, along with Tolstoy in Russia, he was widely regarded as the world's greatest living writer and seer. His fame was spread by writers like William Archer and George Bernard Shaw. Journalists came thousands of miles to interview him in his gloomy apartment in Viktoria Terrace. His daily appearances at the café of the Grand Hotel, where he sat alone, facing a mirror so he could see the rest of the room, reading the newspaper and drinking a beer with a cognac chaser, were one of the sights of the capital. When he entered the café each day, punctual to the minute, the entire room stood up and raised their hats. None dared to sit down again until the great man was seated. The English writer Richard le Gallienne, who like many people came to Norway expressly to witness this performance, as others went to Yasnaya Polyana to see Tolstoy, described his entrance: 'A forbidding, disgruntled, tight-lipped presence, starchily dignified, straight as a ramrod ... no touch of human kindness about his parchment skin or fierce badger eyes. He might have been a Scotch elder entering the Kirk.'[8]

As Le Gallienne hinted, there was something not quite right about this great humanist writer, already embalmed in popular esteem and public honours in his own lifetime. Here was the Great Liberator, the man who had studied and penetrated mankind, wept for it, and whose works taught it how to free itself from the fetters of convention and stuffy prejudice. But if he felt so strongly for humanity, why did he seem to repel individual people? Why did he reject their advances and prefer to read about them only through the columns of his newspaper? Why always alone? Whence his fierce, self-imposed isolation?

The closer one looked at the great man, the odder he appeared. For a man who had stamped on convention and had urged the freedoms of bohemian life, he now struck a severely orthodox figure himself; orthodox, perhaps, to the point of caricature. Queen Victoria's granddaughter, Princess Marie Louise, observed that he had a little mirror glued into the inside crown of his hat, which he used to comb his hair. The first thing many people noticed about Ibsen was his extraordinary vanity, well brought out in Max Beerbohm's famous cartoon. It was not always thus. Magdalene Thoresen, his wife's stepmother, wrote that when she first saw the young Ibsen in Bergen, 'he looked like a shy little marmot ... he had not yet learned to despise his fellow human beings and there-fore lacked self-assurance.'[9] Ibsen first became a fussy dresser in 1856 after the success of *Solhaug*. He adopted the poet's frilly cuffs, yellow gloves and an elaborate cane. By the mid-1870s, his attention to dress had grown but in a more sombre mode, which fitted in well with the increasingly shuttered facade he presented to the world. The young writer John Paulsen described him in the Austrian Alps in 1876 thus: 'Black tailcoat with order ribbons, dazzling white linen, elegant cravat, black gleaming silk hat, gold spectacles ... fine, pursed mouth, thin as a knife-blade ... I stood before a closed mountain wall, an impenetrable riddle.'[10] He carried a big walnut stick with a huge gold head. The following year he got his first honorary doctorate from Uppsala University; thereafter he not only indicated his wish to be addressed as 'Doktor' but wore a long black frock coat, so formal that the Alpine peasant girls thought him a priest and knelt to kiss his hand on his walks.[11]

His attention to dress was unusually detailed. His letters contain elaborate instructions about how his clothes are to be hung in the wardrobes and his socks and underpants put away in chests. He always polished his own boots and would even sew on his own buttons, though he allowed a servant to thread the needle. By 1887, when his future biographer Henrik Jaegar visited him, he was spending an hour each morning dressing.[12] But his efforts at elegance failed. To most people he looked like a bosun or sea-captain; he had the red, open-air face of his ancestors, especially after drinking. The journalist Gottfried Weisstein thought his habit of pronouncing truisms with impressive certitude made him resemble 'a small German professor' who 'wished to inscribe on the tablets of our memory the information, "Tomorrow I shall take the train to Munich." '[13]

There was one aspect of Ibsen's vanity which verged on the ludicrous. Even his most uncritical admirers found it hard to defend. He had a lifelong passion for medals and orders. In fact, he went to embarrassing lengths to get them. Ibsen had a certain skill in drawing and often sketched these tempting baubles. His first surviving cartoon features the Order of the Star. He would draw the 'Order of the House of Ibsen' and present it to his wife.[14] What he really wanted however were decorations for himself. He got his first in the summer of 1869 when a conference of intellectuals – a new and, some would argue, sinister innovation on the international scene – was held in Stockholm to discuss language. It was the first time Ibsen had been lionized: he spent an evening drinking champagne at the royal palace with King Carl xv, who presented him with the Order of Vasa. Later, Georg Brandes, on his first meeting with Ibsen (they had long corresponded) was amazed to find him wearing it at home.

He would have been still more astonished to discover that Ibsen, by the next year, was already soliciting for more. In September 1870 he wrote to a Danish lawyer who dealt with such matters, asking his help in getting him the Order of Danneborg: 'You can have no idea of the effect this kind of thing has in Norway ... A Danish decoration would much strengthen my standing there ... the matter is important to me.' Two months later he was writing to an Armenian honours-broker who operated from Stockholm but had links with the Egyptian court, asking for an Egyptian medal which 'would be of the greatest help to my literary standing in Norway'.[15] In the end he got a Turkish one, the Medjidi Order, which he delightedly described as 'a handsome object'. The year 1873 was a good one for medals: he got an Austrian gong and the Norwegian Order of St Olaf. But there was no relaxation in his efforts to amass more. To a friend he denied he had 'any personal longing' for them but 'when these orders come my way I do not refuse them.' This was a lie, as his letters testify. It was even said that, in the 1870s, in his quest for medals, he would sweep off his hat when a carriage passed by bearing royal or noble arms on its side, even if there was no one in it.[16]

That particular story may be a malicious invention. But there is ample evidence for Ibsen's passion since he insisted on displaying his growing galaxy of stars on every possible occasion. As early as 1878 he is reported to have worn all of them, including one like a dog-collar round his neck, at a club dinner. The Swedish painter Georg Pauli came across Ibsen sporting his medals (not the ribbons alone but the actual stars) in a Rome street. At times he seems to have put them on virtually every evening. He defended his practice by saying that, in the presence of 'younger friends', it 'reminds me that I need to keep within certain limits'.[17] All the same, people who had invited him to dinner were always relieved when he arrived without them, as they attracted smiles and even open laughter as the wine circulated. Sometimes he wore them even in broad daylight. Returning to Norway by ship, he put on formal dress and decorations before going on deck when it docked in Bergen. He was horrified to see four of his old drinking companions, two carpenters, a sexton and a broker, waiting to greet him with shouts of 'Welcome old Henrik!' He returned to his cabin and cowered there until they had gone.[18] He was still at his tuft-hunting even in old age. In 1898 his anxiety to get the Grand Cross of Danneborg was so great that he bought one from a jeweller before it was formally awarded to him; the King of Denmark sent him a jewelled specimen in addition to the one actually presented, so he ended up with three, two of which had to be returned to the court jeweller.[19]

Yet this international celebrity, glittering with his trophies, gave an ultimate impression not so much of vanity, let alone of foolishness, but of malevolent power and barely suppressed rage. With his huge head and thick neck he seemed to radiate strength, despite his small stature. Brandes said 'he looked as though you would need a club to overpower him.' Then there were his terrifying eyes. The late-Victorian period seems to have been the age of the fierce eye. Gladstone had it to the point where he could make a Member of Parliament forget what he was trying to say when it was turned upon him. Tolstoy likewise used his basilisk eye to strike critics dumb. Ibsen's gaze reminded people of a hanging judge. He instilled fear, said Brandes: 'there lay stored within him twenty-four years of bitterness and hatred.' Anyone who knew him at all well was uneasily aware of a volcanic rage simmering just below the surface.

Drink was liable to detonate the explosion. Ibsen was never an alcoholic or even, except briefly, a drunkard. He never drank during periods of working and would sit down at his desk in the morning not only sober and un-hung-over but wearing a freshly pressed frock coat. But he drank socially, to overcome his intense shyness and taciturnity, and the spirits which loosened his tongue might also inflame his rage. At the Scandinavian Club in Rome his post-prandial outbursts were notorious. They frightened people. They were particularly liable to occur at the endless testimonial and celebratory banquets which were a feature of the nineteenth century all over Europe and North America but were particularly beloved of Scandinavian man. Ibsen appears to have attended hundreds of them, often with disastrous results. Frederick Knudtzon, who knew him in Italy, tells of one friendly dinner at which Ibsen attacked the young painter August Lorange, who was suffering from tuberculosis (one reason why so many Scandinavians were in the South). Ibsen told him he was a bad painter: 'You are not worthy to walk on two feet but ought to crawl on four.' Knudtzon adds: 'We were all left speechless at such an attack on an unoffending and defenceless man, an unfortunate consumptive who had enough to contend with without being banged on the head by Ibsen.' When they finally rose from dinner, Ibsen was unable to stand and had to be carried home.[20] Unfortunately the drink which knocked his legs from under him did not necessarily still his savage tongue. When Georg Pauli and the Norwegian painter Christian Ross carried Ibsen home, wearing all his medals, after another celebration dinner in Rome, he 'showed his gratitude by incessantly giving us his confidential opinion on our insignificance. I, he said, was "a frightful puppy" and Ross "a very repulsive character".'[21] In 1891 when Brandes gave a big dinner in his honour at the Grand Hotel in Christiania, Ibsen created an 'oppressive atmosphere', shook his head ostentatiously during Brandes' generous speech in praise of him, refused to reply to it, saying merely, 'One could say much about that speech,' and finally insulted his host by declaring he 'knew nothing' about Norwegian literature. At other receptions at which he was the chief guest he would turn his back on the company. Sometimes he was so drunk he would just say, repeatedly, 'What, what, what?'

It is true that Ibsen was sometimes, in his turn, the victim of Viking intoxication. Indeed a book could be written describing Scandinavian banquets which went wrong during this period. At a particularly solemn one given for Ibsen at Copenhagen in 1898, the principal speaker, Professor Sophus Schandorph, was so drunk that his two neighbours, a bishop and a count, had to hold him up, and when one guest giggled he shouted: 'Shut your ——— mouths while I speak.' On the same occasion Ibsen was bear-hugged by an appreciative but drunken painter and shouted angrily, 'Take this man away!' When sober, he extended no latitude to behaviour of which he was habitually guilty. Indeed he could be very censorious. When a girl, dressed as a man, was illicitly smuggled into the Rome Scandinavian Club, he insisted the member responsible be expelled. Any kind of behaviour, whether pompous or antinomian, was liable to unleash his fury. He was a specialist in anger, a man to whom irascibility was a kind of art form in itself. He even treasured its manifestations in nature. While he was writing his ferocious play *Brand*, he later recorded, 'I had on my table a scorpion in an empty beer glass. From time to time the brute would ail. Then I would throw a piece of ripe fruit into it, on which it would cast itself in a rage and inject its poison into it. Then it was well again.'[22]

Did he see in the creature an echo of his own need to get rid of the rage within himself? Were his plays, in which anger usually simmers and sometimes boils over, a vast therapeutic exercise? No one knew Ibsen intimately, but many of his acquaintances were aware that his early life and struggles had left him with a huge burden of unappeasable resentment. In this respect he was like Rousseau: his ego bore the bruises for the rest of his life and he was a monster of self-centredness in consequence. Quite unfairly, he held his father and mother responsible for his unhappy youth; his siblings were guilty by association. Once he left Skien he made no effort to keep in touch with his family. On the contrary: on his last visit to Skien in 1858, to borrow money from his wealthy uncle, Christian Paus, he deliberately did not visit his parents. He had some contact with his sister, Hedvig, but this may have had to do with unpaid debts. In a terrifying letter he wrote in 1867 to Bjornstjerne Bjornson, his fellow writer, whose daughter later married Ibsen's son, he wrote: 'Anger increases my strength. If there is to be war, then let there be war! ... I shall not spare the child in its mother's womb, nor any thought nor feeling that may have motivated the actions of any man who shall merit the honour of being my victim ... Do you know that all my life I have turned my back on my parents, on my whole family, because I could not bear to continue a relationship based on imperfect understanding?'[23] When his father died in 1877, Ibsen had not been in touch with him for nearly forty years. Defending himself in a letter to his uncle, he cited 'impossible circumstances from a very early stage' as 'the principal cause'. By this he really meant that they had fallen, he was rising, and he did not want them to drag him down. He was ashamed of them; he feared their possible financial demands. The richer he became, the more able to help them, the less inclined he was to make any contact. He made no effort to assist his crippled younger brother, Nicolai Alexander, who eventually went to the United States and died in 1888, aged fifty-three: his tomb recorded, 'By strangers honoured and by strangers mourned.' He likewise ignored his youngest brother, Ole Paus, by turns sailor, shopkeeper, lighthouse-keeper. Ole was always poor but was the only one to help their wretched father. Ibsen once sent him a formal testimonial for a job but never gave him a penny nor left him anything in his will: he died in an old people's home in 1917, destitute.[24]

Behind the formal family there was a yet more carefully concealed and painful tale. It might have come from one of Ibsen's own plays – indeed in a sense the whole of Ibsen's life is a furtive Ibsenesque drama. In 1846, when he was eighteen and still living over the chemist's shop, he had an affair with the housemaid employed there, Elsie Sofie Jensdatter, who was ten years older. She conceived and bore a son, born 9 October 1846, whom she called Hans Jacob Henriksen. This girl was not an illiterate peasant, like Marx's Lenchen, but from a distinguished family of yeoman farmers; her grandfather, Christian Lofthuus, had led a famous revolt of farmers against Danish rule, and had died, chained to the rock, in the Akershus Fortress. The girl, like Lenchen, behaved with the greatest discretion. She went back to her parents to have the child and never sought to get anything out of the father.[25] But under Norwegian law, and by order of the local council, Ibsen was forced to pay maintenance until Hans Jacob was fourteen.[26] Poor already, he resented bitterly this drain on his meagre salary and never forgave either the child or the mother. Like Rousseau, like Marx, he never acknowledged Hans Jacob, took any interest in him or gave him the smallest voluntary assistance, financial or otherwise. The boy became a blacksmith and lived with his mother until he was twenty-nine. She went blind, and when her parents' house was taken away from them, she went to live in a hut. The son scrawled on the rock 'Syltefjell' – Starvation Hill. Elsie too died destitute, aged seventy-four, on 5 June 1892, and it is unlikely that Ibsen heard of her death.

Hans Jacob was by no means a savage. He was a great reader, especially of history, and travel books. He was also a skilled carver of fiddles. But he was drunken and shiftless. He came sometimes to Christiania, where those who knew his secret were struck by his extraordinary resemblance to his famous father. Some of them planned a scheme to dress Hans Jacob in clothes similar to Ibsen's own, and sit him down, early, at the table in the Grand Hotel which the great man habitually occupied, so that when he arrived for his morning beer he would be confronted by the ocular evidence of his own sin. But their courage failed them. Francis Bull, the great authority on Ibsen, says that Hans

Jacob met his father only once. This was in 1892 when the son, penniless, went to his father's apartment to ask for money. Ibsen himself answered the door, apparently seeing his son, then forty-six, for the first time. He did not deny their relationship but handed Hans Jacob five crowns, saying: 'This is what I gave your mother. It should be enough for you' – then slammed the door in his face.[27] Father and son never met again, and Hans Jacob got nothing in Ibsen's will, dying destitute on 20 October 1916.

Fear that his family, both lawful and illegitimate, would make demands on his purse was undoubtedly one reason why Ibsen fought them off. The penury of his early life left him with a perpetual ache for security which only the constant earning, amassing and conserving of money could soothe. It was one of the great driving forces of his existence. He was mean, as he was everything else, on a heroic scale. For money he was quite prepared to lie: considering that he was an atheist who secretly hated the monarchy, his petition to Carl xv begging for a £100 pension is remarkable: 'I am not fighting for a sinecure existence but for the calling which I inflexibly believe and know God has given me ... It rests in Your Majesty's royal hands whether I must remain silent and bow to the bitterest deprivation that can wound a man's soul, the deprivation of having to abandon one's calling in life, having to yield when I know I have been given the spiritual armoury to fight.' By this time (1866), having earned a little from *Brand*, he was beginning to save. It started with silver coins in a sock, then progressed to purchases of government stock. In Italy, fellow exiles noted that he set down even the smallest purchase in a notebook. From 1870 until his first stroke in 1900 he kept two black notebooks, one recording his earnings, the other his investments, which were all in ultra-safe government securities. Until his last two decades his earnings were not large, at any rate by Anglo-Saxon standards, since his plays were slow to achieve worldwide performances and were in any case ill-protected by copyright. But in 1880 he earned, for the first time, over £1000, an enormous income by current Norwegian levels. The total continued to grow steadily. So did his investments. In fact it is unlikely that any other author ever invested so large a proportion of his earnings, between one-half and two-thirds during the last quarter-century of his life. What was it all for? When asked by his legitimate son, Sigurd, why they lived so frugally, he replied: 'Better to sleep well and not eat well, than eat well and not sleep well.' Despite his growing wealth, he and his family continued to live in drab furnished rooms. He said he envied Bjornson because he had a house and land. But he never himself attempted to purchase any property or even his own furniture. The last Ibsen apartments in

Viktoria Terrace and Arbiens Street were just as impersonal and hotel-like as the others.

All Ibsen's apartments, however, had one unusual characteristic: they appeared to be divided into two halves, with husband and wife each arranging a separate fortress, for defensive and offensive operations against the other.[28] In a curious way this fulfilled a youthful vow, since he told his earliest friend, Christopher Due, that 'his wife, if he ever acquired one, would have to live on a separate floor. They would see each other [only] at mealtimes, and not address each other as *Du*.'[29] Ibsen married Suzannah Thoresen, daughter of the Dean of Bergen, in 1858, after a chilly two-year engagement. She was bookish, determined and plain, but with fine hair. Her bluestocking stepmother said of Ibsen scornfully that, next to Sören Kierkegaard, she had never known anyone with 'so marked a compulsion to be alone with himself'. The marriage was functional rather than warm. In one sense it was crucial to Ibsen's achievement because, at a time of great despondency in his life, when his plays were turned down or failures and he was seriously thinking of developing his other talent, painting, she forbade him to paint at all and forced him to write every day. As Sigurd later put it: 'The world can thank my mother that it has one bad painter the fewer and got a great writer instead.'[30] Sigurd, who was born in 1859, always portrayed his mother as the strength behind Ibsen: 'He was the genius, she the character. His character. And he knew it, though he would not willingly have admitted it until towards the end.'

Naturally, Sigurd portrayed the marriage as a working partnership. Others at the time saw it, and him, differently. There is a harrowing picture of the Ibsens, during the Italian years, in the diary of a young Dane, Martin Schneekloth. Ibsen, he noted, was in 'the desperate situation' of finding himself married to a woman he did not love and 'no reconciliation is possible.' He found him 'a domineering personality, egocentric and unbending, with a passionate masculinity and a curious mixture of personal cowardice, compulsively idealistic yet totally indifferent to expressing those ideals in his daily life ... She is womanly, tactless, but a stable, hard character, a mixture of intelligence and stupidity, not deficient in feeling but lacking humility and feminine love ... They wage war on each other, ruthlessly, coldly, and yet she loves him, if only through their son, their poor son, whose fate is the saddest that could befall any child.' He went on: 'Ibsen himself is so obsessed with his work that the proverb "Humanity first, art second" has practically been reversed. I think his love for his wife has long vanished ... His crime now is that he cannot discipline himself to correct the situation but rather asserts his moody and despotic nature over her and their poor, spiritually warped, terrified son.'[31]

Suzannah was by no means defenceless in the face of Ibsen's granitic egotism. Bjornson's wife quotes her as saying, after the birth of Sigurd, that there would be no more children, which then meant no more sex. (But she was a hostile witness.) From time to time there were rumours of a parting. Ibsen certainly loathed marriage as such: 'It sets the mark of slavery on everyone,' he noted in 1883. But, prudent and loving security, he kept his own together. There survives a curious letter from him to his wife, dated 7 May 1895, in which he hotly denies rumours that he intended to leave her for Hildur Andersen, blaming them on her stepmother Magdalene Thoresen, whom he hated.[32] Ibsen was often harsh and unpleasant to his wife. But she knew how to get her own back. When he grew angry, she simply laughed in his face, aware of his inherent timidity and fear of violence. Indeed, she played on his fears, combing the newspapers for accounts of horrible but everyday catastrophes, which she passed on to him.[33] They cannot have been an agreeable couple to observe together.

Ibsen had equally chilly, and often stormy, relationships with his friends. Perhaps friends is not the right word. His correspondence with his fellow writer Bjornson, whom he knew as well as anyone, and for longer, makes painful reading. He saw Bjornson as a rival, and was jealous of his early success, his extrovert nature, his cheery, kindly ways, his manifest ability to enjoy life. In fact Bjornson did everything in his power to bring Ibsen to public recognition and Ibsen's bleak ingratitude strikes one as pitiful. Their relationship resembles Rousseau's with Diderot, Ibsen like Rousseau doing the taking, Bjornson the giving, though there was no final, spectacular quarrel.

Ibsen found reciprocity difficult. In view of all Bjornson had done for him, the congratulatory telegram he was finally induced to send on Bjornson's sixtieth birthday is a minimalist masterpiece: 'Henrik Ibsen sends good wishes for your birthday.' Yet he expected Bjornson to do a great deal for him. When the critic Clemens Petersen published a hostile review of *Peer Gynt*, Ibsen wrote a furious letter to Bjornson, who had had nothing to do with it. Why had he not knocked Petersen down? 'I would have struck him senseless before allowing him to commit so calculated an offence against truth and justice.' The next day he added a postscript: 'I have slept on these words and read them in cold blood . . . I shall send them nevertheless.' He then worked himself up again and continued: 'I reproach you merely with inactivity. It was not good of you to permit, by doing nothing, such an attempt to be made in my absence to put my reputation under the auctioneer's hammer.'[34]

But, while expecting Bjornson to fight his battles, Ibsen regarded him as fair game for satire. He figures as the unpleasant character of Stens-

gaard in Ibsen's play *The League of Youth*, a savage attack on the progressive movement. In this monument of ingratitude, Ibsen went for all the
people who had helped him with money and had signed the petition
for his state grant. He took the view that anyone of prominence was
a legitimate target. But he bitterly resented any similar references to
himself. When John Paulsen published a novel about a domineering
father with a passion for medals, Ibsen seized one of his visiting cards,
wrote the one word 'Scoundrel' on the back and sent it, open, addressed
to Paulsen at his club – the same technique the Marquess of Queensberry
was to apply to Oscar Wilde in the next decade.

Virtually all Ibsen's relationships with other writers ended in rows.
Even when there was no quarrel, they tended to die of inanition. He
could not follow Dr Johnson's advice: 'Friendships must be kept in constant repair.' He kept them in constant tension, interspersed by periods
of silence: it was always the other party who had to make the effort.
He came in fact close to articulating a philosophy of anti-friendship.
When Brandes, who was living in sin with another man's wife and so
was ostracized in Copenhagen, wrote Ibsen a letter complaining he was
friendless, Ibsen replied: 'When one stands, as you do' – and, by implication, 'as I do also' – 'in so intensely personal a relationship to one's
lifework, one cannot really expect to keep one's friends.... Friends are
an expensive luxury, and when one invests one's capital in a calling
or mission in this life, one cannot afford to have friends. The expensive
thing about friends is not what one does for them but what, out of
consideration for them, one leaves undone. Many spiritual ambitions
have been crippled thus. I have been through this, and that is why
I had to wait several years before I succeeded in being myself.'[35] This
bleak and revealing letter exposes, as with the other intellectuals we
have been examining, the intimate connection between the public doctrine and the private weakness. Ibsen was saying to humanity: 'Be yourselves!' Yet in this letter he was in effect admitting that to be oneself
involved the sacrifice of others. Personal liberation was at bottom self-
centred and heartless. In his own case he could not be an effective playwright without ignoring, disregarding and if necessary trampling on
others. At the centre of Ibsen's approach to his art was the doctrine
of creative selfishness. As he wrote to Magdalene Thoresen: 'Most criticism boils down to a reproach to the writer for being himself ... The
vital thing is to protect one's essential self, to keep it pure and free
from all intrusive elements.'

Creative selfishness was Ibsen's attempt to turn the vulnerability of
his own character into a source of strength. As a boy he had been horrify-
ingly alone: 'an old man's face', said his schoolmaster, 'an inward-look-

ing personality'. A contemporary witnessed: 'We small boys didn't like him as he was always so sour.' He was only once heard to laugh 'like other human beings'. Later as a young man his poverty dictated further solitude: he would go out for long walks by himself, so that other guests and the servants in his lodging house would think he was out to dinner. (Pitifully, Ibsen's meanness later forced his son to similar subterfuge; unwilling to invite other little boys to his grim home, he would tell them that his mother was a giant Negress who kept his young brother, who did not actually exist, imprisoned in a box.) Ibsen's long, solitary walks became a habit: 'I have,' he wrote, 'wandered through most of the Papal States at various times on foot with a knapsack on my back.' Ibsen was a natural exile: he saw the surrounding community as alien at best, but often as hostile. In his youth, he wrote, 'I found myself in a state of war with the little community in which . . . I sat imprisoned.'[36]

So it is not surprising that Ibsen chose actual exile for the longest and most productive period of his life. As with Marx, this reinforced his sense of alienation and locked him up in an intensely parochial expatriate group with its quarrels and animosities. Ibsen began by recognizing the shortcomings of his isolation. In a letter of 1858 he described himself as 'walled about with a kind of off-putting coldness which makes any close relationship with me difficult . . . Believe me, it is not pleasant to see the world from an October standpoint.' Six years later, however, he was becoming reconciled to his inability to reach out to others, writing to Bjornson in 1864: 'I cannot make close contact with people who demand that one should give oneself freely and unreservedly . . . I prefer to shut up [my true self] within me.' His solitude became creative, a subject in itself. From his earliest surviving poem, 'Resignation', written in 1847, until he ceased to write poetry in 1870–71, it is the underlying theme of his verse. As Brandes said, 'it is the poetry of loneliness, portraying the lonely need, the lonely strife, the lonely protest.'[37] His writing, reflecting his solitude, became a defence, refuge and weapon against the alien world; 'all his mind and passion', as Schneekloth said of his life in Italy, was given to 'the demonic pursuit of literary fame'. Gradually he came to see his egotistic isolation and self-concealment as a necessary policy, even a virtue. The whole of human existence, he told Brandes, was a shipwreck, and therefore 'the only sane course is to save oneself.' In old age he advised a young woman: 'You must never tell everything to people. . . . To keep things to oneself is the most valuable thing in life.'[38]

But naturally it was unrealistic to suppose that such a policy could be kept under control. It degenerated into a general hostility towards mankind. Brandes was forced to conclude: 'His contempt for humanity knew no bounds.' The searchlight of his hatred moved systematically over all aspects of human societies, pausing from time to time, almost lovingly, on some idea or institution which evoked his particular loathing. He hated the conservatives. He was perhaps the first writer – the scout of what was to become an immense army – to persuade a conservative state to subsidize a literary life devoted to attacking everything it held dear. (When he came back for more money, one member of the grants board, the Reverend H. Riddervold, said that what Ibsen deserved was not another grant but a thrashing.) He came to hate the liberals even more. They were 'poor stuff with which to man the barricades'. Most of them were 'hypocrites, liars, drivellers, curs'. Like his contemporary Tolstoy, he had a particular dislike for the parliamentary system, which he saw as the source of bottomless corruption and humbug; one reason he liked Russia was because it had none. He hated democracy. His *obiter dicta* as recorded in the diaries of Kristofer Janson, make grim reading.[39] 'What is the majority? The ignorant mass. Intelligence always belongs to the minority.' Most people, he said, were 'not entitled to hold opinions'. He likewise told Brandes: 'Under no circumstances will I ever link myself with any party which has the majority behind it.' He saw himself, if anything, as an anarchist, foolishly believing (as many then did) that anarchism, communism and socialism were all essentially the same. 'The state must be abolished,' he told Brandes, who liked to collect his views. 'Now *there's* a revolution to which I will gladly lend my shoulder. Abolish the concept of the state, establish the principle of free will.'

Ibsen undoubtedly thought he possessed a coherent philosophy of public life. His favourite saying, which he gave to his character Dr Stockmann, was: 'The minority is always right.' By minority, he explained to Brandes, he meant 'the minority which forges ahead in territory which the majority has not yet reached'. To some extent he identified himself with Dr Stockmann, telling Brandes:

an intellectual pioneer can never gather a majority around him. In ten years the majority may have reached the point where Dr Stockmann stood when the people held their meeting. But during those ten years the Doctor has not stood stationary: he is at least ten years ahead of the others. The majority, the masses, the mob, will never catch him up; he can never rally them behind him. I myself feel a similarly unrelenting compulsion to keep pressing forward. A crowd now stands where I stood when I wrote my earlier books. But I myself am there no longer. I am somewhere else – far ahead of them – or so I hope.[40]

The difficulty with this view, which was typically Victorian in its way, was that it assumed that humanity, led by the enlightened minority, would always progress in a desirable direction. It did not occur to Ibsen that this minority – what Lenin was later to call 'the vanguard elite' and Hitler 'the standard bearers' – might lead mankind into the abyss. Ibsen would have been surprised and horrified by the excesses of the twentieth century, the century whose mind he did so much to shape.

The reason Ibsen got the future, which he claimed to foresee, so badly wrong sprang from the inherent weakness of his personality, his inability to sympathize with people, as opposed to ideas. When individuals or groups were mere embodied ideas, as in his plays, he could handle them with great insight and sympathy. The moment they stepped into his life as real people, he fled or reacted with hostility. His last group of plays, with their powerful grasp of human psychology, coincided with rows, outbursts and misanthropy in his own life, and a steady deterioration in the few personal relationships he possessed. The contrast between idea and reality is reflected in most of his public attitudes. On 20 March 1888 he sent a cable to the Christiania Workers Union: 'Of all the classes in my country, it is the working class which is nearest to my heart.'[41] This was humbug. Nothing was near to his heart except his wallet. He never paid the slightest attention to working men in real life or had anything but contempt for their opinions. There is no evidence he ever did anything to help the workers' movement. Again, he found it politic to ingratiate himself with the students. They in turn liked to honour him with torchlight processions. But his actual dealings with them ended in a furious row, reflected in a childish and absurdly long letter he wrote to the Norwegian Students Union, on 23 October 1885, denouncing 'the preponderance of reactionary elements' among them.[42]

It was the same story in his relations with women. In theory he was on their side. It could be argued that he did more, in the long run, to improve the position of women than any other nineteenth-century writer. *A Doll's House*, with its clear message – marriage is not sacrosanct, the husband's authority is open to challenge, self-discovery matters more than anything else – really started the women's movement. He has never been excelled at putting the woman's case, and, as *Hedda Gabler* showed, few have equalled him at presenting a woman's feelings. To do him justice, he occasionally tried to help woman, as an embodied idea, in real life too. One of his drunken banquet speeches was in favour of admitting women to the Scandinavian Club in Rome: it was particularly ferocious and may not have done their cause much good – one countess in the audience fainted in terror. However, he had no patience at all with women who actually participated in the cause, especially if they were writers too. At the disastrous dinner Brandes gave to him in the Grand Hotel in 1891, he was incensed to discover he had been put next to the middle-aged woman painter and intellectual Kitty Kielland. When she ventured to criticize the character of Mrs Elvstead in *Hedda Gabler*, he snarled: 'I write to portray people and I am completely indifferent to what fanatical bluestockings like or do not like.'[43] His idea of hell was to attend a protracted banquet at which he was seated besides an elderly suffragette or authoress – and there were large numbers of both in all the Scandinavian capitals by the 1890s. He tried hard to get out of a big formal dinner given in his honour at Christiania on 26 May 1898 by the Norwegian League for Women's Rights. When it proved unavoidable he made a characteristically curmudgeonly speech.[44] He was equally bad-tempered at a dinner in Stockholm given for him jointly by two women's societies; but disaster was avoided when the ladies had the sense to put on a display of folk-dancing by pretty young girls, of which Ibsen, it was known, was passionately fond.[45]

One of the dancers was Rosa Fitinghoff, daughter of a woman who wrote children's stories. She became the last of a long succession of girls with whom Ibsen had complex and in some ways vertiginous relationships. Ibsen seems always to have had a taste for extreme youth, which he associated bitterly with the unattainable. The first time he fell seriously in love, when he was working at the Bergen theatre, was with a fifteen-year-old, Henrikke Holst. But he had no money, the father objected, and that was that. By the time he had his first success, he felt himself too old and ugly and would risk a rebuff if he bid for a girl many years younger than himself. But he continued to contract *liaisons dangereuses*. In 1870 it was the brilliant young women's rightist Laura Petersen. Four years later it was Hildur Sontum, a mere ten-year-old, granddaughter of his old landlady. The taste did not diminish with age: on the contrary. He was fascinated by the story of Goethe's elderly feelings for the delicious Marianne von Willemer, which gave his art renewed youth. It became accepted that actresses, if young and pretty, could usually persuade Ibsen to do what they wanted, especially if they introduced other young girls to him. When he visited the Scandinavian capitals, girls would hang about his hotel; he sometimes agreed to talk to them, and would give them a kiss and a photograph of himself. He liked young girls in general but his interest usually centred on one in particular. In 1891 it was Hildur Andersen. Rosa Fitinghoff was the last.

The two most significant were Emilie Bardach and Helene Raff, whom he met on an Alpine holiday in 1889. Both kept diaries and a number of letters have survived. Emilie, an eighteen-year-old Austrian girl (Ibsen was forty-three years older) recorded in her diary: 'His ardour ought to make me feel proud ... He puts such strong feelings into what he says to me ... Never in his whole life, he says, has he felt so much joy in knowing anyone. He never admired anyone as he admires me.' He asked her 'to be absolutely frank with him so that we may become fellow workers together'. She thought herself in love with him, 'but we both feel it is best outwardly to remain as strangers.'[46] The letters he wrote to her after they parted were fairly harmless, and forty years later she told the writer E. A. Zucker that they had not even kissed; but she also said that Ibsen had spoken of the possibility of his divorce – then they would marry and see the world.[47] Helene, a more sophisticated city girl from Munich, allowed him to kiss her but was clear their relationship was romantic and literary rather than sexual, let alone serious. When she asked him what he saw in her he replied: 'You are youth, child, youth personified, and I need that for my writing.' That of course explained what he meant by the term 'fellow workers'. Helene wrote forty years afterwards: 'His relations with young girls had in them nothing whatever of infidelity in the usual sense of the term but arose solely from the needs of his imagination.'[48] Such girls were archetypes, ideas-made-flesh to be exploited in his dramas, not real women with feelings whom he wished to like or love for their own sakes.

Hence it is unlikely that Ibsen ever seriously considered having an affair with any of these girls, let alone marrying one. He had deep inhibitions about sex. His physician, Dr Edvard Bull, said he would not expose his sexual organ even for the purpose of medical examination. Was there something wrong with it – or did he think there was? One is tempted to call Ibsen, who theoretically at least had a profound understanding of female psychology, the male equivalent of a flirt. He certainly led Emilie on. She was over-imaginative and no doubt silly, and she had no idea Ibsen was using her. In February 1891 he broke off their correspondence, having got what he wanted. The same month, the critic Julius Elias related that over a lunch in Berlin Ibsen told him that:

he had met in the Tyrol ... a Viennese girl of very remarkable character, who had at once made him her confidant ... she was not interested in the idea of marrying some decently brought-up young man ... What tempted, fascinated and delighted her was to lure other women's husbands away from them. She was a demonic little wrecker ... a little bird of prey, who would gladly have included him among her victims. He had studied her very very closely. But she had had no great success with him. 'She did not get hold of me but I got hold of her – for my play.'[49]

In short, Ibsen simply used Emilie to get the idea for one of his characters, Hilde Wangel in *The Master Builder*, transforming her in the process and turning her into a reprehensible character. Not only Elias's account but in due course Ibsen's letters were published and poor Emilie was identified with Hilde.[50] For more than half of her long life (she remained unmarried and lived to be ninety-two) she was branded as a wicked woman. This was characteristic not only of the way in which Ibsen pitched real people into his fictional brews but of his cruel disregard for their feelings in carelessly exposing them. The worst case of all was Laura Kieler, an unhappy young Norwegian woman whom Ibsen had met a few times. She was very much under the influence of her husband and in order, as she thought, to help him she stole; when she was detected, he treated her as an embarrassment and disgrace and had her put in a lunatic asylum for a time. Ibsen saw her as a symbol of the oppression of woman – another idea-made-flesh rather than a real person – and used her to create his fictional character of Nora in *A Doll's House*. The immense, worldwide publicity this brilliant play attracted naturally cast a fierce spotlight on Laura, who was widely identified as the original. She was distressed and wanted Ibsen to state publicly that Nora was not her. It would have cost him nothing to do so and the letter in which he refused is a masterpiece of mean-spirited humbug: 'I don't quite understand what Laura Kieler really has in mind in trying to drag *me* into these squabbles. A statement from me, such as she proposes, to the effect that "she is not Nora" would be both meaningless and absurd, since I have never suggested that she is ... I think you will agree that I can best serve our mutual friend by remaining silent.'

Ibsen's ruthless character-exploitation embraced both those closest to him and virtual strangers. The play which wrecked Emilie's life also damaged and hurt his wife, since Suzannah was understandably identified as the wife of Solness in *The Master Builder*, the co-architect and victim of an unhappy marriage. Yet another character in this play, Kaja Fosli, was an act of human larceny. A woman was surprised to get several invitations to dine with Ibsen, happily did so, was again mildly surprised when they abruptly ceased – then understood all when she saw the play and recognized bits of herself in Kaja. She had been used.

Ibsen often wrote about love and it was, after all, the principal theme of his poetry, if only in a negative sense of expressing the ache of loneliness. But it is doubtful if he ever did, or could, feel love for a particular person as opposed to an idea or a person-as-idea. Hate was a far more genuine emotion for him. Behind the hate was a still more fundamental feeling – fear. In the innermost recesses of Ibsen's personality was an all-pervasive, unspoken, unspeakable dread. It was probably the most important thing about him. His timidity he inherited from his mother, who at every opportunity would lock herself in her room. Ibsen, too, as a child, would bolt himself in. Other children noticed his fear – that he was afraid to cross the ice on a sleigh, for instance – and 'cowardice', both physical and moral, was a word constantly applied to him by observers throughout his life.

There was one particularly dark incident in his life, which occurred in 1851 when he was twenty-three and writing anonymous articles for the radical newspaper *Arbejderforeningernes Blad*. In July that year the police raided its offices and arrested two of his friends, Theodor Abild-gaard and the workers' leader Marcus Thrane. Happily for Ibsen, the police did not find anything in the office papers to link him to the articles. Terrified, he lay low for many weeks. The two men were sentenced and spent seven years in jail. Ibsen was too cowardly either to come forward on their behalf or to protest against the savage punishment.[51] He was a man of words, not deeds. He was incensed when Prussia invaded Denmark in 1864 and annexed Schleswig-Holstein, and fur-iously denounced Norway's pusillanimity in failing to come to Den-mark's aid – 'I had to get away from all that swinishness up there to become cleansed,' he wrote.[52] But he did not actually do anything to help Denmark. When a young Danish student, Christopher Bruun, who had volunteered and fought in the army, asked Ibsen – having heard his vociferous views – why he too had not volunteered, he got the lame reply: 'We poets have other tasks to perform.'[53] Ibsen was cowardly in personal as well as political matters. His relationship with his first love, Henrikke Holst, broke up simply because, when her formidable father found them sitting together, Ibsen literally ran off, terrified. Many years later when she was married, the following conversation occurred between them: Ibsen: 'I wonder why nothing ever came of our relation-ship?' Henrikke: 'Don't you remember? – you ran away.' Ibsen: 'Yes, yes, I never was a brave man face to face.'[54]

Ibsen was an elderly, frightened child who became an old woman early in life. The list of things he feared was endless. Vilhelm Bergsoe describes him on Ischia, in 1867, petrified that the cliffs or rocks would collapse and scared of the height, screaming: 'I want to get out, I want to go home.' Walking the streets, he was always worried that a tile would fall on his head. Garibaldi's rebellion upset him dreadfully as he feared blood in the streets. He worried about the possibility of earth-quakes. He was scared of going in a boat: 'I won't go out with those Neopolitans. If there's a storm they'll lie flat in the boat and pray to the Virgin Mary instead of reefing the sails.' Another fear was a cholera outbreak – indeed, contagious disease was always a prime worry. He wrote to his son Sigurd, on 30 August 1880: 'I much dislike the idea of your luggage being deposited at Anna Daae's hospital. The children she attends are from a class of people among whom one might expect epidemics of smallpox to be rampant.'[55] He worried about storms, both on sea and land, about bathing ('can easily bring on a fatal attack of cramp'), about horses ('well known for their habit of kicking') and anyone with a sporting gun ('keep well away from people carrying such weapons'). He was particularly scared of carriage accidents. He was so obsessed by the danger of hailstones that he took to measuring their circumference. To the annoyance of children, he insisted on blowing out the candles on Christmas trees because of the fire risk. His wife had no need to frighten him by reading out tales of disasters from the newspapers because he scoured the press himself – it was his chief source of material for plots – and fearfully studied accounts of horrors, both natural and man-made. His letters to Sigurd are extraordinary catalogues of warnings – 'I read in almost every Norwegian paper of accidents caused by careless handling of loaded firearms' – and pleas for circumspection: 'Wire if there is the slightest accident.' 'The least carelessness can have the gravest consequences.' 'Be cautious and careful in every way.'[56]

His greatest horror was dogs. Bergsoe relates that on one occasion in Italy he became frightened of a harmless dog and suddenly began to run. The dog then chased and bit him. Ibsen shouted: 'The dog is mad and must be shot, otherwise I shall go mad too.' He was 'foaming with rage and it was several days before his fear departed'. Knudtzon records a more striking, indeed sinister, incident, also in Italy. Ibsen and other Scandinavians lunched together in a restaurant and drank a lot of wine: 'There was thunder in the air. Ibsen seemed from the start to have some worm of indignation in the depths of his soul. [It] weighed on him and demanded an outlet.' When they rose to go Ibsen could not stand and two of them had to help him to walk. His attention was caught by an iron gate and 'a huge dog behind it which barked angrily at us'. Then:

> Ibsen had a stick in his hand which he now began to poke at the dog, one of those gigantic brutes which resemble small lions. It came closer and Ibsen poked and struck at it, trying in every way to madden it, and succeeding. It rushed at the gate, Ibsen prodded and struck it anew, and worked it into such a rage that, without doubt, had not the solid iron gate stood between it and us, it would have torn us apart . . . Ibsen must have stood teasing the dog for six or eight minutes.[57]

As this incident suggests, Ibsen's life-long rage and his perpetual fears were closely linked. He raged because he feared. Alcohol anaesthetized the fear but unleashed the fury too; inside the angry man a fearful one cowered. Ibsen lost his faith early, or so he said, but he carried the fear of sin, and punishment, to the grave. He hated jokes about religion: 'Some things one doesn't make fun of.' He claimed that Christianity 'demoralizes and inhibits both men and women' but he remained intensely superstitious himself. He may not have believed in God but he feared devils. He wrote in a copy of *Peer Gynt*: 'To live is to war with trolls in heart and soul.' Bjornson wrote to him: 'There are many goblins in your head which I think you ought to placate ... a dangerous army to have around for they turn on their masters.' Ibsen knew this well enough. He spoke of his 'super-devil' – 'I lock my door and bring him out.' He said: 'There must be troll in what I write.' In his desk he kept a collection of small rubber devils with red tongues.[58] There were times, after a few glasses of spirits, when his reasoned critique of society collapsed into incoherence and fury, and he seemed a man possessed by devils. Even William Archer, his greatest advocate, thought his political and philosophical views, when closely examined, not so much radical as merely chaotic: 'I am becoming more and more convinced,' he wrote in 1887, 'that as a many-sided thinker, or rather a systematic thinker, Ibsen is nowhere'. Archer thought he was simply contrary, against every established idea on principle. Ingvald Undset, father of the novelist Sigrid Undset, who listened to his half-tipsy rantings in Rome, recorded: 'he is a complete anarchist, wants to wipe everything out ... mankind must start from the foundations to rebuild the world ... Society and everything else must be wiped out ... the great task of our age is to blow the existing fabric into the air.' What did it all mean? Very little, really: just the fallout from fear and hate contending for mastery in a heart which did not know, or could not express, love. The bars of the northern world are full of men holding forth in similar fashion.

In his last years, which began with an apoplectic fit in 1900, recurring on a smaller scale at intervals, Ibsen continued the alternating pattern of worry and rage, watched by his sardonic wife. His chief anxiety now was insurance, while his main source of irritation was physical debility and an intense dislike of being helped. Fury, as usual, got the upper hand. The resident nurse was told to disappear as soon as she had helped Ibsen into the street. When she failed to do so, 'Ibsen swung his stick at her so that she fled back into the house.' A barber came to shave him every day. Ibsen never spoke a word to him except once when he suddenly hissed: 'Ugly devil!' He died on 23 May 1906. Suzannah later claimed that, just before he did so, he said: 'My dear, dear wife, how good and kind you have been to me!' This seems totally out of character. In any case Dr Bull's diary makes it clear he was in a coma that afternoon and incapable of speech. Another, and far more plausible, account gives his last words as: 'On the contrary!'

5

Tolstoy: God's Elder Brother

*O*F all the intellectuals we are examining, Leo Tolstoy was the most ambitious. His audacity is awe-inspiring, at times terrifying. He came to believe that by the resources of his own intellect, and by virtue of the spiritual force he felt welling within him, he could effect a moral transformation of society. His aim, as he put it, was 'To make of the spiritual realm of Christ a kingdom of this earth.'[1] He saw himself as part of an apostolic succession of intellectuals which included 'Moses, Isaiah, Confucius, the early Greeks, Buddha, Socrates, down to Pascal, Spinoza, Feuerbach and all those, often unnoticed and unknown, who, taking no teaching on trust, thought and spoke sincerely upon the meaning of life'. But Tolstoy had no intention of remaining 'unnoticed and unknown'. His diaries reveal that, as a young man of twenty-five, he was already conscious of special power and a commanding moral destiny. 'Read a work on the literary characterization of genius today, and this awoke in me the conviction that I am a remarkable man both as regards capacity and eagerness to work.' 'I have not yet met a single man who was morally as good as I, and who believed that I do not remember an instance in my life when I was not attracted to what is good and was not ready to sacrifice anything to it.' He felt in his own soul 'immeasurable grandeur'. He was baffled by the failure of other men to recognize his qualities: 'Why does nobody love me? I am not a fool, not deformed, not a bad man, not an ignoramus. It is incomprehensible.'[2] Tolstoy always felt a certain apartness from other men, however much he tried to sympathize and identify himself with them. In a curious way he felt himself sitting in judgment over them, exercising moral jurisdiction. When he became a novelist, perhaps the greatest of all novelists, he effortlessly assumed this godlike power. He told Maxim Gorky: 'I myself, when I write, suddenly feel pity for some character, and then I give him some good quality, or take a good quality away from someone else, so that in comparison with the others he may not appear too black.'[3] When he became a social reformer, the identification with God became stronger, since his actual programme was co-extensive with divinity as he defined it: 'The desire for universal welfare ... is that which we call God.' Indeed, he felt himself divinely possessed, noting in his diary: 'Help, father, come and dwell within me. You already dwell within me. You are already "me".'[4] But the difficulty about both Tolstoy and God dwelling in the same soul was that Tolstoy was extremely suspicious of his Creator, as Gorky noted. It reminded him, he said, of 'two bears in one den'. There were times when Tolstoy seemed to think of himself as God's brother, indeed his elder brother.

How did Tolstoy come to feel about himself in this way? Perhaps the largest single element in his sense of majesty was his own birth. Like Ibsen he was born in 1828, but as a member of the hereditary ruling class in a vast country which, for the next thirty years, was to retain the form of slavery called serfdom. Under this, serf families, men, women and children, were bound by law to the land they tilled, and ownership of them went with the title deeds. Some noble families had as many as 200,000 serfs when the institution was abolished in 1861. The Tolstoys were not rich by these standards; Tolstoy's father and grandfather had both been spendthrifts, and the father saved himself only by marrying the plain daughter of Prince Volkonsky. But the Volkonskys were of the very highest rank, co-founders of the realm, on a social level with the Romanovs when their dynasty emerged in 1613. Tolstoy's maternal grandfather had been Catherine the Great's Commander-in-Chief. His mother's dowry included the Yasnaya Polyana estate near Tula, and Tolstoy inherited it from her, with its 4000 acres and 330 serfs.

In his young days Tolstoy thought little of landed responsibilities and indeed sold off portions of his estate to pay gambling debts. But he was proud, indeed vain, of his title and lineage and of the entry it gave him to fashionable salons. He appalled his literary friends with his posing and snobbery. 'I cannot understand,' wrote Turgenev, 'this ridiculous affection for a wretched title of nobility.' 'He disgusted us all,' was Nekrasov's comment.[5] They resented the way he tried to get the best of both worlds, high society and bohemia. 'Why do you come here among us?' asked Turgenev angrily. 'This is not the place for you – go to your princess.' As he grew older, Tolstoy abandoned the more meretricious aspects of his caste but developed instead a land-hunger which went much deeper, using his literary earnings to buy land, piling hectare on hectare with all the grim cupidity of a dynastic founder himself. Until the moment came when he decided to give it all up, he not merely owned land but ruled it. His spirit was authoritarian, springing directly from hereditary title to earth and souls. 'The world was divided into two parts,' his son Ilya wrote, 'one composed of ourselves and the other of everyone else. We were special people and the others were not our equals... [My father] was responsible to a considerable degree for the groundless arrogance and self-esteem that such an upbringing inculcated into us, and from which I found it so hard to free myself.'[6] To the last Tolstoy retained the belief he was born to rule, in one way or another. In old age, wrote Gorky, he remained the Master, the *barin*, expecting his wishes to be obeyed instantly.

Together with this fundamental desire to rule came a fierce unwillingness to be ruled by others. Tolstoy had an adamantine will which circumstances helped to harden. Both his parents died when he was young. His three elder brothers were weak, unfortunate, dissolute. He was brought up by his Aunt Tatiana, a penniless second cousin, who did her best to teach him duty and unselfishness, but she had no authority over him. His account of his early years, 'Boyhood', and his diaries mislead the reader, like Rousseau's, by their apparent honesty but really conceal more than they reveal. Thus he describes being beaten by a ferocious tutor, Monsieur de Saint-Thomas, 'one reason for that horror and aversion for every kind of violence which I have felt throughout my whole life'.[7] In fact there were many kinds of violence, including his own violent nature, which did not dismay Tolstoy until late in life. As for Saint-Thomas, Tolstoy had got the better of him by the age of nine and thereafter his life was as indisciplined as he chose to make it. At school he read what he wanted and worked when he felt like it (often very hard). By the age of twelve he was writing poetry. At sixteen he went to Kazan University on the Volga and for a time studied Oriental languages with a view to a diplomatic career. Later he tried law. At nineteen he gave up university and returned to Yasnaya Polyana to study by himself. He read fashionable fiction – de Kock, Dumas, Eugène Sue. He also read Descartes and, above all, Rousseau. In a number of important respects he was a posthumous pupil of Rousseau; at the end of his life he said that Rousseau had had more influence on him than anyone else, except the Jesus Christ of the New Testament. He saw in Rousseau a fellow spirit, another gigantic ego, conscious of superlative goodness, anxious to impart it to the world. Like Rousseau he was essentially self-educated, with all the pride, insecurity and intellectual touchiness of the autodidact. Like Rousseau he tried many things before settling down to a career as a writer – diplomacy, law, educational reform, agriculture, soldiering, music.

Tolstoy found his *métier* almost by accident, while serving as an apprentice officer in the army. In 1851, aged twenty-two, he went to the Caucasus, where his elder brother Nikolai was on active service. He had no real motive in going there, other than to do something, to fill in time, and to win medals which would serve him well in the salons. He was in the army the best part of five years, first in mountain frontier-warfare, then in the Crimea against the British, French and Turks. He had the assumptions and attitudes of a Russian imperialist. On being accepted by the army and assigned to a gun-battery – the natives had no artillery – he wrote to his brother Sergei: 'With all my strength I shall help with my guns in the destruction of the predatory and turbulent Asiatics.'[8] Indeed, he never repudiated his Russian imperialism or the chauvinist spirit, the conviction that the Russians were a special race, with unique moral qualities (personified in the peasant) and a God-ordained role to perform in the world.

These were the simple, unspoken beliefs of his fellow officers. Tolstoy reflected them. But in other ways he felt himself different. 'Once and for all,' he noted in his diary, 'I must accustom myself to the thought that I am an exception, that either I am ahead of my age or that I am one of those incongruous, unadaptable natures that will never be content.'[9] Army opinion about him differed. Some thought him modest. Others saw in him 'an incomprehensible air of importance and self-satisfaction'.[10] They all noticed his fierce, implacable gaze, his at times terrible eyes; he could stare down anyone. No one disputed his bravery, in or out of action. It was a function of his huge will. As a boy he had forced himself to ride. He had overcome shyness. He had likewise made himself hunt, including the dangerous sport of bear-hunting; as a result of his own arrogant carelessness he was badly mauled and nearly killed in his first bear-hunt. In the army he showed himself brave under fire, and this eventually brought him promotion to full lieutenant. But his efforts to get himself medals came to nothing. He was three times recommended for one but at some level the award was blocked. An eagerness for gongs is easily spotted in armies, and disliked. The fact is, Tolstoy was not a satisfactory officer; he lacked not only humility and willingness to obey and learn, but solidarity with his comrades. He was a loner, out for himself, and if there was nothing going on which helped his career, he simply left the front, often without permission or telling anyone. His colonel noted: 'Tolstoy is eager to smell powder, but only fitfully.' He tended to avoid 'the difficulties and hardships incidental to war. He travels about to different places like a tourist, but as soon as he hears firing he at once appears on the field of battle. When it is over he is off again at his own discretion whenever the fancy takes him.'[11]

Tolstoy, then and always, liked drama. He was willing to make a sacrifice, of comfort, pleasure, even life, provided it could be done as a grand, theatrical gesture, and everyone noticed. As a student, to stress his Russian fortitude, he had made himself a combined poncho-sleeping bag; it was a gesture and it aroused comment. In the army he was willing to perform but not, as it were, to serve. The routine discomforts and hardships, the aspects of army life which had no potential celebrity-value and went unobserved, did not interest him. So it was to be always: his heroism, his virtue, his sanctity were for the public stage, not for the dull, unrecorded routine of everyday life.

But in one respect his army career was truly heroic. During it he made himself a writer of prodigious power. It is obvious in retrospect that Tolstoy was a born writer. Obvious too, from his later descriptions, that from a very early age he observed nature and people with an accuracy of detail which has never been surpassed. But born writers do not always become them. The point at which Tolstoy's two outstanding gifts came together occurred when he first saw the Caucasus mountains on his way to join the army. The almost supernatural splendour of the sight not only excited his intense visual appetite and stirred a still-dormant urge to set it down in words but evoked his third outstanding characteristic – his sense of God's majesty and his desire in some way to mingle himself with it. He was soon writing *Childhood*, then stories and sketches of army life: 'The Raid', 'The Cossacks', 'The Woodfelling', 'Notes of a Billiard Marker', three 'Sebastapol Sketches', 'Boyhood' (part of *Youth*), 'A Landlord's Morning', 'Christmas Eve'. *Childhood* was sent off in July 1852 and published with considerable success. 'The Cossacks' was not finished for another ten years, 'Christmas Eve' never and some material, the campaign against the Chechen chief Shamyl, Tolstoy saved for his last, brilliant story, 'Hadji Murad', which he wrote as an old man. But the remarkable thing was that this considerable body of work was produced in brief intervals of active soldiering or even in the front line, and at a time when Tolstoy, by his own account, was also chasing Cossack women, gambling and drinking. The drive to write must have been overpowering, the industry and will required to satisfy it awesome.

Yet this drive to write was intermittent, and therein lay Tolstoy's tragedy. Sometimes he wrote with exhilaration, proudly conscious of his power. Thus, in October 1858: 'I will spin such a yarn there will be no head or tail to it.' Early 1860: 'I am working at something that comes as naturally to me as breathing and, I confess with culpable pride, enables me to look down on what the rest of you are doing.'[12] Not that his writing was ever easy. He set himself high standards; the work was exacting and arduous. Most of the vast bulk of *War and Peace* went through at least seven drafts. There were even more drafts and revisions of *Anna Karenina*, and the changes were of fundamental importance – we see, in these successive revisions, the metamorphosis of Anna from a disagreeable courtesan into the tragic heroine we know.[13] From the enormous trouble Tolstoy took with his work at its best it is clear he was conscious of his high calling as an artist. How can he not have been? There are times when he writes better than anyone who has ever lived, and surely no one has depicted nature with such consistent truth and thoroughness. 'The Snowstorm', written in 1856, which records his near-death in a blizzard while returning from the Caucasus to Yasnaya, an early example of his mature technique, has an almost mesmeric power. This is achieved directly, by the selection and accuracy of detail. He does not use overtones or undertones, poetry or suggestion. As Edward Crankshaw pointed out, he is like a painter who disdains shadows and chiaroscuro, employing only perfect clarity and visibility.[14] Another critic has compared him to a Pre-Raphaelite painter: shapes, textures, tones and colours, sounds, smells, sensations are all conveyed with crystalline transparency and directness.[15] Here are two examples, both passages evolved from many revisions. First, the extrovert Vronsky:

'Good, splendid!' he said to himself, crossing his legs and, taking a leg in his hand, felt the springy muscle of his calf where he had bruised it the day before in his fall . . . He enjoyed the slight ache in his strong leg, he enjoyed the muscular sensation of movement in his chest as he breathed. The bright, cold August day, which had made Anna feel so hopeless, seemed exhilarating to him . . . Everything he saw through the carriage window was as fresh and jolly and vigorous as he was himself: the roofs of the houses shining in the setting sun, the sharp outlines of fences and angles of buildings, even the fields of potatoes: everything was beautiful, like a lovely landscape fresh from the artist's brush and lately varnished.

Then here is Levin shooting snipe with his dog Laska:

The moon had lost all its lustre and was like a white cloud in the sky. Not a single star could be seen. The sedge, silvery before, now shone like gold. The stagnant pools were all like amber. The blue of the grass had changed to yellow-green . . . A hawk woke up and settled on a haystack, turning its head from side to side and looking discontentedly at the marsh. Crows were flying about the field and a bare-legged boy was driving the horses to an old man who had got up from under his coat and was combing his hair. The smoke from the gun was white as milk over the green of the grass.[16]

It is clear that Tolstoy's writing power sprang directly from his veneration for nature, and that he retained both the capacity and the excitement, if spasmodically, to the end. In his diary for 19 July 1896 he records seeing a tiny shoot of burdock, still living, in a ploughed field, 'black from dust but still alive and red in the centre... It makes me want to write. It asserts life to the end, and alone in the midst of the whole field, somehow or other had asserted it.'[17] When Tolstoy was seeing nature, with that cold, terrible, exact eye of his, and setting it down in words with his precise, highly-calibrated pen, he was as close to happiness, or at any rate peace of spirit, as his character permitted.

Unfortunately, writing alone did not satisfy him. He had a will to power. The authority he exercised over his characters was not enough. For one thing, he did not feel part of them. They were a different race, almost a different species. Just occasionally, above all in the character of Anna, he works himself by prodigies of effort into the mind of the person he is describing, and the fact that he does so with such success in this case reminds one of the dangers of generalizing about this extraordinary man. But as a rule he sees from outside, from afar, most of all from above. His serfs, his soldiers, his peasants are brilliantly rendered animals; he describes horses – Tolstoy had great knowledge and understanding of horses – just as well and in the same spirit. He sees for us, as he takes us through the course of a great battle, almost as though he were observing it from another planet. He does not feel for us. We do feel, as a result of his selective sight on our behalf, and therefore he controls our feelings: we are in the grip of a great novelist. But he does not feel himself. He remains disengaged, aloof, Olympian. Compared to his older contemporary Dickens, his near-contemporary Flaubert – both novelists moving on the same high plane of creation – Tolstoy invested comparatively little of his emotional capital in his fiction. He had, or he thought he had, better things to do with it.

We think of Tolstoy as a professional novelist, and of course in a sense this is true. In both his major works he exercised what can only be called genius: organizing masses of detail into the purposeful pattern of great themes, carried through to their relentless conclusions. Being a true artist, he did not repeat himself: *War and Peace* surveys a whole society and epoch, *Anna Karenina* focuses closely on a particular group of people. These books made him a national hero, brought him worldwide fame, wealth and a reputation for moral sagacity which perhaps no other novelist has ever enjoyed. But most of his life he was not writing fiction at all. There were three creative periods: the early stories in the 1850s; the six years he spent producing *War and Peace* in the 1860s; the creation of *Anna Karenina* in the 1870s. The rest of his long life he was doing and being a multitude of other things, which in his view had higher moral priority.

Aristocrats under the old order found it difficult to shake off the notion that writing was for their inferiors. Byron never regarded poetry as his most important work, which was to assist the subject peoples of Europe to achieve their independence. He felt himself called to lead, as befitted his class. So did Tolstoy. Indeed he felt called to do more than lead: to prophesy, at times to play the Messiah. What, then, was he doing, spending his time writing? 'To write stories,' he told the poet Fet, 'is stupid and shameful.' Note the second adjective. This was an intermittent theme, that art was an outrageous misuse of God's gifts, which Tolstoy orchestrated in ever more sonorous terms when the iconoclastic mood came upon him. So from time to time, and increasingly as he grew older, he would renounce art and exert moral leadership.

Now here was a disastrous case of self-deception. It is remarkable that Tolstoy, who thought about himself as much as any man who ever lived – including even Rousseau – who wrote about himself copiously, and much of whose fiction revolves around himself in one way or another, should have been so conspicuously lacking in self-knowledge. As a writer he was superlatively qualified; and while he was writing he was least dangerous to those around him and to society generally. But he did not wish to be a writer, at any rate of profane matter. Instead he wanted to lead, for which he had no capacity at all, other than will; to prophesy, to found a religion, and to transform the world, tasks for which he was morally and intellectually disqualified. So great novels remained unwritten, and he led, or rather dragged, himself and his family into a confused wilderness.

There was a further reason why Tolstoy felt impelled to set himself to great moral tasks. Like Byron, he knew himself to be a sinner. Unlike Byron, he felt an overwhelming sense of guilt about it. Tolstoy's guilt was a selective and inaccurate instrument – some of his worst failings, even crimes, the atrocious products of his overweening ego, he did not see as sins at all – but it was a very powerful one. And to be sure there was a very great deal in his youthful life to feel guilty about. He seems to have learned to gamble heavily in Moscow and St Petersburg early in 1849. On 1 May he wrote to his brother Sergei: 'I came to St Petersburg for no good reason, I've done nothing worth doing here, simply run through a lot of money and fallen into debt.' He told Sergei to sell part of the estate at once: 'While I am waiting for the money to come through I must absolutely have 3500 roubles straight away.' He added: 'You can commit this kind of idiocy once in a lifetime. I have had to pay for my freedom (there was no one to thrash me; that was my chief misfortune) and for philosophizing, and now I have paid.'[18] In fact he went on to gamble, intermittently, sometimes heavily and disastrously, for the next ten years, in the process selling much of his estates and accumulating debts to relations, friends and tradesmen, many of which were never repaid. He gambled in the army. At one stage he planned to start an army paper, to be called *The Military Gazette*, and sold the central block of Yasnaya Polyana to finance it; but when the cash, 5000 roubles, arrived he used it for gambling and promptly lost it. After he left the army, and travelled in Europe, he gambled again, with the same result. The poet Polonsky, who observed him in Stuttgart in July 1857, recorded: 'Unfortunately roulette attracted him violently... [He] has been completely plucked at play. He dropped 3000 francs and is left without a sou.' Tolstoy himself wrote in his diary: 'Roulette till six. Lost everything.' 'Borrowed 200 roubles from the Frenchman and lost it.' 'Borrowed money from Turgenev and lost it.'[19] Years later his wife was to note that, while he felt guilty about his gambling as such, and had renounced it, he seemed to feel no compunction about his failure to settle the debts he acquired at this time, some of them owed to poor men. There was nothing dramatic about paying an old debt.

Tolstoy had an even stronger sense of guilt about his sexual desires and their satisfaction, though here again his self-castigations were curiously selective and even indulgent to himself. Tolstoy believed himself to be very highly sexed. Diary entries record: 'Must have a woman. Sensuality gives me not a moment's peace' (4 May 1853). 'Terrible lust amounting to a physical illness' (6 June 1856).[20] At the end of his life he told his biographer Aylmer Maude that, so strong were his urges, he was unable to dispense with sex until he was eighty-one. In youth he was extremely shy with women and so resorted to brothels, which disgusted him and brought the usual consequences. One of his earliest diary entries in March 1847 notes he is being treated for 'gonorrhoea, obtained from the customary source'. He records another bout in 1852 in a letter to his brother Nikolai: 'The venereal sickness is cured but the after-effects of the mercury have caused me untold suffering.' But he continued to patronize whores, varied by gypsies, Cossack and native girls, and Russian peasant girls when available. The tone in his diary entries is invariably self-disgust blended with hatred for the temptress: 'something pink... I opened the back door. She came in. Now I can't bear to look at her. Repulsive, vile, hateful, causing me to break my rules' (18 April 1851). 'Girls have led me astray' (25 June 1853). The following day he made good resolution but 'the wenches prevent me' (26 June 1853). An entry for April 1856 records, after a visit to a brothel: 'Horrible, but absolutely the last time.' Another 1856 entry: 'Disgusting. Girls. Stupid music, girls, heat, cigarette smoke, girls, girls, girls.' Turgenev, whose house he was then using like a hotel, gives another glimpse of Tolstoy in 1856: 'Drinking bouts, gypsies, cards all night long, and then sleeps like the dead until two in the afternoon.'[21]

When Tolstoy was in the country, especially on his own estate, he took his pick of the prettier serf-girls. These occasionally excited more than simple lust on his part. He wrote later of Yasnaya Polyana, 'I remember the nights I spent there, and Dunyasha's beauty and youth... her strong, womanly body.'[22] One of Tolstoy's motives in travelling in Europe in 1856 was to escape what he saw as the temptations of an attractive serf-girl. His father, as he knew, had had such an affair, and the girl had given birth to a son, who was simply treated as a male estate serf, being employed in the stables (he became a coachman). But Tolstoy, after his return, could not keep his hands off the women, especially a married one called Aksinya. His diary for May 1858 records: 'Today, in the big old wood. I'm a fool, a brute. Her bronze flesh and her eyes. I'm in love as never before in my life. Have no other thought.'[23] The girl was 'clean and not bad-looking, with bright black eyes, a deep voice, a scent of something fresh and strong and full breasts that lifted the bib of her apron'. Probably in July 1859, Aksinya gave birth to a son, called Timofei Bazykin. Tolstoy brought her into the house as a domestic and allowed the little boy to play at her heels for a time. But, like Marx and Ibsen, and like his own father, he never acknowledged the child was his, or paid the slightest attention to him. What is even more remarkable is that, at a time when he was publicly preaching the absolute necessity to educate the peasants, and indeed ran schools for their children on his estate, he made no effort to ensure that his own illegitimate son even learned how to read and write. Possibly he feared later claims. He seems to have been pitiless in dismissing the rights of illegitimate offspring. He resented the fact, perhaps because it showed up his own behaviour, that Turgenev not only acknowledged his illegitimate daughter but took pains to bring her up in a suitable manner. On one occasion Tolstoy insulted the poor girl, alluding to her birth, and this led to a serious quarrel with Turgenev, which nearly ended in a duel.[24] So Tolstoy's son Timofei was put to work in the stables; later, on the grounds of bad behaviour, he was demoted to woodsman. There is no further record of Timofei after 1900, when he was forty-three, but we know he was befriended by Tolstoy's son Alexei, who made him his coachman.

Tolstoy knew he was doing wrong in resorting to prostitutes and seducing peasant women. He blamed himself for these offences. But he tended to blame the women still more. They were all Eve the Temptress to him. Indeed it is probably not too much to say that despite the fact that he needed women physically all his life and used them – or perhaps because of this – he distrusted, disliked and even hated them. In some ways he found the manifestation of their sexuality repulsive. He remarked at the end of his life, 'the sight of a woman with her breasts bared was always disgusting to me, even in my youth.'[25] Tolstoy was by nature censorious, even puritanical. If his own sexuality upset him, its manifestations in others brought out his strongest disapproval. In Paris in 1857, at a time when his own philandering was surging in full spate, he noted: 'At the furnished lodgings where I stayed, there were thirty-six ménages, of which nineteen were irregular. That disgusted me terribly.'[26] Sexual sin was evil, and women were the source of it. On 16 June 1847, when he was nineteen, he wrote:

> Now I shall set myself the following rule. Regard the company of women as an unavoidable social evil and keep away from them as much as possible. Who indeed is the cause of sensuality, indulgence, frivolity and all sorts of other vices in us, if not women? Who is to blame for the loss of our natural qualities of courage, steadfastness, reasonableness, fairness, etc if not women?

The really depressing thing about Tolstoy is that he retained these childish, in some respects Oriental, views of women right to the end of his life. In contrast to his efforts to portray Anna Karenina, he never seems to have made any serious attempt in real life to penetrate and understand the mind of a woman. Indeed he would not admit that a woman could be a serious, adult, moral human being. He wrote in 1898, when he was seventy: '[Woman] is generally stupid, but the Devil lends her brains when she works for him. Then she accomplishes miracles of thinking, farsightedness, constancy, in order to do something nasty.' Or again: 'It is impossible to demand of a woman that she evaluate the feelings of her exclusive love on the basis of moral feeling. She cannot do it, because she does not possess real moral feeling, i.e. one that stands higher than everything.'[27] He disagreed strongly with the emancipationist views expressed in J.S. Mill's *The Subjection of Women*, arguing that even unmarried women should be barred from entering a profession. Indeed, he regarded prostitution as one of the few 'honourable callings' for women. The passage in which he justifies the whore is worth quoting:

Should we permit promiscuous sexual intercourse, as many 'liberals' wish to do? Impossible! It would be the ruin of family life. To meet the difficulty, the law of development has evolved a 'golden bridge' in the form of the prostitute. Just think of London without its 70,000 prostitutes! What would become of decency and morality, how would family life survive without them? How many women and girls would remain chaste? No, I believe the prostitute is necessary for the maintenance of the family.[28]

The trouble with Tolstoy was that, while he believed in the family he did not really believe in marriage; at any rate in a Christian marriage between adults with equal rights and duties. No one who ever lived, perhaps, was less suited to such an institution. A neighbouring girl in the country, a twenty-year-old orphan called Valerya Arsenev, had a lucky escape. He conceived an attachment for her while he was in his late twenties and for a time considered himself her fiancé. But he only liked her childish aspects; her more womanly, mature side, as it emerged, repelled him. His diaries and letters tell the tale. 'A pity she has no bone, no fire – a pudding.' But 'her smile is painfully submissive.' She was 'badly educated, ignorant, indeed stupid . . . I began to needle her so cruelly that she smiles uncertainly, tears in her smile.' After dithering for eight months and lecturing her unmercifully, he provoked her into an irritable letter and used this as an excuse for breaking it off: 'We are too far apart. Love and marriage would have given us nothing but misery.' He wrote to his aunt: 'I have behaved very badly. I have asked God to pardon me . . . but to mend this matter is impossible.'[29]

His choice finally fell, when he was thirty-four, on an eighteen-year-old doctor's daughter, Sonya Behrs. He was no great catch: not rich, a known gambler, in trouble with the authorities for insulting the local magistrate. He had described himself, some years before, as possessing 'the most ordinary coarse and ugly features . . . small grey eyes, more stupid than intelligent . . . the face of a peasant, and a peasant's large hands and feet'. Moreover, he hated dentists and would not visit them, and by 1862 he had lost nearly all his teeth. But she was a plain, immature girl, only five feet high and competing with her two sisters; she was glad to get him. He proposed formally by letter, then seems to have had doubts until the last minute. The actual wedding was a premonition of disaster. On the morning he burst into her apartment, insisting: 'I have come to say that there is still time . . . all this business can still be put a stop to.' She burst into tears. Tolstoy was an hour late for the ceremony itself, having packed all his shirts. She cried again. After-wards they had supper and she changed, and they climbed into a travelling carriage called a dormeuse, pulled by six horses. She cried again. Tolstoy, an orphan, could not understand this and shouted: 'If leaving your family means such great sorrow to you, you cannot love me very much.' In the dormeuse he began to paw her and she pushed him away. They had a suite at a hotel, the Birulevo. Her hands trembled as she poured him tea from the samovar. He tried to paw her again, and was again repulsed. Tolstoy's diary relentlessly recorded: 'She is weepy. In the carriage. She knows everything and it is simple. But she is afraid.' He thought her 'morbid'. Later still, having finally made love to her, and she having (as he thought) responded, he added: 'Incredible happiness. I can't believe this can last as long as life.'[30]

Of course it did not. Even the most submissive wife would have found marriage to such a colossal egotist hard to bear. Sonya had sufficient brains and spirit to resist his all-crushing will, at least from time to time. So they produced one of the worst (and best recorded) marriages in history. Tolstoy opened it with a disastrous error of judgment. It is one of the characteristics of the intellectual to believe that secrets, especially in sexual matters, are harmful. Everything should be 'open'. The lid must be lifted on every Pandora's box. Husband and wife must tell each other 'everything'. Therein lies much needless misery. Tolstoy began his policy of *glasnost* by insisting that his wife read his diaries, which he had now been keeping for fifteen years. She was appalled to find – the diaries were then in totally uncensored form – that they contained details of all his sex life, including visits to brothels and copulations with whores, gypsies, native women, his own serfs and, not least, even her mother's friends. Her first response was: 'Take those dreadful books back – why did you give them to me?' Later she told him: 'Yes, I have forgiven you. But it is dreadful.' These remarks are taken from her own diary, which she had been keeping since the age of eleven. It was part of Tolstoy's 'open' policy that each should keep diaries and each should have access to the other's – a sure formula for mutual suspicion and misery.

The physical side of the Tolstoy marriage probably never recovered from Sonya's initial shock at learning her husband was (as she saw it) a sexual monster. Moreover, she read his diaries in ways which Tolstoy had not anticipated, noting faults he had been careful (as he thought) to conceal. She spotted, for instance, that he had failed to repay debts contracted as a result of his gambling. She observed, too, that he failed to tell women with whom he had sex that he had contracted venereal disease and might still have it. The selfishness and egotism the diaries so plainly convey to the perceptive reader – and who more perceptive than a wife? – were more apparent to her than to the author. Moreover, the Tolstoyan sex life so vividly described in his diaries was now inextricably mingled in her mind with the horrors of submitting to his demands and their ultimate consequence in painful and repeated pregnancies. She endured a dozen in twenty-two years; in quick succession she lost her child Petya, while pregnant with Nikolai, who in turn died the same year he was born; Vavara was born prematurely and died immediately. Tolstoy himself did not help with the business of childbearing by taking an intimate though insensitive interest in all its details. He insisted on attending the birth of his son, Sergei (later using it for a scene in *Anna Karenina*), and broke into a frightening rage when Sonya was unable to breast-feed the baby. As the pregnancies and miscarriages proceeded, and his wife's distaste for his sexual demands became manifest, he wrote to a friend: 'There is no worse situation for a healthy man than to have a sick wife.'

Early in the marriage he ceased to love her; her tragedy was that her residual love for him remained. At this time she confided in her diary:

> I have nothing in me but this humiliating love and a bad temper,
> and these two things have been the cause of all my misfortunes, for
> my temper has always interfered with my love. I want nothing but
> his love and sympathy but he won't give it to me, and all my pride
> is trampled in the mud. I am nothing but a miserable crushed worm,
> whom no one wants, whom no one loves, a useless creature with
> morning sickness and a big belly.[31]

It is hard to believe, on the available evidence, that the marriage was ever bearable. During a comparatively calm period in 1900, when they had been married thirty-eight years, Sonya wrote to Tolstoy: 'I want to thank you for the former happiness you gave me and to regret that it did not continue so strongly, fully and calmly throughout our whole life.' But this was a gesture of appeasement. Sonya, from the start, tried to keep the marriage functioning by making herself the manager of his affairs, in some ways an obsessive one, by rendering him indispensable services, by becoming his rebellious slave. She took on the fearsome task of making fair copies of his novels from his appalling handwriting.[32] This was drudgery but in a way she enjoyed it because she early on grasped that Tolstoy was least unbearable and destructive when he was exercising his true metier. As she wrote to her sister Tatiana, they were all happiest when he was writing his fiction. For one thing, it made money whereas his other activities wasted it. But 'it is not so much the money. The main thing is that I love his literary works, I admire them and they move me.' She learned from bitter experience that once

Tolstoy stopped writing fiction he was capable of filling the vacuum in his life with great folly, certain to hurt the family she was trying to hold together.

Tolstoy saw things quite differently. Raising and maintaining a family required money. His novels made money. He came to associate the writing of fiction with the need to earn money, and so to dislike both. In his mind, novels and marriage were linked, and the fact that Sonya was always pressing him to write fiction confirmed the link. And both marriage and novels, he now realized, were preventing him from taking on his real work of prophecy. As he put it in his *Confessions*:

> The new conditions of happy family life completely diverted me from all search for the general meaning of life. At that time my whole existence was centred on my family, my wife, my children, and therefore on concern for the increase of our means of livelihood. My striving after self-perfection, for which I had already substituted a striving for perfection in general, for progress, was ... replaced by the effort simply to secure the best possible conditions for my family.[33]

Hence Tolstoy came to see marriage not only as a source of great unhappiness but as an obstacle to moral progress. He generalized from the particular disaster of his own to inveighing against the institution and marital love itself. In 1897, in a Lear-like outburst, he told his daughter Tanya:

> I can understand why a depraved man may find salvation in marriage. But why a pure girl should want to get mixed up in such a business is beyond me. If I were a girl I would not marry for anything in the world. And so far as being in love is concerned, for either men and women – since I know what it means, that is, it is an ignoble and above all an unhealthy sentiment, not at all beautiful, lofty or poetical – I would not have opened my door to it. I would have taken as many precautions to avoid being contaminated by that disease as I would to protect myself against far less serious infections such as diphtheria, typhus or scarlet fever.[34]

This passage suggests, as indeed does much else, that Tolstoy had not thought seriously about marriage. Take the famous sentence from *Anna Karenina*: 'All happy families are alike, but each unhappy family is unhappy in its own way.' The moment one begins to search one's own observed experience, it becomes clear that both parts of this statement are debatable. If anything, the reverse is closer to the truth. There are obvious, recurrent patterns in unhappy families – where, for instance, the husband is a drunk or a gambler, where the wife is incompetent, adulterous, and so forth; the stigmata of family unhappiness are drearily familiar and repetitive. On the other hand, there are happy families of every kind. Tolstoy had not thought about the subject seriously, and above all honestly, because he could not bring himself to think seriously and honestly about women: he turned from the subject in fear, rage and disgust. The moral failure of Tolstoy's marriage, and his intellectual failure to do justice to half the human race, were closely linked.

However, even Tolstoy's marriage, doomed as it was from the start in some ways, might have fared better had it not been for the additional problem of his inheritance, the estate. After gambling and sex, the estate was the third source of Tolstoy's guilt and by far the most important. It came to dominate and finally to destroy his settled existence. It was the source of his pride and authority, and of his moral unease too. For the land and its peasants were inextricably tied together: in Russia you could not own one without owning the other. Tolstoy inherited the estate from his mother when he was a very young man, and almost from the start he began to consider the great question – part honourable, part self-indulgent – 'What am I to do with my peasants?' If he had been a sensible man, he would have recognized that managing an estate was not for him; that his gift and his duty were to write. He would have sold the estate and so rid himself of the moral problem, exercising leadership through his books. But Tolstoy was not a sensible man. He would not relinquish the problem. But neither would he solve it radically. For nearly half a century he wavered, dithered and tinkered with it.

Tolstoy instituted his first 'reform' of the peasants when he inherited the estate in the late 1840s. He claimed later; 'The idea that the serfs should be liberated was quite unheard-of in our circle in the 1840s.'[35] That was false; it had been bandied about everywhere for an entire generation; it was the theme of every petty provincial Philosophy Club; had it not been it would never have occurred to Tolstoy himself. Tolstoy accompanied his 'reform' by other improvements, including a steam threshing machine he designed himself. None of these efforts came to anything. He soon gave up in the face of the intrinsic difficulties and peasant 'swinishness' (as he put it). The only result was the character of Nekhlyudov in 'A Landlord's Morning', who speaks for the disillusioned young Tolstoy: 'I see nothing but ignorant routine, vice, suspicion, hopelessness. I am wasting the best years of my life.' After eighteen months, Tolstoy left the estate and went on to other things – sex, gambling, the army, literature. But he continued to let the peasants, or rather the idea of the peasants – he never saw them as individual human beings – nag at his mind. His attitude to them remained highly ambivalent. His diary records (1852): 'I spent the whole evening talking to Shubin about our Russian slavery. It is true that slavery is an evil, but an extremely pleasant evil.'

In 1856 he had his second attempt at 'reform'. He declared that he would emancipate his serfs in return for payment of thirty years' rent. He did this, characteristically, without consulting any of his acquaintance who had actually had experience of emancipation. The serfs, as it happened, believed rumours then circulating that the new king, Alexander II, intended to liberate them unconditionally. They smelled a rat. They did not spot Count Tolstoy's pretentiousness but feared, rather, his (non-existent) business acumen, and flatly refused his proposal. He, furious, denounced them as ignorant, hopeless savages. He was already displaying a certain emotional disturbance on the subject. He wrote a hysterical letter to the former Interior Minister, Count Dmitri Bludov: 'If the serfs are not free in six months, we are in for a holocaust.'[36] And to members of his own family who thought his schemes foolish and immature – such as his Aunt Tatiana – he showed a frightening hostility: 'I am beginning to develop a silent hatred of my aunt, in spite of all her affection.'

He now turned to education as the once-and-for-all solution to the peasant problem. It is a curious delusion of intellectuals, from Rousseau onwards, that they can solve the perennial difficulties of human education at a stroke, by setting up a new system. He began by teaching the peasant children himself. He wrote to Countess Alexandra Tolstoy: 'When I enter this school and see this crowd of ragged, dirty, thin children with their bright eyes and so frequently their angelic expressions, a sense of alarm and horror comes upon me such as I experienced at the sight of drowning people . . . I desire education for the people only in order to rescue these Pushkins, Ostrograds, Filaretovs, who are drowning there.'[37] For a brief period he enjoyed teaching them. He later told his official biographer, P.I.Biryukov, that this was the best time of his life: 'I owe the brightest time in my life not to the love of women but to the love of people, to the love of children. That was a wonderful time.'[38] It is not recorded how successful his efforts were. There were no rules. No homework was required. 'They bring only themselves,' he wrote, 'their receptive nature and an assurance that it will be as jolly in school today as it was yesterday.' Soon he was setting up a network of schools, and at one time there were seventy. But his own teaching efforts did not last. He became bored and went off on a tour of Germany, ostensibly to examine educational reforms there. But the famous Julius Fröbel disappointed him: instead of listening to Tolstoy he did the talking, and anyway was 'nothing but a Jew'.

This was the situation when, suddenly in 1861, Alexander II emancipated the serfs by imperial decree. Annoyed, Tolstoy denounced it because it was an act of the State, of which he now began to disapprove. The next year he married, and the estate took on a different significance: as the home of his growing family and, together with his novels, as the source of their income. This was the most productive period of his life, the years of *War and Peace* and *Anna Karenina*. As the income from his books rose, Tolstoy bought land and invested in the estate. At one time he had, for instance, four hundred horses on his stud farm. There were five governesses and tutors in the house, plus eleven indoor servants. But the desire to 'reform', not just the peasants but himself, his family – the entire world – never left him. It slumbered just beneath the outward surface of his mind, liable to break into sensational activity at any moment.

Political and social reform, and the desire to found a new religious movement, were closely linked in Tolstoy's mind. He had written as long ago as 1855 that he wanted to create a faith based on 'the religion of Christ but purged of dogmas and mysticism, promising not a future bliss but giving bliss on earth'. This was a commonplace idea, the everyday coin of countless jejune religious reformers through the centuries. Tolstoy was never much of a theologian. He wrote two long tracts, *Examination of Dogmatic Theology* and *Union and Translation of the Four Gospels*, which do nothing to raise one's opinion of him as a systematic thinker. A lot of his religious writing makes little sense except in terms of a vague pantheism. Thus: 'To know God and to live is one and the same thing. God is life. Live seeking God and then you will not live without God' (1878–79).

But the religious notions drifting around inside Tolstoy's head were potentially dangerous because, in conjunction with his political impulses, they formed highly combustible material, liable to burst into sudden flame without warning. By the time he had finished and published *Anna Karenina*, which greatly reinforced his reputation, he was restless, dissatisfied with writing and ready for public mischief: a world-famous figure, a seer, a man to whom countless readers and admirers looked for wisdom and guidance.

The first explosion came in December 1881, when Tolstoy and his family were in Moscow. He went to the Khitrov market in a poor quarter of Moscow where he distributed money to the derelicts there and listened to their life stories. A crowd surrounded him and he took refuge in the neighbouring dosshouse, where he saw things which further distressed him. Returning home, and taking off his fur coat, he sat down to a five-course dinner served by footmen in dress clothes with white gloves and ties. He started to shout: 'One cannot live so! One cannot live so! It is impossible!', frightening Sonya with his arm-waving and threats to give away all their possessions. He immediately began setting up a new system of charity for the poor, using the recently established census as a statistical basis, then hurried down to the country to consult with his current guru, the so-called 'peasant seer' V.K.Syutayev, on further reforms. Sonya was left alone in Moscow with their sick four-month-old Alexei.

This desertion, as she saw it, provoked from the Countess a letter which struck a new note of bitterness in their relationship. It sums up not only her own difficulties with Tolstoy but the anger most ordinary people come to feel in coping with a great humanitarian intellectual: 'My little one is still unwell, and I am very tender and pitying. You and Syutayev may not especially love *your own* children, but we simple mortals are neither able nor wish to distort our feelings or to justify our lack of love for a *person* by professing some love or other *for the whole world.*'[39]

Sonya was raising the question, as a result of observing Tolstoy's behaviour over many years, not least to his own family, whether he ever really loved any individual human being, as opposed to loving mankind as an idea. His wretched brother Dimitri, for instance, was surely an object of compassion: he sank into the gutter, married a prostitute and died young of tuberculosis in 1856. Tolstoy could barely bring himself to spend an hour at his deathbed and refused to attend the funeral at all – he wanted to go to a party instead – though he later put both episodes, the deathbed and the refusal, to good fictional use.[40] His brother Nikolai, likewise dying of tuberculosis, was another object of pity. But Tolstoy refused to visit him, and in the end Nikolai had to come to him, dying in Tolstoy's arms. He did little to help his third brother, Sergei, when he lost his entire fortune gambling. They were all, to be sure, feeble creatures. But it was one of Tolstoy's principles that the strong should come to the aid of the weak.

The record of his friendships is revealing. He was unselfish and submissive only in one case, to his fellow student at Kazan University, Mitya Dyakov, an older man. But this soon faded. As a rule, Tolstoy took, his friends gave. Sonya, copying his early diaries, wrote: '[His] self-adoration comes out in every one [of them]. It is amazing how people existed for him [only] in so far as they affected him personally.'[41] Even more striking is the willingness of those who knew him, not just hangers-on, dependants and flatterers, but highly critical men of independent personality, to put up with his egotism and to revere him despite it. They quailed before that terrible eye, they bowed before the massive strength of his will, and of course they worshipped at the shrine of his genius. Anton Chekhov, a subtle and sensitive man, well aware of Tolstoy's many faults, wrote: 'I dread Tolstoy's death. If he should die there would be a big empty place in my life... I have never loved any man as I have loved him... As long as there is a Tolstoy in literature, then it is easy and agreeable to be a writer; even the realization that one has done nothing and will do nothing is not so dreadful, since Tolstoy will do enough for all.'

Turgenev had even more reason to be aware of Tolstoy's selfishness and cruelty, having experienced both in good measure. He had been generous and thoughtful in helping the young writer. In return he received coldness, ingratitude and Tolstoy's brutal habit of insulting, often brilliantly, the ideas which he knew his friends cherished. Turgenev was a giant of a man, soft-hearted and mild, incapable of paying Tolstoy back in his own coin. But he confessed himself exasperated by Tolstoy's behaviour. He had 'never experienced anything so disagreeable as that piercing look which, coupled with two or three venomous remarks, was enough to drive a man mad'.[42] When he gave Tolstoy his own novel, *Fathers and Sons*, over which he had struggled so hard, to read, Tolstoy promptly fell asleep over it and the returning Turgenev found him snoring. When, after the quarrel over Turgenev's daughter and the threat of a duel, Turgenev handsomely apologized, Tolstoy (according to Sonya) sneered: 'You are afraid of me. I despise you and want no more to do with you.' The poet Fet, who tried to make peace between them, was told: 'Turgenev is a scoundrel who deserves to be thrashed. I beg you to transmit that to him as faithfully as you transmit his charming comments to me.'[43] Tolstoy wrote many unpleasant things, often quite untrue, about Turgenev in his diaries, and their correspondence reflects the lack of symmetry of their friendship. Knowing himself to be dying, Turgenev wrote his last letter to Tolstoy in 1883: 'My friend, great writer of the Russian land, listen to my appeal. Let me know if you receive this scribble and allow me to embrace you once more hard, very hard, you, your wife, and all your family. I cannot go on. I am tired.' Tolstoy never replied to this pathetic request, though Turgenev lingered on another two months. So one is not impressed by Tolstoy's reaction when he got the news of Turgenev's death: 'I think of Turgenev continually. I love him terribly, I pity him, I read him, I live with him.' It has the ring of an actor, playing the public role expected of him. As Sonya noticed, Tolstoy was incapable of the privacy and intimacy needed for person-to-person love, or real friendship. Instead he embraced humanity, because that could be done noisily, dramatically, sensationally on the public stage.

But if he was an actor, he was one who continually changed his role; or, rather, varied the role on the great central theme of service to mankind. His didactic urge was stronger than any other. The moment a subject attracted him, he wanted to write a book about it, or engage in a course of revolutionary reform, usually without taking the trouble to master it himself or to consult real experts. Within months of taking up agriculture he was designing and making farm machinery. He learned to play the piano and instantly began writing *Foundations of Music and Rules for its Study*. Soon after opening a school he was turning educational theory upside down. He believed throughout his life that he could seize upon any discipline, find out what was wrong with it, and then rewrite its rules from first principles. He had at least three shots at educational reform, as he did at land reform, on the last occasion writing his own textbooks, which a disgusted and cynical Sonya was obliged to copy out in legible form, complaining. 'I dispise this *Reader*, this *Arithmetic*, this *Grammar*, and I cannot pretend to be interested in them.'[44]

Tolstoy was always as keen to do as to teach. As with most intellectuals, there came a time in his life when he felt the need to identify himself with 'the workers'. It popped up intermittently in the 1860s and 1870s, then began in earnest in January 1884. He dropped his title (though not his authoritative manner) and insisted on being called 'plain Leo Nikolayevich'. This mood coincided with one of those sartorial gestures intellectuals love: dressing as a peasant. The class transvestism suited Tolstoy's love of drama and costume. It also suited him physically, for he had the build and features of a peasant. His boots, his smock, his beard, his cap became the uniform of the new Tolstoy, the world-seer. It was a prominent part of that instinctive talent for public relations which most of these great secular intellectuals seem to possess. Newspaper reporters came thousands of miles to see him. Photography was now universal, the newsreel just beginning in Tolstoy's old age. His peasant dress was ideally suited to his epiphany as the first media prophet.

Tolstoy could also be photographed and filmed performing manual labour, which from the 1880s he proclaimed 'an absolute necessity'. Sonya noted (1 November 1885): 'He gets up at seven when it is still dark. He pumps water for the whole house and lugs it in an enormous tub on a sledge. He saws long logs, chops them for kindling and stacks the wood. He does not eat white bread and never goes out anywhere.'[45] Tolstoy's own diary shows him cleaning the rooms with his children: 'I was ashamed to do what had to be done, empty the chamberpot' – then, a few days later, he conquered his disgust and did so. He took instruction from a shoemaker in his hut, writing of him: 'How like a light, morally splendid, he is in his dirty, dark corner.' After this instant course in a difficult trade, Tolstoy began to make shoes for the family and boots for himself. He also made a pair for Fet, but it is not recorded whether the poet found them satisfactory. Tolstoy's own sons refused to wear the shoes he made for them. Hammering away, Tolstoy exulted: 'It makes one feel like becoming a worker, for the soul flowers.' But soon the urge to cobble wore off and he turned to farm labour: he carted manure, hauled timber, ploughed and helped to build huts. He fancied carpentry and was photographed, a chisel stuck in his broad leather belt, a saw hanging from his waist. Then that phase too ended, as quickly as it had begun.

Except in writing, his true trade, Tolstoy was not a man for the long haul. He lacked patience, persistence and staying power in the face of difficulty. Even his horse-breeding, about which he did know something, was mismanaged since he soon lost interest in the stud. Sonya had a blazing row with him on this subject, on 18 June 1884. She claimed that the horses were in a deplorable condition: he had bought well-bred mares in Samaria, then let them die from neglect and overwork. It was the same, she said, with everything he undertook, including his charities: no properly thought-out plan, no consistency, no men trained and assigned to specific jobs, the whole philosophy changing from one minute to the next. Tolstoy rushed from the room, shouting he would emigrate to America.

The muddle Tolstoy created on his own estate hurt only his personal circle. His public acts and still more his public preaching held much wider dangers. Not all of them were misguided. Starting from 1865 Tolstoy made valuable and in part successful efforts to draw attention to the regional famines from which Russia periodically suffered. His relief schemes did some good, especially during the great famine of 1890, the magnitude of which the government tried to conceal. Occasionally he came to the rescue of one of Russia's many persecuted minorities. He trumpeted the wrongs of the Doukhobors, the vegetarian pacifists whom the government wanted to round up and destroy. He eventually got permission for them to emigrate to Canada. On the other hand, he was harsh about another persecuted group, the Jews, and his views added to their appalling problems.

Far more serious, however, was Tolstoy's authoritarian view that only he had the solution to the world's distress, and his refusal to take part in any efforts at relief which he did not plan and control personally. His selfishness embraced even his charity. At various times in his life his views on most political problems, land reform, colonization, war, monarchy, the State, ownership, etc., changed radically; the list of his contradictions is endless. But in one thing he was consistent. He refused to participate personally in any systematic scheme to bring about reform in Russia – to tackle the problems at their source – and he denounced, with increasing vehemence, the liberal doctrine of 'improvement' as a delusion, indeed a positive evil. He hated democracy. He despised parliaments. The deputies in the Duma were 'children playing at being grown up'.[46] Russia, he argued, without parliaments, was a much freer country than England with them. The most important things in life were not responsive to parliamentary reform. Tolstoy had a particular hatred for the Russian liberal tradition and in *War and Peace* he pilloried the first of the would-be reformers, Count Speransky. He has Prince Andrew say of Speransky's new Council of State, 'What does it matter to me ... Can all that make me any happier or better?' It is a fact of sombre significance in Russian history, that for half a century her greatest writer set his face like flint against any systematic reform of the Tsarist system and did his best to impede and ridicule those who tried to civilize it.

But what was Tolstoy's alternative? If he had argued, as Dickens, Conrad and other great novelists did, that structural improvements were of only limited value and that what was required were changes in human hearts, he would have made some sense. But Tolstoy, while stressing the need for individual moral improvement, would not let matters rest there: he constantly hinted at the need for, the imminence of, some gigantic moral convulsion, which would turn the world upside down and install a heavenly kingdom. His own utopian efforts were designed to adumbrate this millenarian event. But there was no serious thinking behind this vision. It had something of the purely theatrical quality of the cataclysm which, as we have seen, was the poetic origin of Marx's theory of revolution.

Moreover Tolstoy, again like Marx, had a defective understanding of history. He knew very little history and had no conception of how great events came to happen. As Turgenev lamented, the embarrassing history-lectures he inserted into *War and Peace* bore the hallmarks of the autodidact; they were 'farcical', sheer 'trickery'. Flaubert too, writing to Turgenev, noted with dismay '*il philosophise!*'[47] We read this great novel despite, not because of, its theory of history. Tolstoy was a determinist and an anti-individualist. The notion that events were shaped by the deliberate decisions of powerful men was to him a colossal illusion. Those who appear to be in charge do not even know what is happening, let alone make it happen. Only unconscious activity is important. History is the product of millions of decisions by unknown men who are blind to what they are doing. In a way the notion is the same as Marx's, though reached by a different route. What set Tolstoy on this line of thought is not clear. Probably it was his romantic concept of the Russian peasant as the ultimate arbiter and force. At all events, he believed that hidden laws really govern our lives. They are unknown and probably unknowable, and rather than face this disagreeable fact we pretend that history is made by great men and heroes exercising free will. At bottom Tolstoy, like Marx, was a gnostic, rejecting the apparent explanations of how things happen, looking for knowledge of the secret mechanism which lay beneath the surface. This knowledge was intuitively and collectively perceived by corporate groups – the proletariat for Marx, the peasants for Tolstoy. Of course they needed interpreters (like Marx) or prophets (like Tolstoy) but it was essentially their collective strength, their 'rightness', which set the wheels of history in motion. In *War and Peace*, to prove his theory of how history works, Tolstoy distorted the record, just as Marx juggled his Blue Book authorities and twisted his quotations in *Capital*.[48] He refashioned and made use of the Napoleonic Wars, just as Marx tortured the Industrial Revolution to fit his Procrustean bed of historical determinism.

It is not therefore surprising to find Tolstoy moving towards a collectivist solution to the social problem in Russia. As early as 13 August 1865, reflecting on the famine, he set down in his notebook: 'The universal national task of Russia is to endow the world with the idea of a social structure without landed property. *La propriété est le vol* will remain a greater truth than the English constitution as long as the human family exists . . . The Russian revolution can be based on this only.'[49] Forty-three years later he came across this note and marvelled at his prescience. By then Tolstoy had formed links with Marxists and proto-Leninists such as S.I. Muntyanov, who corresponded with him from Siberian exile, refusing Tolstoy's plea to renounce violence: 'It is difficult, Leo Nikolayevich, to remake me. This socialism is my faith and my God. Of course you profess almost the same thing, but you use the tactic of "love"; and we use that of "violence", as you express it.' The argument, then, was about tactics, not strategy; means, not ends. The fact that Tolstoy spoke of 'God' and called himself a Christian made much less difference than one might suppose. The Orthodox Church excommunicated him in February 1901, not surprisingly in view of the fact that he not only denied the divinity of Jesus Christ but asserted that to call him God or pray to him was 'the greatest blasphemy'. The truth is he selected from the Old and New Testaments, the teachings of Christ and the Church, only those bits he agreed with and rejected the rest. He was not a Christian in any meaningful sense. Whether he believed in God is more difficult to determine since he defined 'God' in different ways at various times. At bottom, it would seem, 'God' was what Tolstoy wanted to happen, the total reform. This is a secular, not a religious concept. As for the traditional God the Father, he was at best an equal, to be jealously observed and criticized, the other bear in the den.[50]

In old age Tolstoy turned against patriotism, imperialism, war and violence in any form, and this alone prevented any alliance with the Marxists. He guessed, too, that the Marxists in power would not, in practice, renounce the State, as they said they would. If the Marxist eschatology actually took place, he wrote in 1898, 'the only thing that will happen is that despotism will be transferred. Now the capitalists rule. Then the directors of the workers will rule.'[51] But this did not worry him very much. He had always assumed that the transfer of property to the masses would take place under some kind of authoritarian system – the Tsar would do as well as any. In any case, he did not regard the Marxists as the enemy. The real enemy were the Western style democrats, the parliamentary liberals. They were corrupting the whole world with the spread of their ideas. In his late writings, *A Letter to the Chinese* and *The Significance of the Russian Revolution* (both 1906) he identifies himself, and Russia, firmly with the East. 'Everything,' he wrote, 'that the Western peoples do can and ought to be an example for the peoples of the East not of what should be done but of what ought not to be done in any circumstances. To pursue the path of the Western nations is to pursue the direct path to destruction.' The greatest danger to the world was the 'democratic system' of Britain and the United States. It was inextricably bound up with the cult of the State and the institutionalized violence which the State practised. Russia must turn her face away from the West, renounce industry, abolish the State and embrace non-resistance.

These ideas strike us as bizarre in the light of later events and hope-lessly incongruous even at the time with what was actually happening in Russia. By 1906 Russia was industrializing herself more rapidly than any other nation on earth, using a form of state capitalism which was to be a stepping-stone to Stalin's totalitarian state. But by this stage of his life Tolstoy was no longer in touch with, or even interested in, the real world. He had created, was inhabiting and to some extent ruling, a world of his own at Yasnaya Polyana. He recognized that state power corrupted, and that is why he turned against the State. What he failed to see, though it is obvious enough – it was obvious to Sonya, for instance – was that the corruption by power takes many forms. One kind of power is exercised by a great man, a seer, a prophet, over his followers, and he is corrupted by their adulation, subservience and, not least, flattery.

Even by the mid-1880s, Yasnaya Polyana had become a kind of court-shrine, to which all kinds of people resorted for guidance, help, reassur-ance and miracle-wisdom, or to impart strange messages of their own

– vegetarians, Swedenborgians, supporters of breast-feeding and Henry George, monks, holy men, lamas and bonzes, pacifists and draft-dodgers, cranks, crazies and the chronically ill. In addition there was the regular, though constantly changing, circle of Tolstoy's acolytes and disciples. All in one way or another regarded Tolstoy as their spiritual leader, part pope, part patriarch, part Messiah. Like the pilgrims to Rousseau's tomb in the 1780s, visitors left inscriptions scrawled or carved on the summer-house in the park at Yasnaya Polyana: 'Down with capital punishment!' 'Workers of the world unite and render homage to a genius!' 'May the life of Lev Nikolayevich be prolonged for as many years again!' 'Greetings to Count Tolstoy from the Tula Realists!' and so on. Tolstoy in celebrated old age set a pattern which (as we shall see) was to recur among leading intellectuals who enjoy world fame: he formed a kind of pseudo-government, taking up 'problems' in various parts of the world, offering solutions, corresponding with kings and presidents, dispatching protests, publishing statements, above all signing things, lending his name to causes, sacred and profane, good and bad.

From the 1890s Tolstoy, as ruler of this chaotic regime, even acquired a prime minister in the shape of a wealthy former guards officer, Vladimir Grigorevich Chertkov (1854–1936), who gradually insinuated himself into a dominant position at the court. He appears in photographs taken with the Master: thin mouth, slitty, pouchy eyes, a short beard, an air of assiduous devotion and apostleship. He soon began to exercise a growing influence over Tolstoy's actions, reminding the old man of his vows and prophecies, keeping him up to the mark of his ideals, always pushing him in more extreme directions. Naturally he made himself chorus-master of the flattering choir, whose voice Tolstoy heard with complacency.

Visitors, or members of the inner circle, noted down Tolstoy's *obiter dicta*. They are not impressive. They remind one of *Napoleon's Sayings in Exile* or *Hitler's Table-Talk* – eccentric generalizations, truisms, ancient, threadbare prejudices, banalities. 'The longer I live the more I am convinced that love is the most important thing.' 'Ignore literature written during the last sixty years. It is all confusion. Read anything written before that time.' 'That One which is within us, every one, brings us all closer to one another. As all lines converge at the centre, so we all come together in the One.' 'The first thing that strikes you about the introduction of these airplanes and flying projectiles is that new taxes are being levied on the people. This is an illustration of the fact that in a certain moral state of society no material improvement can be beneficial but only harmful.' On smallpox inoculation: 'There is no point in trying to escape death. You will die anyway.' 'If the peasants had land, we should not have those idiotic flowerbeds.' 'It would be a much better world if women were less talkative ... It is a kind of naive egoism, a desire to put themselves forward.' 'In Shanghai the Chinese quarter gets along very nicely without police.' 'Children need no education whatsoever ... It is my conviction that the more learned a man is, the stupider.' 'The French are a most sympathetic people.' 'Without religion there will always be debauchery, frippery and vodka.' 'That is how one should live, working for the common cause. It's the way birds live, and blades of grass.' 'The worse it is, the better.'[52]

Trapped at the centre of this prophet's court were Tolstoy's family. Since their father chose to live his life in public, they too were scorched by the glare of publicity. They were forced to share in the drama he created and bore its scars. I have already quoted his son Ilya on the dangers of being 'special' people. Another son, Andrei, suffered from nervous collapses, deserted his wife and family and joined the anti-Semitic Black Hundred. The daughters felt the force of their father's growing detestation of sex. Like Marx, he did not approve of them having followers and disliked the men they chose. In 1897 Tanya, already thirty-three, fell for a widower with six children; he was, it seems, a decent man but he was a liberal and Tolstoy was furious. He gave Tanya a hair-raising lecture on the evils of marriage. Masha, who also fell in love and wanted to marry, got the same treatment. The youngest daughter, Alexandra, was more inclined to be one of his disciples, because she got on badly with her mother.

It was Sonya who had to bear the brunt of Tolstoy's moral cataclysms. For a quarter of a century he forced his sexual demands on her and subjected her to repeated pregnancies. Then he suddenly insisted they should both renounce sex and live 'as brother and sister'. She objected to what she saw as an insult to her status as his wife, especially since he was bound to talk and write about it, being incapable of privacy. She did not want the world peering into her bedroom. He demanded that they sleep in separate rooms. She insisted on a double bed, as a symbol of their continuing marriage. At the same time he showed himself jealous, for no reason. He produced a sinister story, 'The Kreutzer Sonata,' about the murder of a wife by an insanely jealous husband who resents her relationship with a violinist. She copied it (as she copied all his writings) with growing distaste and alarm, realizing that people might think it was about her. Publication was held up by the censors, but the story circulated in manuscript and rumours spread. She then felt obliged to demand publication, thinking her attitude would convince people she was not the subject of the story. As a counterpoint to this quasi-public dispute there were gruesome quarrels behind the scenes arising from Tolstoy's inability to stick to his vow of chastity and his periodic sexual assaults upon his wife. At the end of 1888 his diary records: 'The Devil fell upon me... The next day, the morning of the 30th, I slept badly. It was so loathsome as after a crime.' A few days later: 'Still more powerfully possessed, I fell.' As late as 1898 he was telling Aylmer Maude: 'I was myself a husband last night, but that is no reason for abandoning the struggle. God may grant me not to be so again.'[53]

The fact that Tolstoy could so discuss his marital sex life with an outsider is an indication of the extent to which Sonya felt her most intimate secrets were being exposed to the world's gaze. It was during these years of rising tension that the folly of Tolstoy's policy of *glasnost* became apparent. She did not like reading his diaries at first – no normal, sensible person would – but got used to it. In fact, as his handwriting was so bad, she developed the habit of copying out his diaries in fair, the old ones and the current one. But it is a habit of intellectuals, who write everything with an eye to future publication, to use their diaries as *pièces justificatives*, instruments of propaganda, defensive and offensive weapons against potential critics, not least their loved ones. Tolstoy was a prime example of this tendency. As his relations with Sonya deteriorated, his diaries became more critical of her and he, accordingly, less anxious for her to see them. As early as 1890 she noted: 'It is beginning to worry him that I have been copying his diaries... He would like to destroy his old diaries and appear before his children and the public only in his patriarchal robes. His vanity is immense!'[54] Soon he began to conceal his current diary. So the policy of *glasnost* collapsed and was succeeded by furtiveness on both sides. He used his diary – by now, as he thought, private – to record, for instance, the row with Sonya over 'The Kreutzer Sonata' blow by blow. 'Lyova has broken off all relations with me... I read his diaries secretly, and tried to see what I could bring into our life which would unite us again. But his diaries only deepened my despair. He evidently discovered I had been reading them for he hid them away.' Again: 'In the old days he gave me the job to copy out what he wrote. Now he keeps giving it to his daughters [she does not say 'our'] and carefully hides it from me. He makes me frantic with his way of systematically excluding me from his personal life, and it is unbearably painful.' As the final twist to his abandoned policy of openness, Tolstoy began to keep a 'secret' diary, which he hid in one of his riding boots. She, finding nothing in his usual diary, began to suspect the existence of the secret one, searched for it and eventually found it, bearing it away in triumph for her own secret perusal. She then pasted a sheet of paper over it on which she had written:

'With an aching heart I have copied this lamentable diary of my husband's. How much of what he says about me, and even about his marriage, is unjust, cruel and – God and Levochka forgive me – untrue, distorted and fabricated.'

The background to this nightmarish battle of the diaries was Tolstoy's growing conviction that his wife was preventing his spiritual fulfilment by insisting on a 'normal' way of life he now found morally abhorrent. Sonya was not, as he made out, a gross materialist; she did not deny the moral truth of much of what he preached. As she wrote to him: 'Together with the crowd I see the light of the lantern. I acknowledge it to be *the light*, but I cannot go faster, I am held back by the crowd and by my surroundings and habits.' But Tolstoy, as he grew older, became more impatient and more repelled by the luxury of a life which he associated with Sonya. Thus: 'We sit outside and eat ten dishes. Ice cream, lackeys, silver service – and beggars pass.' To her he wrote: 'The way you live is the very way that I have just been saved from, as from a terrible horror, almost leading me to suicide. I cannot return to the way I lived, in which I found destruction... Between us there is a struggle to the death.'

The tragic and pitiful climax to this struggle began in June 1910. It was precipitated by the return from exile of Chertkov, whom she had learned to hate, and who clearly regarded her as a rival to his power over the prophet. We have an intimate and to a great extent objective record of what happened, as Tolstoy's new secretary, Valentin Bulgakov, kept a diary. It is an indication of the obsession with diaries in Tolstoy's circle that Bulgakov was originally ordered by Chertkov to send a copy of his daily entries to Chertkov's secretary. However, Bulgakov relates, when Chertkov returned from exile and 'appeared on the Yasnaya Polyana scene and the events taking place in the Tolstoy family assumed a dramatic character, I realized how restricted I was by this "censorship" and on various pretexts ceased sending [Chertkov] copies of my diary, his demands notwithstanding.' He says he arrived with a bias against the Countess, being 'warned' that she was 'thoroughly unsympathetic, not to say hostile'. In fact he found her 'gracious and hospitable'; 'I liked the direct look of her sparkling brown eyes, I liked her simplicity, affability and intelligence.'[55] His diary entries indicate that he slowly began to see she was more sinned against than sinning; Tolstoy, his idol, began to topple over.

Chertkov's return was first marked by his taking possession of Tolstoy's diaries. Unknown to Tolstoy, he secretly took photographs of them. On 1 July Sonya insisted that 'objectionable passages' be struck out, so they could not be published. There was a scene. Later she rode in the carriage with Bulgakov, imploring him to persuade Chertkov to return the diaries: 'she wept the whole way and was exceedingly pitiful... I could not look at this weeping, unfortunate woman without feelings of deep compassion.' When he spoke to Chertkov about the diaries, he became 'exceedingly agitated', accused Bulgakov of telling the Countess where they were hidden and 'to my utter amazement... made a hideous grimace and stuck out his tongue at me'. He clearly complained to Tolstoy, who wrote Sonya a letter (14 July) insisting that 'your disposition in recent years has become more and more irritable, despotic and lacking in self-control'; they both now had 'an absolutely contrary understanding of the meaning and purpose of life'. To resolve the dispute the diaries were put under seal and locked in the bank.[56]

A week later, on 22 July, Tolstoy observed: 'Love is the joining of souls separated from each other by the body.' But the same day he went secretly to a nearby village, Grumont, in order to sign a new will, leaving all his copyrights to his youngest daughter, with Chertkov as administrator. Chertkov arranged all this and drew up the instrument himself, and Bulgakov was kept out of the business because it was felt he might tell Sonya. He complained he was not sure Tolstoy knew what he was signing. 'And so an act has been committed which [she] had dreaded above everything: the family, whose material interests she had so jealously guarded, was deprived of the literary rights to Tolstoy's works after his death.' He added that Sonya felt instinctively 'that something awful and irreparable had just happened'. On 3 August there were 'nightmarish scenes' during which Sonya apparently accused Chertkov of having a homosexual relationship with her husband. Tolstoy was 'frozen with indignation'.[57] On 14 September there was another terrible scene, Chertkov saying to Tolstoy in her presence: 'If I had a wife like yours I would shoot myself.' Chertkov said to her: 'If I had wanted to, I could have dragged your family through the mud, but I haven't done it.' A week later, Tolstoy discovered Sonya had found his secret diary in the boot and read it. The next day, contrary to a previous agreement, he rehung Chertkov's photograph in his study. While he was out riding, she tore it up and flushed it down the lavatory. Then she fired off a toy pistol and ran into the park. These rows frequently involved the youngest, Alexandra, also; she formed a habit of striking a boxing posture, goading her mother to say, 'Is that a well-bred young lady or is it a coachman?' – referring, doubtless, to dark family secrets.[58]

On the night of 27–28 October Tolstoy discovered Sonya, at midnight, going through his papers, apparently looking for the secret will. He woke Alexandra and announced: 'I am leaving at once – for good.' He caught a train that night. The next morning, Bulgakov was told the news, by a triumphant Chertkov: 'His face expressed joy and excite-ment.' When Sonya was informed, she threw herself into the pond, and there were further, though unconvincing, suicide attempts. By 1 November Tolstoy, having become ill with bronchitis and pneumonia, had to leave the train and was put to bed at Astapovo Station on the Ryazan–Ural line. Sonya and the family went to join him by special train two days later. On the 7th came the news of the prophet's death. What makes the last months of his life so heartbreaking, especially to those who admire his fiction, is that they were marked, not by any ennobling debate over the great issues the quarrel in theory embodied, but by jealousy, spite, revenge, furtiveness, treachery, bad temper, hys-teria and petty meanness. It was a family dispute of the most degrading kind, envenomed by an interfering and self interested outsider and end ing in total disaster. Tolstoy's admirers later tried to make a scene of Biblical tragedy from the deathbed at Astapovo Station, but the truth is his long and stormy life ended not with a bang but a whine.

Tolstoy's case is another example of what happens when an intellectual pursues abstract ideas at the expense of people. The historian is tempted to see it as a prolegomenon, on a small, personal scale, of the infinitely greater national catastrophe which was soon to engulf Russia as a whole. Tolstoy destroyed his family, and killed himself, by trying to bring about the total moral transformation he felt imperative. But he also yearned for and predicted – and by his writings greatly encouraged – a millenarian transformation of Russia herself, not by gradual and painstaking reforms of the kind he despised, but in one volcanic convulsion. It finally came in 1917, as a result of events he could not foresee and in ways he would have shuddered to contemplate. It made nonsense of all he wrote about the regeneration of society. The Holy Russia he loved was destroyed, seemingly for ever. By a hateful irony, the principal victims of the New Jerusalem thus brought about were his beloved peasants, twenty million of whom were led to mass slaughter on the sacrificial altar of ideas.

6

The Deep Waters of
Ernest Hemingway

ALTHOUGH the United States grew in numbers and strength throughout the nineteenth century, and by the end of it had already become the world's largest and richest industrial power, it was a long time before its society began to produce intellectuals of the kind I have been describing. For this there were several reasons. Independent America had never possessed an *ancien régime*, a privileged establishment based on prescriptive possession rather than natural justice. There was no irrational and inequitable existing order which the new breed of secular intellectual could scheme to replace by millenarian models based on reason and morality. On the contrary: the United States was itself the product of a revolution against the injustice of the old order. Its constitution was based on rational and ethical principles, and had been planned, written, enacted and, in the light of early experience, amended by men of the highest intelligence, of philosophical bent and moral stature. There was thus no cleavage between the ruling and the educated classes: they were one and the same. Then too, as de Tocqueville noted, there was in the United States no institutionalized clerical class, and therefore no anti-clericalism, the source of so much intellectual ferment in Europe. Religion in America was universal but under the control of the laity. It concerned itself with behaviour, not dogma. It was voluntary and multi-denominational, and thus expressed freedom rather than restricted it. Finally, America was a land of plenty and opportunity, where land was cheap and in ample supply, and no man need be poor. There was none of the ocular evidence of flagrant injustice which, in Europe, incited clever, well-educated men to embrace radical ideas. No sins cried out to heaven for vengeance – yet. Most men were too busy getting and spending, exploiting and consolidating, to question the fundamental assumptions of their society.

Early American intellectuals, like Washington Irving, took their tone and manners, their style and content, from Europe, where they spent much of their time; they were a living legacy of cultural colonialism. The emergence of a native and independent American intellectual spirit was itself a reaction to the cringing of Irving and his kind. The first and most representative exponent of this spirit – the archetypal American intellectual of the nineteenth century – was Ralph Waldo Emerson (1803–82), who proclaimed that his object was to extract 'the tape-worm of Europe' from America's body and brain, to 'cast out the passion for Europe by the passion for America'.[1] He too went to Europe but in a critical and rejecting mood. But his insistence on the Americanism of his mind led to a broad identification with the assumptions of his own society which became closer as he grew older, and which was the exact antithesis of the outlook of Europe's intelligentsia. Emerson was born in Boston in 1803, the son of a unitarian minister. He became one himself but left the ministry because he could not conscientiously administer the Lord's Supper. He travelled in Europe, discovered Kant, returned and settled in Concord, Massachusetts, where he developed the first indigenous American philosophical movement, known as Transcendentalism, encapsulated in his first book, *Nature*, published in 1836. It is neo-Platonic, somewhat anti-rational, a little mystical, a touch Romantic, above all vague. Emerson noted in one of his many notebooks and journals:

For this was I born and came into the world to deliver the self of myself to the Universe from the Universe; to do a certain benefit which nature could not forgo, nor I be discharged from rendering, and then emerge again into the holy silence and eternity, out of which as a man I arose. God is rich and many more men than I he harbours in his bosom, biding their time and the needs and beauty of all. Or, when I wish, it is permitted to me to say, these hands, this body, this history of Waldo Emerson are profane and wearisome, but I, I descend not to mix myself with that or any man. Above his life, above all creatures I flow down forever a sea of benefit into races of individuals. Nor can the stream ever roll backwards or the sin or death of man taint the immutable energy which distributes itself into men as the sun into rays or the sea into drops.[2]

This does not make much sense, or, in so far as it does, constitutes a truism. But in an age which admired Hegelianism and the early Carlyle, many Americans were proud that their young country had produced an undoubted intellectual of their own. It was later observed that his appeal rested 'not on the ground that people understand him, but that they think such men ought to be encouraged'.[3] A year after he published *Nature*, he delivered a Harvard address, entitled 'The American Scholar', which Oliver Wendell Holmes was to call 'our intellectual declaration of independence'.[4] His themes were taken up by America's burgeoning press. The paper which published Marx's dispatches from Europe, Horace Greeley's *New York Tribune*, by far the most influential in the country, promoted Emerson's Transcendentalism in a sensational manner, as a kind of national public property, like Niagara Falls.

Emerson is worth examining because his career illustrates the difficulty experienced by American intellectuals in breaking away from their native consensus. In many ways he remained the product of his New England background, especially in his naive, puritanical and etiolated approach to sex. When he descended on the Carlyles at Craigenputtock in August 1833, he seemed to Jane Carlyle a bit etherial, coming 'out of the clouds, as it were'; Carlyle himself noted he left 'like an angel, with his beautiful, transparent soul'.[5] On a subsequent visit, in 1848, Emerson described in his diary how he was obliged to defend American standards of morals at a dinner party at John Foster's house, attended by Dickens, Carlyle and others:

> I said that, when I came to Liverpool, I enquired whether the
> prostitution was always as gross in that city, as it then appeared,
> for to me it seemed to betoken a fatal rottenness in the state, and
> I saw not how any boy could grow up safe. But I had been told,
> it was not worse or better for years. Carlyle and Dickens replied that
> chastity in the male sex was as good as gone in our times, and in
> England was so rare that they could name all the exceptions. Carlyle
> evidently believed that the same things were true in America . . . I
> assured him it was not so with us; that, for the most part, young
> men of good standing and good education with us, go virgins to their
> nuptial bed, as truly as their brides.[6]

As Henry James later wrote of Emerson, 'his ripe unconsciousness of evil . . . is one of the most beautiful signs by which we know him'; though he adds, cruelly, 'We get the impression of a conscience gasping in the void, panting for sensations, with something of the movements of the gills of a landed fish.'[7] Evidently the sexual drive in Emerson was not powerful. His young first wife called him 'Grandpa'. His second, who had to put up with Emerson's much-adored mother living in the household until she died, occasionally gave vent to bitter remarks, which

Emerson naively recorded in his journal: 'Save me from magnificent souls. I like a small, common-sized one'; or again, 'There is no love to restrain the course of, and never was, that poor God did all he could but selfishness fairly carried the day.'[8] Emerson's poem 'Give All to Love' was thought daring but there is no evidence he ever gave much himself. His one great extra-marital friendship with a woman was strictly platonic, or perhaps neo-platonic, and not by her choice. He remarked, cautiously: 'I have organs also and delight in pleasure, but I have experience also that this pleasure is the bait of a trap.'[9] His journal, which constantly tells us more about him than he evidently intended, records a dream, in 1840–41, in which he attends a debate on Marriage. One of the speakers suddenly turned on the audience 'the spout of an Engine which was copiously supplied ... with water, and whisking it vigorously about', drove everyone out, finally turning it fully on Emerson 'and drenched me as I gazed. I woke up relieved to find myself quite dry.'[10]

Emerson married both his wives for prudential reasons and thereby acquired a capital which gave him a measure of literary independence. Soundly invested, it also brought a growing measure of affinity with the fast-expanding entrepreneurial system. He made what eventually became an unrivalled national reputation as a sage and prophet not so much by his books as through the lecture circuit, which was part of that system. First came his course 'Human Life' in Boston (1838), then 'The Times' in New York, (1842), followed by his study of great minds, 'Representative Men' (1845). Emerson's emergence as a highbrow, but popular, lecturer, whose discourses were widely reported in the local, regional and even national press, coincided with the development of the Lyceum movement, founded by Josiah Holbrook in 1829, to educate the expanding nation.[11] Lyceums were opened in Cincinnati in 1830, in Cleveland in 1832, in Columbus in 1835, and then throughout the expanding Midwest and Mississippi Valley. By the end of the 1830s almost every considerable town had one. They were accompanied by Young Men's Mercantile Libraries and lecture and debating societies especially aimed at young, unmarried men – bank clerks, salesmen, bookkeepers and so forth – who then made up an astonishingly high proportion of the population in the new towns.[12] The idea was to keep them off the streets and out of the saloons, and to promote their commercial career and moral welfare.

Emerson's views fitted neatly into this concept. He disapproved of cultural and intellectual elites. He thought America's own culture must be truly national, universal and democratic. Self-help was vital. The first American who read Homer in a farmhouse, he said, performed a great service to the United States. He said that if he found a man out west, reading a good book in a train, he wanted to hug him. His personal economic and political philosophy was identical with the public philosophy pushing Americans across the continent to fulfill their manifest destiny:

> The only safe rule is found in the self-adjusting meter of demand and supply. Do not legislate. Meddle, and you snap the sinews with your sumptuary laws. Give no bounties, make equal laws, secure life and property, and you need not give alms. Open the doors of opportunity to talent and virtue and they do themselves justice, and property will not be in bad hands. In a free and just commonwealth, property rushes from the idle and imbecile to the industrious, brave and persevering.[13]

It would be difficult to think of anything more diametrically opposed to the doctrines Marx was developing and preaching at exactly the same time. And Emerson's actual experience in the field repeatedly contradicted the way in which Marx said capitalism not only did but must behave. Far from opposing this quest for enlightenment, owners and managers positively promoted it. When Emerson came to Pittsburgh in 1851, firms closed early so the young clerks could go to hear him. His courses were not obviously designed to reinforce the entrepreneurial spirit: 'Instinct and Inspiration', 'The Identity of Thought with Nature', 'The Natural History of Intellect', and so on. But he tended to argue that knowledge, plus moral character, promoted business success. Many who came expecting to be bewildered by this eminent philosopher found he preached what they thought common sense. The *Cincinnati Gazette* reported him as 'unpretending . . . as a good old grandfather over his Bible'. Many of his *obiter dicta* – 'Every man is a consumer and ought to be a producer,' '[Man] is by constitution expensive and ought to be rich,' 'Life is the search after power' – struck his listeners as true, and when simplified and taken out of context by the newspapers passed into the common stock of American popular wisdom. It did not seem odd that Emerson was often associated in the same lecture series with P.T. Barnum, whose subjects were 'The Art of Money Getting' and 'Success in Life'. To listen to Emerson was a sign of cultural aspiration and elevated taste: he became the embodiment of Thinking Man. At his last lecture in Chicago in November 1871, the *Chicago Tribune* reported: 'The applause . . . bespoke the culture of the audience.' To a nation which pursued moral and mental improvement with the same enthusiasm as money, and regarded both as essential to the creation of its new civilization, Emerson was by the end of the 1870s a national hero and mentor, as Hugo was to France or Tolstoy to Russia. He had set an American pattern.

It is against this background, in which the nation's economic development and its cultural and intellectual life were seen as in broad harmony, that we should place Ernest Hemingway. At first glance he is not easily recognized as an intellectual at all. On closer inspection he is not only seen to exhibit all the chief characteristics of the intellectual but to possess them to an unusual degree, and in a specifically American combination. He was, moreover, a writer of profound originality. He transformed the way in which his fellow Americans, and people throughout the English-speaking world, expressed themselves. He created a new, personal, secular and highly contemporary ethical style, which was intensely American in origin, but translated itself easily into many cultures. He fused a number of American attitudes together and made himself their archetypal personification, so that he came to embody America at a certain epoch rather as Voltaire embodied France in the 1750s or Byron England in the 1820s.

Hemingway was born in 1899 in the salubrious suburb of Oak Park near Chicago, which had applauded Emerson so heartily a quarter-century before. His parents, Grace and Edmunds ('Ed') Hemingway, and he himself were all outstanding products of the civilization which Emerson and his lectures, and the economic dynamism they upheld, had helped to bring into being. The parents were, or they certainly seemed to be, healthy, industrious, efficient, well-educated, many-talented and well adjusted to their society, grateful for their European cultural inheritance but proudly conscious of the way America had triumphantly improved upon it. They feared God and lived a full life, indoors and outdoors. Dr Hemingway was an excellent physician who also hunted, shot, fished, sailed, camped and pioneered; he possessed, and taught his son, all the wilderness skills of the woodsman. Grace Hemingway was a woman of strong intelligence, powerful will and many accomplishments. She was widely read, wrote excellent prose and skilful verse, painted, designed and made furniture, sang well, played various instruments and wrote and published original songs.[14] Both did everything in their power to transmit to their children, of whom Ernest as the eldest son was the most favoured, all their cultural inheritance and add to it. In many ways they were model parents and Hemingway grew up well-read and highly literate, a skilled sportsman and an all-round athlete.

Both parents were strongly religious. They were Congregationalists, and Dr Hemingway was a strict Sabbatarian too. They not only went to church on Sunday and said grace at meals but, according to

Hemingway's sister Sunny, 'We had morning family prayers accompanied by a Bible reading and a hymn or two.'[15] The moral code of broadstream Protestantism was minutely enforced by both parents and any infringements severely punished. Grace Hemingway spanked the children with a hairbrush, the Doctor with a razor-strop. Their mouths were washed out with bitter soap when they were detected lying or swearing. After punishment they were made to kneel down and beg God for forgiveness. Dr Hemingway made it clear at all times that he identified Christianity with male honour and gentlemanly conduct: 'I want you to represent,' he wrote to Hemingway, 'all that is good and noble and brave and courteous in Manhood, and fear God and respect Woman.'[16] His mother wanted him to be a conventional Protestant hero, non-smoking, non-drinking, chaste before marriage, faithful within it and at all times to honour and obey his parents, especially his mother.

Hemingway rejected his parents' religion in toto and with it any desire to be the sort of son they wanted. In his teens he seems to have decided, quite firmly, that he was going to pursue his genius and his inclination in all things, and to create for himself a vision both of the man of honour and of the good life which was his reward. This was a Romantic, literary and to some extent an ethical concept, but it had no religious content at all. Indeed Hemingway seems to have been devoid of the religious spirit. He privately abandoned his faith at the age of seventeen when he met Bill and Katy Smith (the latter to become the wife of John Dos Passos), whose father, an atheist don, had written an ingenious book 'proving' Jesus Christ had never existed. Hemingway ceased to practise religion at the earliest possible moment, when he went to work at his first job on the *Kansas City Star* and moved into unsupervised lodgings. As late as 1918, when he was nearly 20, he assured his mother: 'Don't worry or cry or fret about my being a good Christian. I am just as much as ever, and pray every night and believe just as hard.'[17] But this was a lie, told for the sake of peace. He not only did not believe in God but regarded organized religion as a menace to human happiness. His first wife, Hadley, said she only saw him on his knees twice, at their wedding and at the christening of their son. To please his second wife, Pauline, he became a Roman Catholic, but he had no more conception of what his new faith meant than did Rex Mottram in *Brideshead Revisited*. He was furious when Pauline tried to observe its rules (e.g. over birth control) in ways which inconvenienced him. He published blasphemous parodies of the Our Father in his story 'A Clean, Well-Lighted Place' and of the Crucifixion in *Death in the Afternoon*; there is a blasphemous spittoon-blessing in his play *The Fifth Column*. In so far as he did understand Roman Catholicism, he detested it. He raised not the slightest protest when, at the beginning of the Civil War in Spain, a place he knew and said he loved, hundreds of churches were burnt, altars and sacred vessels desecrated, and many thousands of priests, monks and nuns slaughtered. He abandoned even the formal pretence of being a Catholic after he left his second wife.[18] All his adult life he lived, in effect, as a pagan, worshipping ideas of his own devising.

Hemingway's rejection of religion was characteristic of the adolescent intellectual, and still more characteristic in that it was part of a rejection of his parents' moral culture. He later sought to differentiate between his mother and father, in a way which exonerated the latter. When his father committed suicide, he tried to hold his mother responsible, though it was clearly a case of a doctor anticipating what he knew would be a painful, terminal illness. Dr Hemingway was the weaker of the two parents but he supported his wife entirely in their disputes with their son, whose quarrel was with both rather than the mother alone. But Grace was the person on whom Hemingway's resistance concentrated, probably, in my view, because he recognized in her the chief source of his egotistical will and his literary power. She was a formidable woman as he was becoming a formidable man. There was not room in the same circle for both.

Their dispute came to a head in 1920 when Hemingway, who had spent the latter part of the Great War in an ambulance unit on the Italian front, and had returned something of a war hero, not only failed to find himself a job but offended his parents by his idle and (by their standards) vicious conduct. In July that year Grace wrote him a Grand Remonstrance. Every mother's life, she said, was like a bank. 'Every child that is born to her enters the world with a large and prosperous bank account, seemingly inexhaustible.' The child draws and draws – 'no deposits during *all* the early years'. Then, up to adolescence, 'while the bank is heavily drawn upon,' there are 'a few deposits of pennies, in the way of some services willingly done, some thoughtfulness and "thank yous"'. With manhood, while the bank goes on handing out love and sympathy:

> The account needs some deposits by this time, some good-sized ones in the way of gratitude and appreciation, interest in Mother's ideas and affairs. Little comforts provided for the home; a desire to favour any of Mother's peculiar prejudices, on no account to outrage her ideas. Flowers, fruit or candy, or something pretty to wear, brought home to Mother with a kiss and a squeeze . . . A surreptitious paying of bills, just to get them off Mother's mind . . . deposits which keep the account in good standing. Many mothers I know are receiving

these and much more substantial gifts and returns from sons of less abilities than my son. Unless you, my son, Ernest, come to yourself, cease your lazy loafing and pleasure seeking . . . stop trading on your handsome face . . . and neglecting your duties to God and your Saviour, Jesus Christ . . . there is nothing before you but bankruptcy: *You have overdrawn.*[19]

She brooded on this document for three days, polishing it as carefully as Hemingway was ever to do his own prize passages, then presented it personally. It indicates whence he got the strong sense of moral outrage, not unmixed with self-righteousness, which is so important a part of his fiction.

Hemingway reacted as might have been expected, with slow, mounting and prolonged fury, and from then on he treated his mother as an enemy. Dos Passos said Hemingway was the only man he had ever come across who really hated his mother. Another old acquaintance, General Lanham, testified: 'From my earliest days with Ernest Hemingway he always referred to his mother as "that bitch". He must have told me a thousand times how much he hated her and in how many ways.'[20] This hatred was reflected repeatedly, and variously, in the fiction. It spilled over into a related detestation of his elder sister, 'my bitch sister Marcelline', 'a bitch complete with handles'. It broadened into a general hatred of families, often expressed in irrelevant contexts, as in the discussion of bad painters (his mother painted) in his autobiography, *A Moveable Feast*: 'they do not do terrible things and make intimate harm, as families do. With bad painters all you need to do is not look at them. But even when you have learned not to look at families nor listen to them and have learned not to answer letters, families have many ways of being dangerous.' His hatred of his mother was so intense that to a considerable extent it poisoned his life, not least because he always felt a residual guilt about it, which nagged at him and kept the hatred evergreen. He was still hating her in 1949 when she was nearly eighty, writing to his publisher from his house in Cuba: 'I will not see her and she knows she can never come here.'[21] His loathing for her exceeded the purely utilitarian dislike that Marx felt for his mother, and was emotionally akin to Marx's attitude to the capitalist system itself. For Hemingway, mother-hatred attained the status of a philosophical system.

The family breakup drove Hemingway to the *Toronto Star* and thence to Europe as a foreign correspondent and novelist. He repudiated not merely his parents' religion but his mother's view of an optimistic, Christianized culture, expressed in her powerful but conventional – and to him detestable – prose. One of the forces which drove Hemingway towards the literary perfectionism which became his outstanding characteristic was the overwhelming urge not to write like his mother, using the stale rhetoric of an over-elaborate literary inheritance. (A sentence of hers which he particularly hated, as epitomizing her prose style, came from one of her letters to him: 'You were named for the two finest and noblest gentlemen I have ever known.')

From 1921 Hemingway led the life of a foreign correspondent, using Paris as his base. He covered warfare in the Middle East and international conferences, but the main focus of his attention was on the expatriate *literati* of the Left Bank. He wrote poetry. He was trying to write prose. He read ferociously. One of the many habits he inherited from his mother was carrying books around with him, shoved into his pockets, so that he could read at any time or place during a pause in the action. He read everything, and all his life he bought books, so that any Hemingway habitation had stacks running along the walls. At his house in Cuba he was to build up a working library of 7400 volumes, characterized by expert studies on all the subjects in which he was interested and by a wide range of literary texts, which he read and re-read. He arrived in Paris having read virtually all the English classics but determined to broaden his range. He was never chippy about having missed a university education, but he regretted it and was anxious to fill any gaps its absence might have left. So he settled down to Stendhal, Flaubert, Balzac, Maupassant and Zola, the major Russian novelists, Tolstoy, Turgenev and Dostoyevsky, and the Americans, Henry James, Mark Twain and Stephen Crane. He read the moderns, too: Conrad, T.S. Eliot, Gertrude Stein, Ezra Pound, D.H. Lawrence, Maxwell Anderson, James Joyce. His reading was wide but also dictated by a growing urge to write. Since the age of fifteen he had made a cult of Kipling, and continued to study him all his life. To this was now added close attention to Conrad, and Joyce's brilliant collection, *Dubliners*. Like all really good writers, he not only devoured but analysed and learned from the second-rate, such as Marryat, Hugh Walpole and George Moore.

Hemingway moved right to the centre of the Paris intelligentsia in 1922 with the arrival there of Ford Madox Ford. Ford was a great unearther of literary talent, helping to bring out Lawrence, Norman Douglas, Wyndham Lewis, Arthur Ransome and many others. In 1923 he published the first issue of *Transatlantic Review* and, on the recommendation of Ezra Pound, hired Hemingway as a part-time assistant. Hemingway admired Ford as a literary entrepreneur but had many complaints about him: he ignored most of the younger writers, he was not sufficiently interested in new styles and literary forms, his taste was too close to that of the mainstream magazines, above all he assumed most good literary things came from France and England, and largely ignored America's output, rapidly growing in quantity and quality. Hemingway saw himself as the impresario of the American avant-garde. 'Ford,' he grumbled, 'is running [the] whole damn thing as a compromise.'[22] Once installed in the tiny Ile St Louis offices over the Three Mountain Press, Hemingway began to tilt the *Review* in an adventurous, American direction, so that in addition to sixty British and forty French pieces it carried ninety by Americans, among others Gertrude Stein, Djuna Barnes, Lincoln Steffens, Natalie Barnard, William Carlos Williams and Nathan Asch. When Ford left Paris for a trip to the United States, Hemingway ruthlessly turned the July and August issues into a triumphant parade of young American talent, so that Ford, on his return, felt he had to apologize for the 'unusually large sample of the work of that Young America whose claims we have so insistently – but not with such efficiency – forced upon our readers'.[23]

But Hemingway had his own intense drive for literary fame and power, and in the long run was less concerned with the parties and intrigues of the Left Bank intelligentsia than in developing his own talent. Pound had introduced Hemingway to Ford with words: 'He writes very good verse and he's the finest prose stylist in the world.'[24] Made in 1922, the remark is highly perceptive, for Hemingway had by no means developed his mature method. But he was working on it, as his early notebooks, with their infinite erasures and amendments, testify. Probably no writer of fiction has ever struggled so hard and so long to fashion a personal manner of writing exactly suited to the work he wished to do. A study of Hemingway during these years is a model of how a writer should acquire his professional skills. It is comparable, in nobility of aim and persistence of effort, with Ibsen's arduous efforts to become a playwright. It also had the same revolutionary impact on the craft.

It was Hemingway's belief that he had inherited a false world, symbolized by his parents' religion and moral culture, and that it must be replaced by a truthful one. What did he mean by truth? Not the inherited, revealed truth of his parents' Christianity – that he rejected as irrelevant – or the truth of any other creed or ideology derived from the past and reflecting the minds of others, however great, but the truth as he himself saw it, felt it, heard, smelt and tasted it. He admired Conrad's literary philosophy and the way he summed up his aim – 'scrupulous fidelity to the truth of my own sensations'. That was his starting point. But how do you convey that truth? Most people when they write, including most professional writers, tend to slip into seeing events through the eyes of others because they inherit stale expressions and combinations of words, threadbare metaphors, clichés and literary conceits. This is particularly true of journalists, covering at speed occasions which are often repetitive and banal. But Hemingway had had the advantage of an excellent training on the *Kansas City Star*. Its successive editors had compiled a house-style book of 110 rules designed to force reporters to use plain, simple, direct and cliché-free English, and these rules were strictly enforced. Hemingway later called them 'the best rules I ever learned for the business of writing'.[25] In 1922, covering the Genoa Conference, he was taught the ruthless art of cablese by Lincoln Steffens, which he acquired with rapidity and growing delight. He showed Steffens his first successful effort, exclaiming: 'Steffens, look at this cable: no fat, no adjectives, no adverbs – nothing but blood and bones and muscle . . . It's a new language.'[26]

On this journalistic basis, Hemingway built his own method, which was both theory and practice. At one time or another he put down a lot about how to write – in *A Moveable Feast*, in *The Green Hills of Africa*, in *Death in the Afternoon*, and in *By-line* and elsewhere.[27] The 'basic principles of writing' he set down for himself are well worth study.[28] He once defined the art of fiction, following Conrad, as 'find what gave you the emotion; what the action was that gave you the excitement. Then write it down making it clear so that the reader can see it too.'[29] All had to be done with brevity, economy, simplicity, strong verbs, short sentences, nothing superfluous or for effect. 'Prose is architecture,' he wrote, 'not interior decoration, and the Baroque is over.'[30] Hemingway paid particular attention to exactitude of expression and ransacked dictionaries for words. It is important to remember that, during the formative period of his prose style, he was also a poet, and strongly under the influence of Ezra Pound, who he said taught him more than anyone else. Pound was 'the man who believed in the *mot juste* – the one and only correct word to use – the man who taught me to distrust adjectives'. He also closely studied Joyce, another writer whose nose for verbal precision he respected and imitated. Indeed, in so far as Hemingway had literary progenitors, it might be said he was the offspring of a marriage between Kipling and Joyce.

But the truth is Hemingway's writing is *sui generis*. His impact on the way people not only wrote but saw, in the quarter-century 1925–50, was so overwhelming and conclusive, and his continuing influence since so pervasive, that it is now impossible for us to subtract the Hemingway factor from our prose, especially in fiction. But in the early 1920s he found it difficult to win approval, or even to get published at all. His first work, *Three Stories and Ten Poems*, was a typical avant-garde venture, locally published in Paris. The big magazines would not look at his fiction, and as late as 1925, *The Dial*, itself regarded as adventurous, was still rejecting his stories, including that superb tale 'The Undefeated'. What Hemingway did is what all really original great writers do – he created his own market, he infected readers with his own taste. The method, which brilliantly combined bare, exact depiction of events with subtle hints of the emotional response to them, emerged in the years 1923–25, and it was in 1925 that the breakthrough came with the publication of *In Our Time*. Ford felt able to hail him as America's leading writer: 'the most conscientious, the most master of his craft, the most consummate'. To Edmund Wilson the book revealed prose 'of the first distinction', which was 'strikingly original' and of impressive 'artistic dignity'. This first success was quickly followed by two vivid and tragic novels, *The Sun Also Rises* (1926) and *A Farewell to Arms* (1929), the latter perhaps the best thing he ever wrote. These books sold hundreds of thousands of copies and were read and re-read, digested, regurgitated, envied and quarried by writers of every kind. As early as 1927, Dorothy Parker, reviewing his collection *Men Without Women* in the *New Yorker*, referred to his influence as 'dangerous' – 'the simplest thing he does looks so easy to do. But look at the boys who try to do it.'[31]

The Hemingway manner could be parodied but not successfully imitated because it was inseparable from the subject matter of the books and especially their moral posture. Hemingway's aim was to avoid explicit didacticism of any kind, and he denounced it in others, even the greatest. 'I love *War and Peace*,' he wrote, 'for the wonderful, penetrating and true descriptions of war and of people, but I have never believed in the great count's thinking ... He could invent more and with more insight and truth than anyone who ever lived. But his ponderous and messianic thinking was no better than many another evangelical professor of history and I learned from him to distrust my own Thinking with a capital T, and to try to write as truly, as straightly, as objectively and as humbly as possible.'[32] In his best work he always avoided preaching at the reader, or even nudging his elbow by drawing attention to the way his characters behaved. Nevertheless, his books are suffused throughout with a new secular ethic, and this springs directly from the way Hemingway describes events and actions.

It is the subtle universality of the Hemingway ethic which makes him so archetypically an intellectual, and the nature of the ethic which reflects his Americanism. Hemingway saw the Americans as a vigorous, active, forceful, even violent people, doers, achievers, creators, conquerors and pacifiers, hunters and builders. He was a vigorous, active, forceful, even violent person himself. Talking to Pound and Ford about literature, he would break off from time to time to shadow-box round Ford's studio.

He was a big, strong man, skilled in a vast range of physical activities. It was natural for him, as an American and a writer, to lead a life of action, and to describe it. Action was his theme.

There was nothing new in that, of course. Action had been the theme of Kipling, whose heroes, or subjects, had been soldiers, dacoits, engineers, sea-captains and rulers big and small – anyone or thing, indeed, periodically subject to the strain and motion of violent activity, even animals and machinery. But Kipling was not an intellectual. He was a genius, he had a 'daemon' but he did not believe he could refashion the world by his own unaided intelligence, he did not reject the vast corpus of its inherited wisdom. On the contrary, he fiercely upheld its laws and customs as unalterable by puny man and depicted with relish the nemesis of those who defied them. Hemingway is much closer to Byron, another writer who longed for action and described it with enthusiastic skill. Byron did not believe in the utopian and revolutionary schemes of his friend Shelley, which seemed to him abstract ideals rather than workable concepts – his point is made for him by Shelley himself in *Julian and Maddalo* – but he had fashioned for himself a system of ethics, devised in reaction to the traditional code he had rejected when he left his wife and England for good. In this sense, and only in this sense, he was an intellectual. He never set down his system formally, though it was coherent enough, but it emerges strongly in his letters and it saturates every page of his great narrative poems, *Childe Harold* and *Don Juan*. It is a system of honour and duty, not codified but illustrated in action. No one can read these poems without being quite clear how Byron saw good and evil and especially how he measured heroism.

Hemingway worked in a similar manner, by illustration. He once specified his ideal as the ability to exhibit 'grace under pressure' (a curious phrase in view of his mother's name) but he went no further in definition. Probably his ethic was incapable of a precise definition and would have been injured and reduced by attempts to construct one. But it was infinitely capable of illustration and that is the driving force behind Hemingway's entire work. His novels are novels of action and that makes them novels of ideology because to Hemingway there was no such thing as a morally neutral action. To him even a description of a meal is a moral statement since there are the right and the wrong things to eat and drink, and right and wrong ways to eat and drink them. Almost any action can be performed correctly or incorrectly, or to be precise nobly or ignobly. The author himself does not point the moral but he presents everything within an implicit moral framework so that the actions speak for themselves. The framework is personal and pagan; certainly not Christian. His parents, especially his mother, found his stories immoral, often outrageously so, because she at any rate could recognize their strong ethical tone, to her a false and blasphemous one. What Hemingway was saying, or rather implying, was that there were right and wrong ways to commit adultery, to steal and to kill. The essence of Hemingway's fiction is observing boxers, fishermen, bullfighters, soldiers, writers, sportsmen, or almost anyone who has definite and skilled actions to perform, trying to live a good and honest life, according to the values of each, and usually failing. Tragedy occurs because the values themselves turn out to be illusory or mistaken, or because they are betrayed by weakness within or external malice or the intractability of objective facts. But even failure is redeemed by truth-seeing, by having the ability to perceive the truth and the courage to stare it in the face. Hemingway's characters stand or fall by whether they are truthful or not. Truth is the essential ingredient of his prose and is the one thread that runs right through his ethical system, its principle of coherence.

Having created his style and his ethic, Hemingway necessarily found himself living both. He became, as it were, the victim, the prisoner, the slave of his own imagination, forced to enact it in real life. Here again he was not unique. Once Byron had published the first canto of *Childe Harold*, he found himself treading the path it indicated. He might vary the direction a little by writing *Don Juan* but he left himself no real choice but to live as he had sung. But then with Byron it was a matter of taste as well as compulsion: he enjoyed the womanizing, the heroics, the liberator role. It was the same with Hemingway's contemporary, André Malraux, another action-intellectual and novelist, revolutionary, explorer, buccaneering questor of art treasures, resistance hero, who ended his career as a senior Cabinet Minister sitting at President de Gaulle's right hand. With Hemingway one is not so sure. His pursuit of 'real' life, the life of action, was an intellectual activity in the sense that it was vital to his kind of fiction. As the hero Robert Jordan says in the Spanish War novel, *For Whom the Bell Tolls* (1940), he 'liked to know how it really was; not how it was supposed to be'. Hemingway, the intellectual obsessed by violent action, was a real person. A perceptive colleague on the *Toronto Star* summed him up at twenty: 'A more weird combination of quivering sensitiveness and preoccupation with violence never walked the earth.' He enjoyed all his father's outdoor pursuits and more – skiing, deep-sea fishing, big-game hunting and, not least, war. There was no doubt about his courage, on occasion. The *New York Times* reporter Herbert Matthews described how Hemingway, during the Ebro river battle in 1938, saved him from drowning in the rapids, by an extraordinary exhibition of strength: 'He was a good man in a pinch.'[33] The white hunters who took him on safari in East Africa, often a good test, bore similar witness. Moreover Hemingway's courage was not unthinking and instinctive but cerebral. He had an acute sense of danger, as many anecdotes testify. He knew what it was to feel afraid and conquer his fear – no writer ever described cowardice more vividly. He made the reader sense his willingness to live his fiction.

This was why Hemingway's action-man image grew as quickly as his fame. Like many other intellectuals, from Rousseau onwards, he had a striking talent for self-publicity. He created the physical, ocular Hemingway persona, reversing the old, soft-velvet, relaxed image of the Romantics, which had done such yeoman service in its time, in favour of a new he-man appeal: safari suits, bandoliers, guns, peaked cap, a whiff of powder, tobacco, whisky. One of his obsessions was adding a few years to his age. In the 1920s, he quickly promoted himself to 'Papa'; the latest girl became 'daughter'. By the early 1940s 'Papa' Hemingway was already a familiar figure in the picture magazines, as famous as the leading Hollywood males. No writer in history ever gave more interviews and photo-calls. In time his white-bearded face became better known than Tolstoy's.

But in trying to personify his ethic and live up to the legend he created, Hemingway was also mounting a treadmill, from which he would not allow himself to descend till death. Rather as his mother saw maternal love in the shape of a bank account, Hemingway was constantly depositing experience of action to his credit, then drawing on it for his fiction. His Italian War, 1917–18, was his initial capital. During the 1920s he used up most of it, balancing the drain by frenzied sportmanship and bullfighting. In the 1930s he made valuable deposits of big game hunting, and the huge windfall of the Spanish Civil War. But he was slothful at exploiting the opportunities of the Second World War and his belated involvement in it added little to his writing capital. Thereafter his chief deposits were hunting and fishing; his attempts to retrace his steps on the big-game shooting and bullfighting circuits bore more farce than fruit. Edmund Wilson noted the contrast, both in the writing and the activity: 'the young master and the old impostor'. The truth is, Hemingway continued to enjoy some of his violent pursuits, but not quite as much as he claimed. There was a perceptible decline in zest for the wilderness, as though he would willingly, if only he dared, hang up his rifle and settle down in his library. A false, forced, boastful note crept into his situation reports to his publisher, Charles Scribner. Thus in 1949 he wrote to him: 'To celebrate my fiftieth birthday ... I fucked three times, shot ten straight pigeons (very fast ones) at the club, drank with friends a case of Piper Heidsieck brut and looked the ocean for big fish all afternoon.'[34]

True? False? An exaggeration? One does not know. None of Hemingway's statements about himself, and very few he made about other people, can ever be accepted as fact without corroboration. Despite the central importance of truth in his fictional ethic he had the characteristic intellectual's belief that, in his own case, truth must be the willing servant of his ego. He thought, and sometimes boasted, that lying was part of his training as a writer. He lied both consciously and without thinking. He certainly knew he was lying on occasion as he makes clear in his fascinating tale, 'Soldier's Home', with its character of Krebs. 'It is not *unnatural* that the best writers are liars,' he wrote. 'A major part of their trade is to lie or invent . . . They often lie unconsciously and then remember their lies with deep remorse.'[35] But the evidence shows that Hemingway habitually lied long before he worked out a professional apologia for it. He lied when he was five, claiming to have stopped a bolting horse unaided. He told his parents he had become engaged to the movie actress Mae Marsh, though he had never set eyes on her except in *Birth of a Nation*; he repeated this lie to his Kansas City colleagues, down to the detail of a $150 engagement ring. Many of these blatant lies were transparent and embarrassing as when, aged eighteen, he told friends he had caught a fish he had obviously bought in the market. He told an elaborate story about being a professional boxer in Chicago, having his nose broken but nonetheless going on fighting. He invented Indian blood for himself and even claimed he had Indian daughters. His autobiography, *A Moveable Feast*, is quite unreliable and, like Rousseau's *Confessions*, most dangerous when it appears to be frank. He was usually mendacious about his parents and sisters, sometimes for no apparent reason. Thus he said his sister Carol had been raped, aged twelve, by a sex-pervert (quite untrue) and later claimed she was divorced or even dead (she was happily married to a Mr Gardiner, whom Hemingway disliked).[36]

Many of Hemingway's most complicated and reiterated lies concern his First World War service. Of course most soldiers, even brave ones, lie about their wars, and the degree of detailed investigation Hemingway's life has been subjected to was bound to turn up some malpractice with the truth.[37] All the same, Hemingway's inventions about what happened in Italy are unusually brazen. In the first place he said he volunteered for the army but was rejected because of poor eyesight. This does not appear in the records and is most unlikely. He was in fact a non-combatant, and by choice. On many occasions, including newspaper interviews, he said he had served in the Italian 69th Infantry Regiment and had fought in three major battles. He also claimed he had belonged to the crack Arditi regiment, and he told his British military friend, 'Chink' Dorman-Smith, that he had led an Arditi charge on Mount Grappa and had been badly wounded during it. He told his Spanish Civil War friend, General Gustavo Duran, that he had commanded first a company, then a battalion, when he was only nineteen. He had indeed been wounded – there was no doubt about that – but he lied repeatedly about the occasion and nature of the injury. He invented a story about being shot in the scrotum, not once but twice, and said he had had to rest his testicles on a pillow. He said he had been knocked down twice by machine-gun fire and hit thirty-two times by .45 bullets. And, as a bonus, he said he had been baptised a Catholic on what the nurses believed was his deathbed. All these statements were untrue.

War brought out the liar in Hemingway. In Spain, jealous of Matthews's superior skills as a correspondent, he reported home in a letter a tissue of lies about the Teruel front: 'got first story of the battle to New York ten hours ahead of Matthews even, went back, made the whole attack with the infantry, entered town behind one company of dynamiters and three of infantry, filed that, went back and had most godwonderful house-to-house fighting story ready to put on wire ...'[38] He also lied about being the first into liberated Paris in 1944. Sex brought out the liar in him too. One of his choicest Italian tales, often repeated, was being held sexual prisoner by a Sicilian woman hotel owner who hid his clothes so he was forced to fornicate with her for a week. He told Bernard Berenson (the recipient of many mendacious letters) that when he finished *The Sun Also Rises*, he got in a girl, his wife came back suddenly and he was forced to smuggle the girl out through the roof; no truth in it at all. He lied about his famous jealous fight with 'that kike [Harry] Loeb' in Pamplona in 1925, saying that Loeb had a gun and threatened to shoot him (the incident was transfigured in *The Sun Also Rises*). He lied about all his marriages, divorces and settlements, both to the women concerned and to his mother. His lies to, and about, his third wife, Martha Gellhorn, were particularly audacious. She, in turn, dismissed him as 'the biggest liar since Munchausen'. As with some other novelist-liars, Hemingway left false trails: some of his most striking stories, seemingly autobiographical by overwhelming internal evidence, may be pure inventions. All one can say is that Hemingway had little respect for truth.

In consequence he was apt and ready for that 'low, dishonest decade', the 1930s. Hemingway never held any set of political convictions with consistency; his ethic was really about personal loyalties. His one-time friend Dos Passos thought that, as a young man, Hemingway 'had one of the shrewdest heads for unmasking political pretensions I've ever run into'.[39] But it is hard to find much evidence for this assertion. In the 1932 election Hemingway supported the Socialist, Eugene Debs. But by 1935 he had become a willing exponent of the Communist Party line on most issues. In the 17 September 1935 issue of the CP paper *New Masses*, he contributed a violent article, 'Who Killed the Vets?', blaming the government for the deaths, in a Florida hurricane, of 450 ex-service-men railway workers employed on federal projects – a typical exercise in CP agitprop. Hemingway's view, throughout the decade, seems to have been that the CP was the only legitimate and trustworthy conductor of the anti-fascist crusade, and that criticism of it, or participation in activities outside its control, was treachery. He said that anyone who took an anti-CP line was 'either a fool or a knave', and he would not allow his name to appear on the masthead of the new left-wing magazine *Ken*, launched by *Esquire*, when he discovered it was not a CP vehicle.

This approach governed his response to the Spanish Civil War, which he welcomed on professional grounds as a source of material – 'Civil war is the best war for a writer, the most complete.'[40] But, curiously in view of his ethical code, which made elaborate provision for conflicts of loyalties, the power of tradition and different concepts of justice, he accepted, from start to finish, the CP line on the war in all its crudity. He paid four visits to the front (spring and autumn 1937, spring and autumn 1938) but even before he left New York he had decided what the Civil War was all about and was already signed up for a propaganda film, *Spain in Flames*, with Dos Passos, Lillian Hellman and Archibald MacLeish. 'My sympathies,' he wrote, 'are always for exploited working people against absentee landlords even if I drink around with the land-lords and shoot pigeons with them.' The CP were 'the people of this country' and the war was a struggle between 'the people' and 'the absen-tee landlords, the Moors, the Italians and the Germans'. He said he liked and respected the Spanish CP, who were 'the best people' in the war.[41]

It was Hemingway's line, in accordance with CP policy, to play down the role of the Soviet Union, especially in directing the Spanish CP's ferocious conduct in the bloodstained internal politics of Republican Spain. This led him into a shameful breach with Dos Passos. Passos's interpreter was José Robles, a former Johns Hopkins University don who had joined the Republican forces on the outbreak of war and was a friend of Andres Nin, head of the anarchist POUM. He had also been interpreter to General Jan Antonovic Berzin, head of the Soviet military mission in Spain, and therefore knew some of the secrets of Moscow's dealings with the Madrid Defence Ministry. Berzin had been murdered by Stalin, who later gave orders to the Spanish CP to liquidate the POUM too. Nin was tortured to death, hundreds of others were arrested, accused of fascist activities, and executed. It was thought prudent to accuse Robles of spying, and he was secretly shot. Dos Passos became worried about his disappearance. Hemingway, who saw himself as an ultra-sophisticate in political matters and Dos Passos as a naive new-comer, pooh-poohed his anxieties. Hemingway was staying at Gaylord's Hotel in Madrid, then the haunt of the CP bosses, and asked his crony Pepe Quintanilla (who, it later emerged, was responsible for most CP executions) what had happened. He was assured Robles was alive and well, under arrest to be sure, but certain to get a fair trial. Hemingway believed this and told Dos Passos. In fact Robles was already dead, and when Hemingway belatedly found this out – from a journalist who had only just arrived in Madrid – he told Dos Passos that it was clear he had been as guilty as hell and only a fool could think otherwise. Dos Passos, greatly distressed, refused to accept Robles' guilt and publicly attacked the Communists. This brought from Hemingway the rebuke: 'A war is being fought in Spain between the people whose side you used to be on and the fascists. If with your hatred of the Communists you feel justified in attacking, for money, the people who are still fighting that war, I think you should at least try to get your facts right.' But Dos Passos, as it turned out, had got his facts right: Hemingway was the naif, the innocent, the dupe.[42]

As such he remained, until the end of the war and for some time afterwards. His work for the Communists reached its climax on 4 June 1937 when he spoke at the Second Writers Congress, which the American CP, through a front organization, held in New York at Carnegie Hall. Hemingway's point was that writers had to fight fascism because it was the only regime which would not allow them to tell the truth; intellectuals had a duty to go to Spain and do something there themselves – they should stop arguing doctrinal points in their armchairs and start fighting: 'There is now, and there will be for a long time, war for any writer to go to who wants to study it.'[43]

Hemingway was certainly a dupe. But he was also consciously partici-pating in a lie, since it is clear from his novel about the Spanish war, For Whom the Bell Tolls, that he was aware of the dark side of the Republi-can cause, and had probably known some of the truth about the Spanish CP all along. But he did not publish the book until 1940, when it was all over. So long as the Civil War lasted, Hemingway took the same line as those who tried to suppress George Orwell's Homage to Catalonia, that truth came a long way behind political and military expediency. His speech to the Writers Congress was thus completely fraudulent.

It was odd in another way too, since Hemingway showed no inclination to follow his own advice and 'study war'. When America's involvement in the crusade against Nazism began in earnest in 1941, he did not join it. By now he had acquired for himself a home, the Finca Vigia, outside Havana in Cuba, which remained his chief residence for most of his remaining years. The success of *For Whom the Bell Tolls*, which became one of the great best-sellers of the century, brought him an enormous income and he wanted to enjoy it, notably in what was now his preferred sport, deep-sea fishing. The result was another discreditable episode in his life, known as 'The Crook Factory'.[44]

Hemingway had a strong propensity to make friends in the urban underworld, especially in Spanish-speaking countries. He loved the dubious characters who made up the *caudrillos* (squads) of bullfighters, and the waterfront café habitués, pimps, prostitutes, part-time fishermen, police informers and the like, who responded warmly to his free drinks and tips. In 1942 in wartime Havana he became obsessed by what he considered the imminent danger of a fascist takeover. There were, he argued, 300,000 Spanish-born inhabitants of Cuba, of whom 15–30,000 were 'violent Falangists'. They might stage an uprising and turn Cuba into a Nazi outpost on America's doorstep. Moreover he was, he said, reliably informed that German submarines were cruising in Cuban waters, and he produced the calculation that a force of 1000 subs could land a 30,000-strong Nazi army in Cuba to assist the insurgents, Whether he believed these fantastic notions is hard to say: throughout his life Hemingway was a mixture of superficial sophistication concealing an abyss of credulity on almost any subject. He may have been influenced by Erskine Childers's spy-mania novel, *The Riddle of the Sands*. He certainly convinced the US Ambassador, a rich drinking and sporting pal of his called Spruille Braden, that something ought to be done.

What Hemingway proposed was that he should recruit and command a group of agents from among his Loyalist underworld friends. They would keep a watch on fascist suspects and at the same time he would use his deep-sea motor-cruiser, suitably armed, to patrol areas likely to be infested with U-boats in an attempt to lure one to the surface. Braden approved the plan and later claimed credit for it.[45] As a result, Hemingway got $1000 a month to pay six full-time agents and twenty undercover ones, chosen from among his café-*caudrillos*. More important, at a time of acute rationing, he got 122 gallons of gasoline a month to run his boat, which was fitted with a heavy machine gun and loaded with hand grenades.

The existence of his 'Crook Factory', as he called it, raised Hemingway's prestige in Havana drinking circles but there is no evidence it turned up a single fascist spy. For one thing, Hemingway made the elementary mistake of paying more for exciting reports. The FBI, which treated this rival venture with the greatest possible disapproval, told Washington that all Hemingway's gang produced were 'vague and unfounded reports of a sensational character . . . His data were almost without fail valueless.' Hemingway, aware of FBI hostility, retorted that all its agents were of Irish origin, Roman Catholics, Franco supporters and 'draft-dodgers'. There were some absurd incidents, too improbable for any spy story, including a report by one of Hemingway's agents of a 'sinister parcel' in the Bar Basque, which turned out to contain a cheap *Life of St Teresa of Avila*. As for the anti-U-boat patrol, it confirmed the views of Hemingway's critics that he needed the gas for his fishing. An eye witness recorded: 'They didn't do a godamned thing – nothing. Just cruise around and have a good time '

The episode led to one of Hemingway's brutal quarrels. Among the men he most admired in Spain was General Duran, who (as 'Manuel') inspired his hero Robert Jordan in *For Whom the Bell Tolls*. Duran was everything Hemingway wanted to be – the intellectual turned master-strategist. He was a musician, a friend of De Falla and Segovia and a member of Spain's pre-war intellectual elite. But he held the view, which Hemingway endorsed, that 'modern war' demands 'intelligence, it is an intellectual's job . . . War is also poetry, tragic poetry.'[46] He got a reserve commission in the Spanish Army in 1934, was called up at the beginning of the Civil War and rapidly became an outstanding general, eventually commanding XX Army Corps. After the Republic collapsed, Duran volunteered in vain for both the British and the US Army. When Hemingway conceived the idea of the Crook Factory he used his influence to get Duran attached to the US Embassy and put him in charge of the scheme. At the same time the general and his English wife Bonté were his guests at the Finca. But Duran quickly realized that the whole thing was a farce and that he was wasting his time. He applied for different work and at the same time there was a bitter personal quarrel, involving Bonté and Hemingway's then-wife Martha, culminating in an explosion at an Embassy lunch. Hemingway never spoke to Duran again, except in May 1945 when the two met by chance and Hemingway sneered: 'You managed quite well to keep out of the war, didn't you?'

That was characteristic of the tone which Hemingway's disputes with former friends tended to take. For a man whose code and whose fiction exalted the virtues of friendships, he found it curiously difficult to sustain any for long. As with so many intellectuals – Rousseau and Ibsen for instance – his quarrels with fellow writers were particularly vicious. Hemingway was unusually jealous, even by the standards of literary life, of the talent and success of others. By 1937 he had quarrelled with every writer he knew. There was one notable exception, which reflects highly on him. The only writer he did not attack in his autobiography was Ezra Pound, and from first to last he wrote approvingly of him. From their first acquaintance he admired Pound's unselfish kindness to other writers. He took from Pound the sharp criticism he would accept from no one, including the shrewd advice in 1926 that he should get down to a novel rather than publish another volume of stories, expressed characteristically: 'Wotter yer think yer are, a bloomink DILLYtante?' He seems to have admired in Pound a virtue he knew he himself conspicuously lacked, a complete absence of professional jealousy.[47] When Pound was in danger of execution for treason in 1945, having made over three-hundred wartime broadcasts for the Axis, Hemingway effectively saved his life. Two years before, when Pound was formally charged, Hemingway had argued: 'He is obviously crazy. I think you might prove he was crazy as far back as the later *Cantos* ... He has a long history of generosity and unselfish aid to other artists and he is one of the greatest of living poets.' In the event it was Hemingway who was responsible for the successful insanity defence which got Pound incarcerated in hospital and saved him from the gas-chamber.[48]

Hemingway also avoided a quarrel with Joyce, perhaps because of lack of opportunity or perhaps because he continued to admire his work, once calling him 'the only alive writer that I ever respected'. For the rest it was a sad tale. He quarrelled with Ford Madox Ford, Sinclair Lewis, Gertrude Stein, Max Eastman, Dorothy Parker, Harold Loeb, Archibald MacLeish and many others. His literary quarrels brought out a peculiar streak of brutal malice as well as his propensity to lie. Indeed many of his worst lies concerned other writers. There is a monstrously false portrait of Wyndham Lewis in the autobiography ('Lewis did not show evil; he just looked nasty ... The eyes had been those of an unsuccessful rapist'), apparently in revenge for some criticism Lewis had once made of him.[49] In the same book he told a string of lies about Scott Fitzgerald and his wife Zelda. Zelda had punctured Hemingway's ego, but Fitzgerald had admired and liked him and done him no harm; Hemingway's repeated assaults on this fragile and bruised spirit are difficult to understand, except in terms of an unappeasable jealousy. According to Hemingway, Fitzgerald told him: 'You know I never slept with anyone except Zelda ... Zelda said that the way I was built I could never make any woman happy and that was what upset her.' The two then went into a men's room and Fitzgerald took out his penis for inspection; Hemingway generously reassured him: 'You're perfectly fine.' This episode seems to be a piece of fiction.

Hemingway's most malevolent quarrel was with Dos Passos, particularly painful in view of their long acquaintance. Jealousy was clearly the original motive – Dos Passos made the cover of *Time* magazine in 1936 (Hemingway had to wait another year). Then came the Robles incident in Spain, followed by a row in New York with both Dos Passos and his wife Katie, an even older friend. Hemingway called Dos Passos a bum who borrowed money and never repaid it, and his wife a kleptomaniac; and there was a lot of sneering about his Portuguese ancestry and supposed illegitimate birth. Hemingway tried to insert these libels into *To Have and Have Not* (1937) but was obliged by his publishers, on legal advice, to cut them. He told William Faulkner in 1947 that Dos Passos was 'a terrible snob (on account of being a bastard)'. In retaliation Dos Passos portrayed Hemingway as the odious George Elbert Warner in *Chosen Country* (1951), which led Hemingway to inform Dos Passos's brother-in-law, Bill Smith, that in Cuba he kept 'a pack of fierce dogs and cats trained to attack Portuguese bastards who wrote lies about their friends'. He unleashed a last quiver of darts at Dos Passos in *A Moveable Feast* – he was a vicious pilot-fish who led sharks like Gerald Murphy to their prey and he had succeeded in destroying Hemingway's first marriage.[50]

The last assertion was palpably false since Hemingway needed no help in destroying his marriages. In his fiction he often wrote about women with remarkable understanding. He shared with Kipling a gift of varying his habitual masculine approach with unexpected and highly effective presentation of a female viewpoint. There have been all kind of speculations about a feminine, even a transvestite or transsexual streak in Hemingway, arising from his apparent obsession with hair, especially short hair in women, and attributed to the fact that his mother declined to dress him in boy's clothes and kept his hair uncut for an unusually long time.[51] What is clear, however, is that Hemingway found it difficult to form any kind of civilized relationship with a woman, at any rate for long, except one based on her complete subservience. The only female in his own family he liked was his younger sister Ursula, 'my lovely sister Ura' as he called her, because she adored him. He told a friend in 1950 that when he came back from the war in 1919, Ursula, then seventeen, 'always used to wait, sleeping, on the stairway of the third-floor staircase to my room. She wanted to wake when I came in because she had been told it was bad for a man to drink alone. She would drink something light with me until I went to sleep and then she would sleep with me so I would not be lonely in the night. We always slept with the light on except she would sometimes turn it off if she saw I was asleep and stay awake and turn it on if she saw I was waking.'[52]

This may have been an invention, reflecting Hemingway's idealized notion of how a woman should behave towards him; but, true or false, he was not going to find such submission in real, adult life. As it happens, three out of his four wives were unusually servile by twentieth-century American standards, but that was not enough for him. He wanted variety, change, drama as well. His first wife, Hadley Richardson, was eight years older and quite well off; he lived off her money until his books began to sell in large quantities. She was an agreeable, accommodating woman, and attractive until she put on weight while pregnant with Hemingway's first child, Jack ('Bumby'), and failed to get it off afterwards.[53] Hemingway had no scruples about fondling other women in her presence – as, for instance, the notorious Lady Twysden, born Dorothy Smurthwaite, who figures as Brett Ashley in *The Sun Also Rises*, a Montparnasse flirt and the source of his row with Harold Loeb. Hadley put up with this humiliation and later with Hemingway's affair with Pauline Pfeiffer, a sexy, slender girl, much richer than Hadley, whose father was one of the biggest landowners and grain-operators in Arkansas. Pauline fell heavily for Hemingway and in effect seduced him. The loving pair then persuaded Hadley to permit the setting up of a *ménage à trois* – 'three breakfast trays', she wrote bitterly from Juan-les-Pins in 1926, 'three wet bathing suits on the line, three bicycles'. When this did not content them, they pushed her out into a trial separation, then into a divorce. She accepted, writing to Hemingway: 'I took you for better, for worse (and meant it).' The settlement was generous on her part, and a delighted Hemingway wrote to her in fulsome terms: 'perhaps the luckiest thing Bumby will ever have is to have you as a mother ... how I admire your straight thinking, your head, your heart and your very lovely hands and I pray God always that he will make up to you the very great hurt that I have done to you – who are the best and truest and loveliest person I have ever known.'[54]

There was a small element of sincerity in this letter in that Hemingway did think Hadley had behaved nobly. On this proposition he began, almost before he married Pauline, to erect a legend of Hadley's sanctity. Pauline, for her part, noted Hadley's unbusinesslike approach to the divorce and determined Hemingway would not be so lucky the next time. She used her money to make their life more ample, buying and embellishing a fine house in Key West, Florida, which introduced Hemingway to the deep-sea fishing he came to love. She gave him a son, Patrick, but when in 1931 she announced she was having another child (Gregory), the marriage went into decline. By now Hemingway had acquired his taste for Havana and there he took up with a strawberry blonde, Jane Mason, wife of the head of Pan-American Airways in Cuba, fourteen years his junior. She was slim, pretty, hard-drinking, a first-class sportwoman who enjoyed hanging around with Hemingway's barroom chums, then driving sports cars at reckless speed. She was in many ways an ideal Hemingway heroine, but she was also a depressive who could not handle her complicated life. She tried to commit suicide and succeeded in breaking her back, at which point Hemingway lost interest.

In the meantime Pauline had taken desperate steps to win back her husband. Her father, she wrote to Hemingway, had just given her a vast sum of money – did he want some? 'Have no end of this filthy money ... Just let me know and don't get another woman, your loving Pauline.' She built him a swimming pool at Key West and wrote: 'I wish you were here sleeping in my bed and using my bathroom and drinking my whisky ... Dear Papa please come home as soon as you can.' She went to a plastic surgeon: 'Am having large nose, imperfect lips, protruding ears and warts and moles all taken off before coming to Cuba'. She also dyed her dark hair gold-colour, which turned out disastrously. But her trip to Cuba did not work. Hemingway called his boat after her but he would not take her out in it. He had issued a warning in *To Have and Have Not*:'The better you treat a man and the more you show him you love him, the quicker he gets tired of you.' He meant it. Moreover, being a man who felt guilt but who responded by shifting it onto other people, he now held her responsible for breaking up his first marriage and therefore felt she deserved anything that was coming to her.

What came was Martha Gellhorn, a passionately keen reporter and writer, Bryn Mawr educated (like Hadley) and, as with most of Hemingway's women, from a secure, upper-middle-class Midwest background. She was tall, with spectacular long legs, a blue-eyed blonde, nearly ten years his junior. Hemingway first met her in Sloppy Joe's Bar, Key West, in December 1936, and the next year invited her to join him in Spain. She did so, and the experience was an eye-opener, not least because he greeted her with a lie: 'I knew you'd get here, Daughter, because I fixed it up so you could' – this was quite untrue, as she was aware. He also insisted on locking her room from the outside, 'so that no man could bother her'.[55] His own room at the Hotel Ambos Mundos, she discovered, was in a disgusting mess: 'Ernest,' she wrote later, 'was extremely dirty ... one of the most unfastidious men I have ever known.' Hemingway had inherited from his father a fondness for onion sandwiches, and in Spain delighted in making them from the powerful local variety, munching them with periodic swigs from his silver hip-flask of whisky, a memorable combination. Martha was inclined to be squeamish and it is unlikely she was ever in love with him physically. She always refused to have a child by him, and later adopted one ('There's no need to have a child when you can buy one. That's what I did'). She married Hemingway primarily because he was a famous writer, something she was passionately keen to become herself: she hoped his literary charisma would rub off on her. But Pauline fought bitterly to keep her husband, and when she felt she was losing remembered Hadley's easy settlement and insisted on a tough one, which delayed the divorce. By the time it was through Hemingway was already inclined to blame Martha for breaking up his marriage; friends testify to their blazing public rows at an early stage.

Martha was easily the cleverest and most determined of his wives, and there was never any chance of the marriage lasting. For one thing, she objected strongly to his drinking and the brutality it engendered. When, at the end of 1942, she insisted on driving the car home because he had been drinking at a party, and they had an argument on the way, he slapped her with the back of his hand. She slowed down his much-prized Lincoln, drove it straight into a tree then left him in it.[56] Then there was the dirt: she objected strongly to the pack of fierce tomcats he kept in Cuba, which smelled fearfully and were allowed to march all over the dining table. While he was away in 1943 she had them castrated, and thereafter he would mutter fiercely: 'She cut my cats.'[57] She corrected his French pronunciation, challenged his expertise on French wines, ridiculed his Crook Factory and hinted broadly that he ought to be closer to the fighting in Europe. He finally decided to go, cunningly arranging an assignment with *Collier's*, which had been employing her and now, to her fury, dropped her. She followed him to London nonetheless and found him, in 1944, living in his customary squalor at the Dorchester, empty whisky bottles rolling about under his bed.

From then on it was downhill all the way. Back in Cuba, he would wake her in the middle of the night when he came to bed after drinking: 'He woke me when I was trying to sleep to bully, snarl, mock – my crime really was to have been at war when he had not, but that was not how he put it. I was supposedly insane, I only wanted excitement and danger, I had no responsibility to anyone, I was selfish beyond belief. It never stopped and believe me it was fierce and ugly.'[58] He threatened: 'Going to get me somebody who wants to stick around with me and let *me* be the writer of the family.'[59] He wrote an obscene poem, 'To Martha Gellhorn's Vagina', which he compared to the wrinkled neck of an old hot-water bottle, and which he read to any woman he could get into bed with him. He became, she complained, 'progressively more insane each year'. She was leading 'a slave's life with a brute for a slave-owner', and she walked out. His son Gregory commented: 'He just tortured Marty, and when he had finally destroyed all her love for him and she had left him, he claimed she had deserted him.'[60] They broke up at the end of 1944, and under Cuban law, since she had deserted, Hemingway kept all her property there. He said his marriage to her was 'the greatest mistake of my life' and in a long letter to Berenson, he listed her vices, accused her of adultery ('a rabbit'), said she had never seen a man die but had nonetheless made more money out of writing about atrocities than any woman since Harriet Beecher Stowe – all untrue.

Hemingway's fourth and final marriage endured to his death mainly because his protagonist this time, Mary Welsh, was determined to hang on whatever happened. She came from a different class to the earlier wives, a logger's daughter from Minnesota. She can have had no illusions about the man she was marrying, since right at the start of their relationship, at the Paris Ritz in February 1945, he got drunk, came across a photo of her Australian journalist husband, Noel Monks, hurled it down the lavatory, fired at it with his sub-machine-gun, smashed the entire apparatus and flooded the room.[61] Mary was a journalist on *Time*, not an ambitious high-flyer like Martha, but hard-working and shrewd. Realizing that Hemingway wanted a wife-servant rather than a competitor, she gave up her journalism completely to marry him, though she continued to have to endure sneers such as '*I* haven't fucked generals in order to get a story for *Time* magazine.'[62] He called her 'Papa's Pocket Venus' and boasted of the number of times he had intercourse with her: he told General Charles ('Buck') Lanham that, after a period of neglect, it was easy to pacify Mary as he had 'irrigated her four times the night before' (when Lanham asked her about this after Hemingway's death she sighed, 'If only it were true').[63]

Mary was a determined woman, a manager; there was something of Countess Tolstoy about her. By this time, of course, Hemingway was as world-famous as Tolstoy, a seer of manliness, a prophet of the outdoors, with drinks, guns, safari clothes, camping gear of all kinds named after him. Wherever he went, in Spain, in Africa, above all in Cuba, he was attended by a court of cronies and freeloaders, sometimes a travelling circus, in Havana usually static. The courtiers were often as eccentric as Tolstoy's, rather lower-grade morally, but equally devoted in their way. Before she made off, Martha Gellhorn recorded what she called 'a very funny sweet scene in Cuba', with Hemingway 'reading aloud from the *Bell* to a bunch of grown-up, well-off, semi-literate pigeon-shooting and fishing pals, they sitting on the floor spellbound'.[64] However, the reality of Hemingway's life, thanks to his appalling habits, was less decorative, let alone decorous, than Yasnaya Polyana. Durie Shevlin, wife of one of Hemingway's many millionaire friends, left a description of the Cuban set-up in 1947: the boat uncomfortable, small and squalid, the Finca roamed by foul cats and bereft of hot water, Hemingway himself stinking of alcohol and sweat, unshaven, muttering in the weird pidgin English he adopted and addicted to the word 'chicken-shit'. Mary had a lot of managing to do.

There were, too, the repetitive, often deliberate, humiliations, Hemingway loved the attentions of women, particularly if they were glamorous, famous and flattering. There was Marlene 'The Kraut' Dietrich, who sang to him in his bathroom while he shaved, Lauren Bacall ('You're even bigger than I'd imagined'), Nancy 'Slim' Hayward ('Darling, you're so thin and beautiful'). There was Virginia 'Jigee' Viertel, part of Hemingway's Paris circus at the Ritz: 'It is now one and a half hours,' Mary recorded grimly, 'since I left Jigee Viertel's room and Ernest said "I'll come in a minute."' In Madrid there were Hemingway's 'whores de combat' as he called them, in Havana the waterfront tarts; he liked to fondle them in Mary's presence, as he had once fondled Dorothy Twysden under Hadley's worried gaze. As he grew older, the girls he wanted grew younger. Hemingway once told Malcolm Cowley, 'I have fucked every woman I wanted to fuck and many I did not, and fucked them all well I hope.'[65] This was never true, and became less true after the Second War. In Venice he became infatuated with a young woman, both dreadful and pathetic, called Adriana Ivancich, whom he made the heroine of his disastrous post-war novel, *Across the River and Into the Trees* (1950). She was a chilly piece, snobbish and unresponsive, who wanted marriage or nothing, and had (as Hemingway's son Gregory put it) 'a hook-nosed mother in constant attendance'. Hemingway lavished hospitality on what must have been one of the most gruesome couples in literary history and, since Adriana had artistic ambitions, forced his reluctant publisher to accept her jacket designs for both *Across the River* and *The Old Man and the Sea* (1952), the book which to some degree restored his reputation and won him the Nobel Prize. Both jackets had to be redrawn. Adriana sneered at Mary as 'uncultured', a judgment echoed by Hemingway himself, who praised the young woman's breeding and civilized ways, drawing the contrast with Mary whom he termed a 'camp-follower' and a 'scavenger'.[66]

There were further humiliations on Hemingway's last big safari, in the winter of 1953–54. He became dirty even by his standards, his tent a muddle of discarded clothes and empty whisky bottles. For mysterious reasons connected with his personal ethic, he took up native dress, shaved his head, dyed some of his clothes orange-pink like the Masai, and even carried a spear. Worse, he took up with a local Wakamba girl called Debba, described by the safari's gamewarden, Denis Zaphina, as 'an evil-smelling bit of camp trash'. She, her girlfriends and Hemingway held celebrations in his tent, during one of which his cot collapsed. Always, according to the diary Mary kept, there was his 'pounding, repetitive conversation that droned on day and night'.[67] Then there was the last big expedition to Spain in 1959, the Hemingway circus travelling with its eighty or ninety pieces of luggage for a summer of bullfights. A nineteen-year-old teenager called Valerie Danby-Smith, daughter of a Dublin builder, came to interview Hemingway for a Belgian news agency, for which she was a stringer. He fell for her, and may even have wished to marry her, but recognized Mary was a better wife to look after an old man, a natural last wife, 'one for the road'. But Valerie was hired for $250 a month, joined the circus and rode in the front seat of the car, for Hemingway's fondling hand, while Mary was put in the back. She endured this, recognizing that Valerie was harmless and by cheering up Hemingway made him less violent – indeed, after he died, continued her employment (she eventually married Gregory Hemingway). But at the time it helped to make the summer 'horrible and hideous and miserable'.[68]

Did Mary have more to endure than Countess Tolstoy? Probably not, in the sense that Hemingway, unlike Tolstoy, was a homebird, with no intention of pushing off into the wilderness. Mary learned Spanish, ran his home well and took part in most of his sporting jaunts. Hemingway wrote a 'situation report' on her at one stage which listed her qualities: 'an excellent fisherwoman, a fair wing-shot, a strong swimmer, a really good cook, a good judge of wine, an excellent gardener ... can run a boat or a household in Spanish.'[69] But he had no sympathy when, as often happened, she injured herself in his wilderness expeditions. She records a characteristic exchange after a painful injury: 'You could keep it quiet.' 'I'm trying.' 'Soldiers don't do that.' 'I'm not a soldier.'[70] There were shattering rows in public, scenes of frightening violence in private. On one occasion he hurled her typewriter to the ground, broke an ashtray she prized, threw wine in her face and called her a slut. She replied that, if he was trying to get rid of her, she was not going to leave the house: 'So try as you may to goad me to leave it and you, you're not going to succeed ... No matter what you say or do, short of killing me, which would be messy, I'm going to stay here and run your house and your Finca until the day when you come here, sober, in the morning and tell me truthfully and straight out that you want me to leave.'[71] It was an offer he was too prudent to take up.

The children of Hemingway's marriages were usually silent, sometimes fearful, witnesses of his marital life. When young, they were much foisted on nannies and servants as the Hemingway circus trundled the world. We hear of one nanny, Ada Stern, described as a lesbian. Bumby, the eldest, bribed her with stolen drinks, Patrick prayed she would be sent to hell, while Gregory, the youngest, was terrified she would leave.[72] Gregory eventually wrote a revealing and rather bitter book about his father. As a young man he got into some petty trouble with the California police. His mother Pauline, long divorced, phoned Hemingway (30 September 1951) to tell him the news and seek comfort and guidance. He replied that she was to blame – 'See how you brought him up' – and they had a frantic argument, Pauline 'shouting into the phone and sobbing uncontrollably'. That night she woke with a grievous internal pain, and the following day she died, aged fifty-six, on the operating table, from a tumour of the adrenal gland. It might have been aggravated by emotional stress. Hemingway blamed his son's delinquency; the son blamed the father's rage. 'It was not my minor troubles that had upset Mother but his brutal phone conversation with her eight hours before she died.' Gregory noted in his book: 'It's fine to be under the influence of a dominating personality as long as he's healthy, but when he gets dry rot of the soul, how do you bring yourself to tell him he stinks?'[73]

The truth, of course, is that Hemingway did not suffer from dry rot of the soul. He was an alcoholic. His alcoholism was as important, indeed central, to his life and work as drug addiction was to Coleridge. Hemingway was a classic textbook case of progressive alcoholism, provoked by deep-seated, chronic and probably inherited depression, and aggravating it in turn. He once told MacLeish: 'Trouble was, all my life when things were really bad I could take a drink and right away they were much better.'[74] He began to drink as a teenager, the local blacksmith, Jim Dilworth, secretly supplying him with strong cider. His mother noted his habit and always feared he would become an alcoholic (there is a theory that his heavy drinking started with his first big row with Grace). In Italy he progressed to wine, then had his first hard liquor at the officers' club in Milan. His wound and an unhappy love affair provoked heavy drinking: in the hospital, his wardrobe was found to be full of empty cognac bottles, an ominous sign. In Paris in the 1920s, he bought Beaune by the gallon at a wine cooperative, and would and did drink five or six bottles of red at a meal. He taught Scott Fitzgerald to drink wine direct from the bottle which, he said, was like 'a girl going swimming without her swimming suit'. In New York he was 'cockeyed' he said for 'several days' after signing his contract for *The Sun Also Rises*, probably his first prolonged bout. He was popularly supposed to have invented the Twenties' phrase 'Have a drink'; though some, such as Virgil Thomson, accused him of being mean about offering one and Hemingway, in turn, was always liable to accuse acquaintances of free-loading, as he did Ken Tynan in Cuba in the 1950s.[75]

Hemingway particularly like to drink with women, as this seemed to him, vicariously, to signify his mother's approval. Hadley drank a lot with him, and wrote: 'I still cherish, you know, the remark you made that you almost worshipped me as a drinker.'[76] The same disastrous role was played by his pretty Thirties' companion in Havana, Jane Mason, with whom he drank gin followed by champagne chasers and huge jars of iced daiquiris; it was indeed in Cuba in this decade that his drinking first got completely out of hand. One bartender there said he could 'drink more martinis than any man I have ever seen'. At the house of his friend Thorwald Sanchez he became fighting drunk, threw his clothes out of the window and broke a set of precious Baccarat glasses; Sanchez's wife was so frightened she screamed and begged the butler to lock him up. On safari, he was seen sneaking out of his tent at 5 am to get a drink. His brother Leicester said that, by the end of the 1930s, at Key West, he was drinking seventeen scotch-and-sodas a day, and often taking a bottle of champagne to bed with him at night.

At this period, his liver for the first time began to cause him acute pain. He was told by his doctor to give up alcohol completely, and indeed tried to limit his consumption to three whiskies before dinner. But that did not last. During the Second World War his drinking mounted steadily and by the mid-1940s he was reported pouring gin into his tea at break-fast. A. E. Hotchner, interviewing him for *Cosmopolitan* in 1948, said he dispatched seven double-size Papa Doubles (the Havana drink named after him, a mixture of rum, grapefruit and maraschino), and when he left for dinner took an eighth with him for the car drive. He claimed: 'Made a run of sixteen here one night.' He boasted to his publisher that he had begun an evening with absinthe, dispatched a bottle of wine at dinner, switched to a vodka session, then 'battened it down with whiskys and soda till 3 am'. His pre-dinner drinks in Cuba were usually rum-based, in Europe martinis, at fifteen-to-one strength. Once, in the early 1950s, I watched him dispatch six of these in quick succession – there was a strong element of public bravado in his drinking – on the terrace outside the Dôme in Montparnasse. His breakfast drinks might be gin, champagne, Scotch or 'Death in the Gulf Stream', a big glass of Hollands gin and lime, another of his inventions. And, on top of all, there was constant whisky: his son Patrick said his father got through a quart of whisky a day for the last twenty years of his life.

Hemingway's ability to hold his liquor was remarkable. Lillian Ross, who wrote his profile for the *New Yorker*, does not seem to have noticed he was drunk a lot of the time he talked to her. Denis Zaphiro said of his last safari: 'I suppose he was drunk the whole time but seldom showed it.' He also demonstrated an unusual ability to cut down his drinking or even to eliminate it altogether for brief periods, and this, in addition to his strong physique, enabled him to survive. But the effects of his chronic alcoholism were nonetheless inexorable. Drinking was also one of the factors in his extraordinary number of accidents. Walter Benjamin once defined an intellectual (himself) as a man 'with spectacles on his nose and autumn in his heart'. Hemingway certainly had autumn in his heart – often midwinter, indeed – but he kept the spectacles off his nose as long as he dared, despite the fact that he had inherited poor sight in his left eye from his mother (who also refused to wear glasses from vanity).

As a result, and perhaps also because of the awkward shape of his big body, Hemingway was prone to accidents all his life. The list of them is dauntingly long.[77] As a child he fell with a stick in his mouth and gouged his tonsils; caught a fishhook in his back; sustained injuries at football and boxing. The year 1918 saw him blown up in the war and smash his fist through a glass showcase. Two years later, he cut his feet walking on broken glass and started internal bleeding by falling on a boat-cleat. He burned himself badly smashing up a water-heater (1922), tore a foot ligament (1925) and had the pupil of his good eye cut by his son (1927). In spring 1928 came the first of his major drinking accidents when, returning home, he mistook the skylight cord for the lavatory chain and pulled the whole heavy glass structure down on his head, sustaining concussion and needing nine stitches. He tore his groin muscle (1929), damaged an index finger with a punch bag, was hurt by a bolting horse and broke his arm in a car smash (1930), shot himself in the leg while drunk and trying to gaff a shark (1935), broke his big toe kicking a locked gate, smashed his foot through a mirror and damaged the pupil of his bad eye (1938) and got two more concussions in 1944, by driving his car into a water tank in the blackout and jumping off a motorcycle into a ditch. In 1945, he insisted on taking over from the driver to take Mary to Chicago airport, skidded and hit a bank of earth, breaking three ribs and a knee and denting his forehead (Mary went through the windscreen). In 1949 he was badly clawed playing with a lion. In 1950 he fell on his boat, gashing his head and leg, severing an artery and concussing himself for the fifth time. In 1953 he sprained his shoulder falling out of his car, and that winter there was a series of accidents in Africa: bad burns while drunkenly trying to put out a bush fire, and two plane accidents, which produced yet another concussion, a fractured skull, two cracked spinal discs, internal injuries, a ruptured liver, spleen and kidneys, burns, a dislocated shoulder and arm, and paralysed sphincter muscles. The accidents, which usually followed drinking, continued almost to his death: torn ligaments, sprained ankle climbing a fence (1958), another car crash (1959).

Despite his physique, his alcoholism also had a direct impact on his health beginning with his damaged liver in the late 1930s. In 1949, while skiing at Contino d'Ampezzo, he got a speck of dust in his eye, and this, combined with drinking, developed into a very serious case of erysipelas, from which he was still suffering ten years later, with a livid, flaking red scar from the bridge of his nose to his mouth. By this time, following his last big drinking bout in Spain (1959), he was experiencing both kidney and liver trouble and possibly haemochromatosis (cirrhosis, bronzed skin, diabetes), oedema of the ankles, cramps, chronic insomnia, blood-clotting and high blood urema, as well as his skin complaints.[78] He was impotent and prematurely aged; the last, sad photograph taken of him, walking near a house he had bought in Idaho, tells its own tale. Even so, he was still on his feet, still alive; and the thought had become unbearable to him. His father committed suicide because of his fear of mortal illness. Hemingway feared that his illnesses were not mortal: on 2 July 1961, after various unsuccessful treatments for depression and paranoia, he got hold of his best English double-barrelled shotgun, put two canisters in it and blew away his entire cranial vault.

Why did Hemingway long for death? It is by no means unusual among writers. His contemporary Evelyn Waugh, a writer in English of comparable stature during this period, likewise longed for death. But Waugh was not an intellectual: he did not think he could refashion the rules of life out of his own head but submitted to the traditional discipline of his church, dying of natural causes five years later. Hemingway created his own code, based on honour, truth, loyalty. He failed it on all three counts, and it failed him. More seriously, perhaps, he felt he was failing his art. Hemingway had many grievous faults but there was one thing he did not lack: artistic integrity. It shines like a beacon through his whole life. He set himself the task of creating a new way of writing English, and fiction, and he succeeded. It was one of the salient events in the history of our language and is now an inescapable part of it. He devoted to this task immense resources of creative skill, energy and patience. That in itself was difficult. But far more difficult, as he discovered, was to maintain the high creative standards he had set himself. This became apparent to him in the mid-1930s, and added to his habitual depression. From then on his few successful stories were aberrations in a long downward slide. If Hemingway had been less of an artist, it might not have mattered to him as a man; he would simply have written and published inferior novels, as many writers do. But he knew when he wrote below his best, and the knowledge was intolerable to him. He sought the help of alcohol, even in working hours. He was first observed with a drink, a 'Rum St James', in front of him while writing in the 1920s. This custom, rare at first, became intermittent, then invariable. By the 1940s, he was said to wake at 4.30 am, 'usually starts drinking right away and writes standing up, with a pencil in one hand and a drink in another'.[79] The effect on his work was exactly as might be expected, disastrous. An experienced editor can always tell when a piece of writing has been produced with the aid of alcohol, however gifted the author may be. Hemingway began to produce large quantities of unpublishable material, or material he felt did not reach the minimum standard he set himself. Some was published nonetheless, and was seen to be inferior, even a parody of his earlier work. There were one or two exceptions, notably *The Old Man and the Sea*, though there was an element of self-parody in that too. But the general level was low, and falling, and Hemingway's awareness of his inability to recapture his genius, let alone develop it, accelerated the spinning circle of depression and drink. He was a man killed by his art, and his life holds a lesson all intellectuals need to learn: that art is not enough.

7

Bertolt Brecht : Heart of Ice

*T*HOSE who want to influence men's minds have long recognized that the theatre is the most powerful medium through which to make the attempt. On 7 February 1601, the day before the Earl of Essex and his men staged their revolt in London, they paid the company to which Shakespeare belonged to put on a special, unexpurgated performance of his *Richard II*, then regarded as a play subversive to monarchy. The Jesuit-led Counter-Reformation placed dramatic presentations right at the centre of the *propaganda fidei*. The first secular intellectuals were no less aware of the importance of the stage. Both Voltaire and Rousseau wrote for it – and Rousseau warned of its dangerous capacity to corrupt public morals. Victor Hugo used it to destroy the last Bourbons. Byron devoted a high proportion of his energy to verse dramas; and even Marx worked on one. But it was Ibsen, as we have seen, who first deliberately and systematically, and with stunning success, used the stage to bring about a revolution in social attitudes. Bertolt Brecht, a totally different playwright in most ways, was his natural successor in this one. He created the modern, sophisticated propaganda play, exploiting brilliantly one of the twentieth century's new cultural institutions, the large-scale state-subsidized theatre. In the two decades after his death, the 1960s and 1970s, he was probably the most influential writer in the world.

Yet Brecht was in his lifetime and remains to some extent even today a mysterious figure. This was the deliberate choice both of himself and of the Communist Party, the organization he served faithfully for the last thirty years of his existence. He, for many reasons, wanted to deflect all public attention from his life to his work. The communist establishment was equally unwilling to have his origins and background, or indeed his style of life, thoroughly explored.[1] His biography thus contains many lacunae, though the main outlines are clear enough. He was born 10 February 1898 in the dull, respectable town of Augsburg, forty miles from Munich. Contrary to the repeated communist assertion, he was not of peasant stock. His forebears on both sides, going back to the sixteenth century, were solidly middle-class – gentlemen-farmers, doctors, schoolteachers, then stationmasters and businessmen.[2] His mother was the daughter of a civil servant. His father was in the paper trade, as chief clerk then sales director of the papermill at Augsburg. His younger brother Walter later went into the business, becoming professor of paper-making at Darmstadt Technical University. Bertolt had a heart-condition and appeared delicate, becoming (like many other leading intellectuals) his mother's favourite child. She said she found the sheer intensity of his wants impossible to refuse. As an adult, however, Brecht took no interest in his family. He scarcely ever mentioned his father. He did not reciprocate his mother's affection. When his mother died in 1920 he insisted on inviting a group of noisy friends to the house the next day – 'the rest of us there were dumb with grief,' his brother recalled – and ostentatiously left town the day before her funeral; though later, in one of his very rare moments of self-reproach, he criticized his behaviour towards his mother: 'I should be stamped out.'[3]

The Brecht legend relates that at school he not only repudiated religion but publicly burned the Bible and the Catechism, and was nearly expelled for his pacifist views. In fact he seems to have written patriotic poetry and was in trouble not for pacifism but for cheating at exams. He was part of the pre-1914, guitar-playing German youth culture with its pro-nature, anti-city ideology. Most of his middle-class contemporaries were conscripted, went straight to the front and were killed there; or, if they survived, became Nazis. Brecht was not a conscientious objector but was excused army service because of his weak heart, and became a medical auxiliary (he had already studied medicine briefly at Munich University). He later painted a horrifying picture of the butchery he witnessed in military hospitals: 'If the doctor ordered me "Amputate a leg, Brecht!" I would answer "Yes, Your Excellency," and cut off the leg. If I was told "Make a trepanning!" I opened the man's skull and tinkered with his brains. I saw how they patched them up to ship them back to the front as soon as possible.'[4] But Brecht was not in fact called up until October 1918, by which time most of the fighting was over, and his work chiefly concerned venereal disease cases. He also lied when he later claimed (in his speech accepting the Stalin Peace Prize) that, in November 1918, he had 'instantly' rallied to the Bavarian Communist Republic and become a soldiers' deputy. He gave various versions of what he did, but it was certainly not, then or at any other time, heroic.[5]

From 1919 onwards Brecht quickly established himself as a literary figure: first as a critic, feared for his rudeness, savagery and cruelty, then in the theatre itself, thanks to his guitar-playing, his skill at writing songs (from first to last, his poetic talent was his finest and purest) and his ability to sing them in a high-pitched, curiously mesmeric voice, not unlike Paul McCartney's in the 1960s. The mood in the German theatre in the early 1920s was strongly leftist, and Brecht took his cue from it. His first success was *Spartakus*, renamed *Drums in the Night* (1922), which won the Kleist Prize for the best young dramatist. This made the right radical noises but Brecht was at this stage in no sense an ideologue, more an opportunist. He wanted to draw attention to himself and was astonishingly successful at it. His aim was *épater la bourgeoisie*. He denounced capitalism and all middle-class institutions. He attacked the army. He praised cowardice and practised it: Keuner, the autobiographical hero of his celebrated short story, 'Measures Against Violence', is an accomplished coward. His friend Walter Benjamin later noted that cowardice and sheer destructiveness were among his salient characteristics.[6] He liked his work to provoke rows and scandals. Ideally he wanted a play of his to evoke boos and hissing from one half of the audience and aggressive applause from the other. A traditional theatre review, based on careful aesthetic analysis, did not interest him. Indeed he despised traditional intellectuals, especially those of the academic or romantic sort.

He invented, in fact, an intellectual of a new kind, rather as Rousseau or Byron had in their times. Brecht's new-model egghead, for whom he himself was the prototype, was harsh, hard, heartless, cynical, part-gangster, part-sports-hearty. He wanted to bring to the theatre the raucous, sweaty, violent atmosphere of the sports arena. Like Byron, he enjoyed the company of professional boxers. Asked to judge a poetry competition in 1926, he ignored the four hundred entries from poets and gave the prize to a crude piece of verse he found in a bicycling magazine.[7] He rejected the Austro-German musical tradition in favour of a metallic, repetitive sound, finding a kindred spirit in the Jewish composer Kurt Weill, with whom he collaborated. He wanted his stage sets to show, as it were, their bones, the machinery behind the illusion: this was his new kind of truth. Machinery fascinated him, as did the men who made and manipulated it, the engineers. He saw himself as a manipulator, a mental engineer. That, indeed, is how he is portrayed, as the engineer Kaspar Proechl in Lion Feuchtwanger's novel *Erfolg*, of whom another character says: 'You lack the most important human organs: sense capable of pleasure and a loving heart.'

Many of Brecht's attitudes and activities in the 1920s reflected his genius for self-publicity. He shared this gift with Hemingway, almost his exact contemporary (as of course with many other intellectuals), and like Hemingway he developed, as part of it, his own distinctive sartorial style. Hemingway's, however, was all-American and predominantly sporting. Brecht clearly, if secretly, admired Hemingway and became very upset if anyone suggested he stole ideas from 'Papa'. In the 1920s he was open about his admiration for the United States – it was the last epoch at which Europe's intelligentsia felt it acceptable to be pro-American – especially its gangsters and sports heroes: he wrote a poem about the Dempsey–Tunney fight in 1926. So some of his dress ideas derived from across the Atlantic. But others were distinctively European. The belted leather jerkin and cap had been favoured by the violent young men of the Cheka which Lenin created early in 1918. To this Brecht added his own invention, a leather tie and waistcoats with cloth sleeves. He wanted to look half student, half workman and wholly smart. His new rig evoked varied comments. His enemies claimed he wore silk shirts under the proletarian leather gear. Carl Zuckmayer called him 'a cross between a truck-driver and a Jesuit seminarist'.[8] He completed his personal style by devising a special way of combing his hair straight down over his forehead and by maintaining a perpetual three-day beard, never more, never less. These touches were to be widely imitated by young intellectuals thirty, forty, even fifty years later. They also copied his habit of wearing steel-rimmed 'austerity' spectacles. In Brecht's case they were grey, his favourite colour. He wrote on a kind of grey tissue-paper and, as he became well-known, began to publish 'Work in Progress', called *Versuche* (drafts), texts of his plays in grey, deliberately sombre paperbacks like school textbooks, a highly effective form of self-promotion also much imitated later. His car, an open Steyre tourer, was also grey; he got it free by writing advertising jingles for its makers. In short, Brecht had a remarkable talent for visual presentation, a field in which, during the 1920s, the Germans led the world: at almost exactly the same time, Hitler was designing the brilliant sumptuary apparatus of the Nazi Party and SS and inventing the night-display technique later known as *son et lumière*.

The rise of Hitler was one factor which pushed Brecht into a more political posture. In 1926 he read *Capital*, or parts of it, and thereafter was associated with the Communist Party, although on the evidence of Ruth Fischer, a German CP leader and sister of his composer-friend Hanns Eisler, he did not actually join it until 1930.[9] The year 1926 was also notable for the beginning of his collaboration with Weill. In 1928 they produced *The Threepenny Opera*, which had its first night on 31 August and was an immediate smash hit, first in Germany, then all over the world. In many ways it was a characteristic example of the way Brecht operated. The main idea was taken from Gay's *The Beggar's Opera* and entire passages of the writing were simply stolen from K.L. Ammers' translation of François Villon (after protests, Ammers was given a share of the royalties). The work's success was in large part due to Weill's catchy and highly original music. But somehow Brecht managed to acquire most of the credit for its lasting success, and when he finally quarrelled with Weill, he announced contemptuously: 'I'll kick that fake Richard Strauss downstairs.'[10]

One reason Brecht captured the credit was his great skill in public relations and showbiz politics. In 1930 G.W. Pabst, who had got the film rights of *Threepenny Opera*, objected to the shooting script Brecht had written, which changed the plot, moving it sharply in a more communist direction. Brecht refused to change it back and the issue came to court in October. He performed some carefully staged tantrums for the cameras and although the case was bound to go against him – Pabst had bought the original play, not a new Marxist version of it – Brecht extracted a massive financial settlement in return for abandoning the suit and was able, moreover, to pose as a martyr to his artistic integrity at the hands of a brutal capitalist system. He published his own script with an introductory essay pointing up the strictly Marxist moral – 'justice, freedom and character have all become conditional upon the process of production'.[11] He had a brilliant knack of advancing his personal interests while proclaiming his devotion to the public's.

A second reason for Brecht's growing celebrity was that, by 1930, he was accepted by the CP as their own star and had all the advantages of its powerful institutional backing. Brecht never cut much ice in Moscow in Stalin's day, and even the German CP, far more flexible in artistic matters, thought some of his work lightweight and heterodox – for instance *The Rise and Fall of the City of Mahagonny* (1930), which provoked rows, fights and Nazi-organized demonstrations. But Brecht showed himself amenable to Party discipline, attended lectures on Marxism-Leninism at the Workers' College in Berlin and indeed, being at heart a Hegelian who loved the mental fantasy-world of the dialectic method – his mind was very German, like Marx's own – found the system intellectually attractive. His first proper Marxist work, *Die Massnahme*, dates from the summer of 1930 and his adaptation of Gorky's *The Mother* was performed all over Germany in CP-controlled halls. He wrote agitprop film scripts. He developed, again with Weill (who was never, however, a keen Marxist), the new political art form of the *Schulopern* (school-opera or didactic drama). Its object was not so much (as it claimed) to educate the audience politically as to turn it into a well-drilled chorus, not unlike the crowds at Nuremberg. The actors became mere political instruments, men-machines rather than artists, and the characters in the plays were not individuals but types, performing highly formalized actions. Such artistic merit as this art form possessed lay in the brilliance of the staging, at which of course Brecht excelled, but its political uses were obvious and it lived on for decades, reaching its nadir with the grim opera-dramas staged by Madame Mao during the Chinese Cultural Revolution of the 1960s. Brecht also invented the use of the set-piece trial (of witches, Socrates, Galileo, Marx's suppressed newspaper, etc.) for agitprop purposes, and it passed into the left-wing repertoire, surfacing from time to time as in the Russell Vietnam War Crimes Tribunal. Indeed many of Brecht's stage inventions – use of white make-up, skeletons, coffins, floats of giant weapons – are still regularly employed in progressive street theatre, processions and demonstrations.

Brecht had other devices for keeping his name before the public. He had himself photographed writing poetry in the midst of a crowd of workers, to emphasize that the days of romantic political individualism were dead and that poetry was now a collective proletarian activity. He embraced publicly the principles of Marxist self-criticism. He took his school opera *Der Jasager* (*The Yes-Man*) to the Communist-controlled Karl Marx Schule, invited comments from the students and re-wrote it in the light of them (later, having got the publicity, he quietly changed it back again).[12] He repeatedly stressed the element of collaboration in his work, though if a play failed he quickly made it clear his own share in it was modest.

Hitler's coming to power in 1933 brought this successful career to an abrupt end, and Brecht left Germany the morning after the Reichstag fire. The 1930s were a difficult time for him. He had no wish to be a martyr. He tried Vienna, did not like the growing mood of pan-German politics, and left for Denmark. He flatly refused to fight in Spain. He went several times to Moscow, and indeed was co-editor of *Das Wort* (with Feuchtwanger and Willi Bredel), published in Russia, which brought him his only regular income. But he rightly judged that Russia was a dangerous place for such as he and never spent more than a few days at a time there. His writing was mainly political hack-work in the years 1933–38. Then, towards the end of the decade, he suddenly began to produce, in quick succession, work of much higher quality – *The Life of Galileo* (1937), *The Trial of Lucullus* (1938), *The Good Woman of Setzuan* (1938–40) and *Mother Courage* (1939). He decided to tackle the

American market, writing *The Resistable Rise of Arturo Ui*, with Hitler presented as a Chicago gangster. The coming of war in 1939 persuaded him Denmark was too risky; he moved to Sweden, then Finland, then – having got a US visa – across Russia and the Pacific to California and Hollywood (1941).

He had been to America before but had made no impact outside left-wing circles. His young man's idealized, comic-strip vision of America quickly faded and he never got to like the reality; hated it, indeed. He could not work the Hollywood studio system, and grew bitterly jealous of other émigrés who succeeded there (Peter Lorre was an exception).[13] His screenplays were not liked. Some of his projects foundered completely. In 1944–45 W.H.Auden worked with him on an English version of *The Caucasian Chalk Circle* and they collaborated on an adaptation of *The Duchess of Malfi*; but at the last minute their version was dropped in favour of the original text, which had had an unexpected hit in London, so Brecht took his name off it. A production of *Galileo*, starring the great Charles Laughton, flopped. Neither in Hollywood nor on Broadway did he understand the market or steel himself to adapt to it. He could not abide theatrical masters or even co-equals; he had to be in absolute charge to be effective.

Realizing that his theatre could not succeed except under ideal conditions which he personally dominated, Brecht prepared himself for a Faustian bargain. It was precipitated by an appearance before the Congressional House Un-American Activities Committee, on 30 October 1947. The Committee was then investigating communist subversion in Hollywood and Brecht, along with nineteen others, was subpoenaed to appear as a potential 'hostile witness'. The others had collectively agreed to refuse to answer questions about their CP membership, and so were cited for contempt of Congress; ten of them got one-year prison sentences.[14] But Brecht had no intention of serving time in a US jail. When asked about membership he flatly denied it: 'No, no, no, no, no, never.' The interrogation had elements of farce, for his interpreter, David Baumgardt of the Library of Congress, had an even thicker accent than Brecht's own, and the furious Chairman, J.Parnell Thomas, shouted: 'I cannot understand the interpreter any more than I can the witness.' However, the Committee had not done its homework and Brecht, realizing this, lied smoothly and earnestly. 'Haven't many of your writings been based on the philosophy of Lenin and Marx?' 'No. I don't think that is quite correct. But of course I studied. I had to study as a playwright who wrote historical plays.' When asked about songs he had contributed to the *Communist Party Song Book*, he said they were mistranslations. He planned, indeed, to make a submissive statement, asserting, 'My activities . . . have always been purely literary activities of a strictly independent nature,' but was given no chance to read it. But he lied with such conviction, was so punctilious in correcting any factual mistakes, appeared so sincere and anxious to help the Committee in any way he could, that he was publicly thanked as an exceptionally cooperative witness.[15] The other subpoenaed writers were so delighted by the cunning way in which he had hoodwinked the Committee that they ignored the fact that he had betrayed them by agreeing to answer questions at all. So he remained a hero of the left. Safely back in Europe, he struck a defiant posture for the press: 'When they accused me of wanting to steal the Empire State Building, I felt it was time for me to leave.'[16]

Basing himself in Switzerland, Brecht now began a careful survey of the European scene before deciding how to plan his future career. He devised a new uniform for himself, a well-cut, grey 'worker's suit', with a grey cloth cap. He had many well-informed contacts through his CP connections. He quickly discovered a fact of critical importance to himself. The emergent Soviet puppet regime in East Germany, struggling for political recognition and, still more, for cultural self-respect, would go a long way to accommodate a major literary figure who helped to give it legitimacy. Brecht had exactly the right literary and ideological credentials for East Germany's purposes. In October 1948 Brecht carried out a reconnaissance in East Berlin, attending a reception in his honour given by the CP Kulturbund. Wilhelm Pieck, later to be President of East Germany, sat on one side of him, Colonel Tupanov, the Soviet Political Commissar, on the other. Called on to reply to the speeches, Brecht used a characteristic gimmick, which both retained all his options and struck a theatrical note of modesty. He simply shook hands with the men on either side of him, and sat down. Three months later, a lavish, heavily subsidized production of *Mother Courage* opened in East Berlin; it was a huge success, with critics coming from all over Western Europe to see it. This finally persuaded Brecht to make East Germany the base for his theatrical operations.

However, his master plan was more complicated. He discovered that Austria, too, was in quest of post-war legitimacy. Austrians had been among Hitler's most enthusiastic supporters and had run many of his concentration camps (including four out of the six big death camps). For strategic reasons the Allies had found it convenient to treat Austria as an 'occupied country', technically a 'victim of Nazi aggression' rather than an enemy, so after 1945 Austrians had neutral status. An Austrian passport was therefore very convenient to hold. At the same time the Austrian authorities were as anxious as the East Germans to win their way back into civilized hearts by stressing their cultural contribution. They too saw Brecht as a useful recruit. So a deal was struck. Brecht certified that he wanted 'to do intellectual work in a country that offered the appropriate atmosphere for it'. He added: 'Let me emphasize that I consider myself only a poet and do not wish to serve any definite political ideology. I repudiate the idea of repatriating myself in Germany.' He insisted that his links with East Berlin were superficial: 'I have no kind of official function or engagement in Berlin and receive no salary at all ... it is my intention to regard Salzburg as my permanent place of residence.'[17] Most of these statements were lies, and Brecht had no intention of living in Salzburg. But he got his Austrian passport, and this not only enabled him to travel wherever he wanted but gave him a considerable degree of independence *vis-à-vis* the East German government.

There was yet a third element to Brecht's carefully designed strategy. His arrangement with the East Germans was that they would provide him with a company and theatre of his own, backed by considerable resources, in return for his artistic identification with the regime. He calculated, rightly as it turned out, that such an investment would give his plays exactly the right push they needed to penetrate the world repertoire. His copyrights would thus become extremely valuable and he had no intention of allowing the East Germans to benefit from them, nor to subject himself to the control of their publishing houses. In the decade 1922–33 he had always refused to have anything to do with the German CP's publishing cooperatives, preferring sound capitalist firms which paid proper royalties. Now too he put his copyrights into the hands of a West German publisher, Peter Suhrkamp, and forced the East Germans to carry, even in their own editions of his works, the line 'By permission of Suhrkamp Verlag, Frankfurt-on-Main.' All his publishing profits, worldwide, and royalties on international productions, were thus paid in hard West German currency, and in due course transferred to the bank account he had opened in Switzerland.

By the summer of 1949, thanks to a good deal of double-dealing and outright lying, Brecht had exactly what he wanted: an Austrian passport, East German government backing, a West German publisher and a Swiss bank account. He took up residence as 'artistic adviser' to what was in effect his own company, the Berliner Ensemble, with his wife Helene Weigel as director. The first big production, *Mr Puntila*, opened on 12 November 1949. In due course, the Theater am Schiffbauerdamm was made over to him as the permanent home of the company, launched with a poster by Picasso. No artist since Wagner had been given a set-up on this scale for the ideal performance of his works. He had sixty actors, plus costume and set designers and musicians, and dozens of production assistants, a total of 250 employees. All the conceivable luxuries which a playwright had ever dreamed of were his. He could rehearse for up to five months. Indeed he could, and did, cancel an evening performance of a play already in the repertoire in order to continue rehearsing a new one – patrons were simply handed their money back when they arrived. There were no worries about the numbers of actors or production costs. He could rewrite and transform his plays several times over in the light of full rehearsals and thus achieve a degree of polish no other playwright in the world could attain. There was a large travelling budget which enabled him to take the company's production of *Mother Courage* to Paris in 1954, and *The Caucasian Chalk Circle* the following year.

These visits were the true beginning of Brecht's international fame and influence. But he had been preparing for this day for many years, using all his wonderful skills of self-promotion. He polished his proletarian image as well as his plays. Extreme care was taken in tailoring his worker's suits. Interviewers were encouraged but scrupulously vetted. Photographs were allowed but only on condition Brecht could select those for publication. Brecht had always been anxious to give his work a 'serious', even solid, dimension and attract the attention of academics, whom he shrewdly saw were excellent long-term promoters of a writer's reputation. That was why he had begun his 'Work in Progress' series, and this was now resumed but on a much larger scale. In the United States he had kept a 'working journal', not so much a diary as a running account of his work and the functioning of his artistic mind, dressed up with what he liked to call 'documentation', newspaper clippings and the like. In 1945 he began to call these and other working papers an 'archive'. He had them all photographed by the then-equivalent of microfilm and persuaded the New York Public Library to take a complete set. The object was to encourage students to write Ph.D. theses on Brecht's work by making it easy for them. Another set went to a Harvard graduate, Gerhard Nellhaus, who was already at work on such a thesis and in due course became an enthusiastic and effective promoter of Brecht's image in the US. Brecht had already acquired an American academic evangelist in the shape of Eric Bentley, an UCLA English professor, who had been working on Stefan Georg. In 1943 Brecht encouraged him to drop Georg and concentrate on himself. Thereafter Bentley not only translated (with Maja Bentley) *The Caucasian Chalk Circle* and arranged its first US production in 1948, but became Brecht's leading drum-beater across the Atlantic. Brecht was cold towards such disciples and forced them to concentrate relentlessly on his work. Bentley testified: 'he did not try to find out much about me. He did not invite me to find out much about him.'[18] Brecht grasped that raising difficulties, even being curmudgeonly, far from discouraging academic sleuths and would-be acolytes, actually whetted their appetites to do him service. He became systematically awkward and exigent, all in the name of artistic integrity. Rousseau had made exactly the same discovery and exploited it, but in Brecht's case the technique was applied with Germanic efficiency and thoroughness.

By the 1950s these efforts were paying increasing dividends in America. Brecht was also assiduously promoting his reputation in Europe and encouraging others to do the same. In East Berlin his vast power of theatrical patronage attracted a circle of young would-be directors and designers. He ordered them around like a Prussian sergeant-major – indeed he ran the whole company with a fierce and arbitrary authority – and they dutifully revered him. His rehearsals themselves became theatrical occasions and were tape-recorded by his disciples, the results being added to the 'archive' as well as circulated in London, Paris and elsewhere. These young men were one means whereby the Brechtian gospel was spread throughout the world of show-business.[19] But he was also promoted by key intellectuals outside his circle. In France the drum was beaten by Roland Barthes in the magazine *Théâtre populaire*. As one of the founders of the new and fashionable science of semiology – study of the modes of human communication – Barthes was in an ideal position to place Brecht on a pedestal for intellectual admiration. In Britain there was a still more influential fugelman in the shape of Kenneth Tynan, who had been converted to Brecht by Eric Bentley in 1950 and who from 1954 was theatre critic of the *Observer*.

This assiduous promotion of Brecht and his work might have been less effective had it not coincided with a fundamental change in the economics of the Western theatre. In the quarter-century 1950–75, for the first time, virtually every country in Europe accepted the principle of state-subsidized theatre. These new institutions were conceived on a large scale and endowed with ample resources, often partly funded from the private sector. Unlike the state theatres of the *ancien régime*, of which the Comédie Française was the archetype, the new companies were usually placed by their constitution outside government control and indeed prided themselves on their independence. Superficially they resembled the generously financed theatres of Eastern Europe, Brecht's in particular; indeed, they tended to take Eastern Europe as their model, concentrating on lavish, meticulously rehearsed productions. Where they differed, however, was in performing not only the classics but 'significant' new plays from an international repertoire. Brecht's oeuvre was a natural choice for this category. Indeed in London, where the change was most revolutionary – the subsidized theatres quickly displacing the commercial ones as providers of 'quality' plays – the National Theatre itself appointed Kenneth Tynan its first literary manager. Hence, throughout Europe and then all over the world, audiences saw Brecht plays performed under heavily subsidized and so ideal conditions, often directly copied from the standards he had laid down in his own theatre. Not even Wagner had enjoyed this degree of good fortune.

Thus Brecht's Faustian bargain paid off, and even in his own lifetime he was rapidly becoming the most influential figure in world theatre. He had always been prepared to deliver his share, or as much of it as he could not withhold by his cunning. From a very early age Brecht had not merely practised but made a cult of self-interested servility. His philosophy was Schweikian. One of the earliest sayings attributed to him is: 'You mustn't forget that art's a swindle. *Life's* a swindle.' To survive you have to engage in swindles yourself, cautiously, success- fully. His works abound in advice to this effect. In *Drums in the Night*, the cowardly soldier Kragler boasts: 'I am a swine – and the swine goes home [from the war].' His hero Galileo, bowing before the Medici, says: 'You think my letter too subservient? . . . A man like me can get into a moderately dignified position only by crawling on his belly. And you know I despise people whose brains are unable to fill their stomachs.' Brecht reiterated this doctrine off stage too. He told his fifteen-year-old son Stefan that poverty must be avoided at all costs, because poverty precluded generosity. To survive, he said, you had to be egotistic. The most important commandment was 'Be good to yourself.'[20]

Behind this philosophy was the adamantine selfishness which seems to be such a common characteristic of leading intellectuals. But Brecht pursued his egotistical objectives with a systematic and cold-blooded ruthlessness which is very rare, even by their standards. He accepted the grim logic of servility: that is, if he bowed to the strong, he tyrannized over the weak. His attitude to women throughout his life had an awe- some consistency: he made all of them serve his purpose. They were hens in a farmyard of which he was cock. He even devised a sartorial style for his women, complementary to his own: long dresses, dark colours, a hint of puritanism.[21] He appears to have had his first success when seventeen, seducing a girl two years younger. As a young man he concentrated on working-class girls: peasants, farmers' daughters, hairdressers, shopgirls; later it was actresses, by the score. No impresario ever used the casting couch more unscrupulously, and Brecht took parti- cular pleasure in corrupting strictly brought-up Catholic girls. It is not clear why women found him so attractive. One actress girlfriend, Mar- ianne Zoff, said he was always dirty: she had to wash his neck and ears herself. Elsa Lanchester, wife of Charles Laughton, said his teeth were 'little tombstones sticking out of a black mouth'. But his voice, thin, reedy, high-pitched, clearly appealed to some; when he sang, said Zoff, his 'grating metallic voice' sent shivers down her spine; she also liked his 'spider thinness' and 'the dark button eyes' which 'could sting'. Brecht was attentive (in the early stages), a great hand-kisser, persistent, above all demanding; it was not only his mother who found the sheer intensity of his requests hard to resist.

Moreover Brecht, though heartless, clearly saw women as more important to him than men; he gave them responsibilities, even if only on a servile basis. He liked to provide each with a special name, which only he used: 'Bi', 'Mar', 'Muck' and so on. He did not mind jealousy, spitting, scratching, rows; liked them in fact. His aim, like Shelley's, was to run small sexual collectives, with himself as master. Where Shelley failed, he usually succeeded. At all times he ran women in tandem, double- and triple-banking them. In July 1919 he had a son by a young woman called Paula Banholzer ('Bi'), before whom he dangled vague promises of marriage. In February 1921 he took up with Zoff ('Mar'), who also became pregnant. She wanted to keep the baby but he refused: 'A child would destroy all my peace of mind.' The two women found out about each other and ran Brecht to earth in a Munich café. They made him sit down between them and choose: which? He replied; 'Both.' Then he proposed to Bi that he marry Mar, make her baby legitimate, then divorce her and marry Bi, making her son legitimate too. Mar read him an angry lecture and swept out of the café in disgust. Bi, rather more timid, would have liked to do the same. Instead she just left. Brecht followed her, got into her train compartment, proposed marriage and was accepted. A few weeks later he did indeed marry – not Bi but Mar. She lost her first child but produced a daughter, Hanne, in March 1923. Within months Brecht was having an affair with another actress, Helene Weigel. He moved into her flat in September 1924 and their son, Stefan, was born two months later. Gradually other members of the sexual collective were acquired, including his devoted secretary, Elizabeth Hauptmann, and yet another actress, Carola Neher, who played Polly in the *Threepenny Opera*. Brecht and Mar were divorced in 1927, thus making him available for matrimony again. Which would he choose this time? He hesitated two years, finally picking Weigel as the most consistently useful. He presented Neher with a compensatory bunch of flowers, saying: 'It can't be helped but it doesn't mean anything.' She bashed him on the head with the bouquet. Hauptmann tried to commit suicide. These messy and for the women harrowing arrangements left Brecht serene. Not once did he give any indication he was perturbed by the sufferings he inflicted on women. They were to be used and discarded, as and when they served his purpose.

There was the tragic case of Margarete Steffin ('Muck'), an amateur actress to whom he had given a part, then seduced during rehearsals. She followed him into exile and became an unpaid secretary. She was a gifted linguist and dealt with all his foreign correspondence (Brecht found it difficult to cope with any language except his own). She suffered from tuberculosis and her condition slowly worsened during the exile years of the 1930s. When her doctor and friend Dr Robert Lund urged her to go into hospital, Brecht objected: 'That won't do any good, she can't stay in hospital now because I need her.' So she forwent the treatment and continued to work for him, being abandoned in Moscow in 1941 when he left for California; she died there suddenly a few weeks later, a telegram from Brecht in her hand. She was thirty-three.

Another case was Ruth Berlau, with whom he began an affair in 1933, a clever Dane aged twenty-seven, whom he stole from her distinguished doctor husband. As with his other mistresses, he gave her a good deal of secretarial and literary work to do. Indeed, he took a lot of notice of what she said about his plays. This infuriated Weigel, who hated Berlau more than any other of Brecht's girls. Berlau was with him in America, where she complained bitterly: 'I am Brecht's backstreet wife' and 'I am the whore of a classical writer.' She became deranged and had to be treated in New York's Bellevue Hospital. Brecht's comment: 'Nobody is as crazy as a crazy communist.' Discharged from hospital, she began to drink heavily. She followed him to East Berlin, sometimes submissive, sometimes creating scenes, until he finally had her shipped back to Denmark, where she collapsed into alcoholism. Berlau was warm-hearted and gifted, and her sufferings over the years do not bear thinking about.

Weigel was the toughest of Brecht's women, but also the most servile. In effect she replaced his mother. Brecht, like Marx, had a perpetual need to exploit people, and in Weigel he achieved his masterpiece of exploitation, Marx's Jenny and Lenchen rolled into one. In many ways Weigel was a strong-minded woman, with powers of leadership and immense organizing capacity. At a superficial level they seemed equals: he called her 'Weigel', she called him 'Brecht'. But she lacked confidence in herself as a woman, especially in her sex appeal, and he seized on this weakness and made use of it. She served him equally in the home and the theatre. At home she washed and scrubbed with passionate energy, scoured the antique shops for fine things, cooked furiously, sometimes brilliantly, and gave endless parties for his colleagues, friends and girls. She promoted his professional interests with every fibre in her being. When he acquired his own theatre in 1949 she ran it for him: box office, bills, builders, cleaners, staff and catering, all the administrative side. But he made it abundantly, even cruelly, clear she was in charge of the building only and had nothing to do with the creative activities, from which she was pointedly excluded. She often had to write to him for an appointment on theatre business.

Indeed at home they each had distinct apartments with their own doorbells. This was to spare her from the full extent of Brecht's philanderings, which continued relentlessly, almost impersonally, during his Berlin years, when his power and position gave him physical access to scores of young actresses. From time to time Weigel, driven beyond endurance, would leave home. But usually she was grimly, resignedly tolerant. Sometimes she gave the young mistresses good advice: Brecht was a very jealous man; promiscuous himself, he expected his women to be faithful or at least to remain firmly under his direction. He demanded control, and that meant good information. He was quite capable of making several telephone calls to check on the activities of a mistress who was not spending the evening with him. Towards the end he sometimes resembled an old stag, working hard to keep his hinds together.

Brecht's intensive, lifelong womanizing left him no time for his children. He had at least two illegitimate ones. Ruth Berlau bore him a son, born in 1944, who died young. His earlier son by Paula, Frank Banholzer, grew to manhood and was killed on the Russian front in 1943. Brecht did not exactly refuse to acknowledge him, as Marx disowned his Freddy. But he took no interest either, scarcely ever saw him, and never mentions him in his journals. But then his legitimate children did not figure much in his life either. He grudged any time he had to spend on or with them. It was the usual tale of intellectual idealism. Ideas came before people, Mankind with a capital 'M' before men and women, wives, sons or daughters. Oscar Homolka's wife Florence, who knew Brecht well in America, summed it up tactfully: 'in his human relationships he was a fighter for people's rights without being overly concerned with the happiness of persons close to him.'[22] Brecht himself argued, quoting Lenin, that one had to be ruthless with individuals to serve the collective.

The same principle applied to work. Brecht had a highly original and creative presentational style but his matter, as often as not, was taken from other writers. He was a gifted adapter, parodist, refurbisher and updater of other men's plots and ideas. Indeed it is probably true that no other writer ever attained such eminence by contributing so little that was truly his own. And why not, he would ask cynically? What did it matter so long as the proletariat was served? Detected in stealing Ammers' Villon, he conceded what he termed his 'basic laxity in matters of literary property' – an odd admission from a man later so tenacious in guarding his own. His *St Joan of the Stockyards* (1932) is a kind of parody of Schiller's *Maid of Orleans* and Shaw's *St Joan*. He based *Señora Carrar's Rifles* on J.M.Synge's *Riders to the Sea*. His *Herr Puntila und sein Knecht Matti* involved stealing the work of the folklorist Hella Wuolojoki, who had been Brecht's hostess in Finland, a characteristic piece of ingratitude. His *Freiheit und Demokratie* (1947) owed a good deal to Shelley's *Mask of Anarchy*. He stole from Kipling. He stole from Hemingway. When Ernest Bornemann drew Brecht's attention to a curious resemblance between one of his plays and a Hemingway short story – thus touching on a tender spot – he provoked an explosion. Brecht shouted: 'Get out – get out – get out!' Helene Weigel, who was in the kitchen cooking and had not heard the beginning of the row – had no idea what it was about – loyally joined in and rushed into the room screaming 'Yes, go, go, go!' and 'swinging her frying pan like a sword'.[23]

Brecht's 'basic laxity' was one reason why, apart from his satellites and those tied to him by party bonds, he was generally unpopular with other writers. He was much despised by the academic writers of the Frankfurt School (Marcuse, Horkheimer, etc.) as a 'vulgar Marxist'. Adorno said that Brecht spent hours every day putting dirt under his fingernails so he looked like a worker. In America he made enemies of both Christopher Isherwood and W.H.Auden. Isherwood resented the efforts of both Brecht and Weigel to destroy his newly acquired Buddhist faith. He found Brecht 'ruthless', a bully, and the pair of them rather like a Salvation Army couple.[24] Auden, Brecht's former collaborator, praised his poetry but dismissed him as a serious political figure ('He couldn't think') and found his moral character deplorable: 'a most unpleasant man', 'an odious person', one of the few who actually deserved the death sentence – 'In fact I can imagine doing it to him myself.'[25] Thomas Mann, too, disliked Brecht: he was 'a party-liner', 'very gifted – unfortunately', 'a Monster'. Brecht lashed back: Mann was 'that short-story writer', a 'clerico-fascist', 'half-wit', 'reptile'.[26]

One of the reasons why Adorno and his friends disliked Brecht so much was that they resented his identification with 'the workers', which they rightly saw as humbug. Of course their own claim to understand what 'the workers' really wanted, felt and believed was equally without foundation; they led entirely middle-class lives too and, like Marx himself, never met the sons of toil. But at least they did not dress up as proles, in clothes carefully designed by expensive tailors. There was a degree of lying, of systematic deception about Brecht which turned even their stomachs. For instance, there was a tale he spread about himself arriving at the door of an expensive hotel for an appointment (the Savoy in London, the Ritz in Paris, the Plaza in New York – the venue changed), dressed of course in his 'worker's suit', and being refused admission by the uniformed doorman. As Brecht was naturally autocratic and quite capable of behaving like an incensed Junker if anyone tried to prevent him from getting what he really wanted, it is most improbable that such an event ever occurred. But Brecht used it as an emblem of his relations with the capitalist system. In one of his versions, the story goes that he was stopped at the entrance to a lavish Western reception to which he had been invited, and asked to fill out a form. When he had done so, the doorman asked: 'Bertolt Brecht? Are you a relative of Bertolt Brecht?' Brecht replied: 'Yes, I am his own son,' then exited, murmuring: 'In every hole, you still find a Kaiser Wilhelm II.'[27]

Some of his publicity tricks Brecht picked up from Charlie Chaplin, whom he admired and once rated a better director even than himself. Thus, when he arrived in his car at an official party and the commission-aire opened its door, Brecht would pointedly get out at the other side, leaving the commissionaire looking foolish and getting a laugh from the gaping crowd. The car, as it happened, was still the old Steyre. Brecht had noisily declined the privilege of an official East German limousine. But keeping and running the Steyre (including fuel, spare parts, repairs, etc.) was just as much a privilege in practice – it was impossible for anyone not connected with the regime to run a private car – and the Steyre had the added advantage to Brecht of serving as a personal publicity symbol.

Again, there was something intrinsically deceptive about the way Brecht lived. In addition to his fine flat overlooking the cemetery where his beloved Hegel was buried (Weigel's flat was on the floor below), Brecht purchased a superb country property at Buckow on Lake Scharmutzel. It had been confiscated by the government from a 'capitalist' and Brecht used it for summer entertaining under its massive old trees. It was in fact two houses, one smaller than the other, and Brecht let it be known that he lived in what he called 'the gardener's cottage'. In his city flat he kept, for display to visiting officials of the regime, portraits of Marx and Engels; but they were arranged in a slightly 'satirical' manner – undetectable, it was supposed, to the official eye – to evoke a titter from friends.

Brecht's anxiety to preserve his image and present at any rate the appearance of independence arose from the undoubted fact that he had made a Faustian exchange. But there was nothing really new in his identification of his professional interests with the survival and spread of communist power. It was implicit and sometimes explicit in his life since 1930. Brecht was a Stalinist throughout the 1930s, sometimes a fanatical one. The American philosopher Sidney Hook records a chilling conversation in 1935 when Brecht called at his apartment on Barrow Street, Manhattan. The great purges were just beginning and Hook, raising the cases of Zinoviev and Kamenev, asked Brecht how he could bear to work with the American communists, who were trumpeting their guilt. Brecht said that the US communists were no good – nor were the Germans either – and that the only body which mattered was the Soviet Party. Hook pointed out that they were all part of the same movement, responsible for the arrests and imprisonment of innocent former comrades. Brecht: 'As for them, the more innocent they are, the more they deserve to be shot.' Hook: 'What are you saying?' Brecht: 'The more innocent they are, the more they deserve to be shot.' (The conversation was in German, which Hook gives in his account.) Hook: 'Why? Why?' He repeated the question but Brecht did not answer. Hook got up, went into the next room and brought Brecht's hat and coat. 'When I returned he was still sitting in his chair, holding a drink in his hand. When he saw me with his hat and coat, he looked surprised. He put his glass down, rose and with a sickly smile took his hat and coat and left.'[28] When Hook first published this account, it was disputed by Eric Bentley. But according to Hook, when he originally related the incident to Bentley (at the 1960 Berlin Congress for Cultural Freedom), Bentley had said: 'That was just like Brecht' – uncannily recalling Byron's initial reaction to the story of Shelley and his illegitimate child by Claire Clairmont. Moreover, confirmation came from Professor Henry Pachter of City University who testified that Brecht had made 'statements to the same effect in my presence', and adding the still more devastating justification which Brecht had produced at the time: 'Fifty years hence the communists will have forgotten Stalin but I want to be sure that they will still read Brecht. Therefore I cannot separate myself from the Party.'[29]

The truth is, Brecht never protested against the purges even when they struck down his own friends. When his former mistress, Carola Neher, was arrested in Moscow, he commented: 'If she has been condemned, there must have been substantial evidence against her'; the furthest he would go was to add that, in this case, 'one does not have the feeling that one pound of crime has been met with one pound of punishment.'[30] Carola vanished – almost certainly murdered by Stalin. When another friend, Tretiakov, was shot by Stalin, Brecht wrote an elegiac poem; but he would not publish it until many years later. At the time his public comment was: 'With total clarity the trials have proved the existence of active conspiracies against the regime ... All the scum at home and abroad, all the parasites, professional criminals, informers joined them. All this rabble had the same objectives as [the conspirators]. I am convinced this is the truth.'[31]

At the time, indeed, Brecht always, and often publicly, supported all Stalin's policies, including his artistic ones. In 1938–39, for instance, he supported the attack on 'formalism' – that is, in effect, any kind of artistic experiment or innovation. 'The very salutary campaign against formalism,' he wrote, 'has helped the productive development of artistic forms, by proving that social content is an absolutely decisive condition for such development. Any formal innovation which does not serve and derive its justification from its social content will remain utterly frivolous.'[32] When Stalin finally died, Brecht's comment was: 'The oppressed of all five continents ... must have felt their heartbeats stop when they heard that Stalin was dead. He was the embodiment of their hopes.'[33] He was delighted in 1955 to be awarded the Stalin Peace Prize. Most of the 160,000 roubles went straight into his Swiss account. But he went to Moscow to receive it and asked Boris Pasternak, apparently unaware of his vulnerable position, to translate his acceptance speech. Pasternak was happy to do this, but later – the prize having been renamed in the meantime – ignored Brecht's request that he translate a bunch of his poems in praise of Lenin. Brecht was dismayed by the circulation of Khrushchev's Secret Session Speech on Stalin's crimes and strongly opposed to its publication. He gave his reasons to one of his disciples: 'I have a horse. He is lame, mangy and he squints. Someone comes along and says: but the horse squints, he is lame and, look here, he is mangy. He is right, but what use is that to me? I have no other horse. There is no other. The best thing, I think, is to think about his faults as little as possible.'[34]

Not thinking was a policy Brecht had perforce adopted himself, since from 1949 he had in effect been a theatrical functionary of the ultra-Stalinist East German regime. He began as he meant to continue, writing a court poem, 'To My Compatriots', to mark the 'election' of Wilhelm Pieck as President of the new German Democratic Republic, 2 November 1949. He enclosed it in a letter to Pieck expressing his 'delight' at the event. On the whole Brecht was the most consistently loyal of all the writers owned by the Communist Party, if we exclude the absolute hacks. He lent his name to whatever international policy the regime was currently promoting. He protested strongly to the West German intelligentsia for conniving at the rearmament by the Federal Republic, while remaining silent about similar arming by the GDR. It was a habit of his to denounce others for his own sins: a repeated theme in these years was the wickedness of Western intellectuals who 'serve' capitalism for money and privileges. He was at work on a play dealing with this subject when he died. He supplied masses of anti-Adenauer material including a preposterous cantata *Herrnburger Bericht*, with such ditties as

> Adenauer, Adenauer, show us your hand
> For thirty pieces of silver you sell our land, etc.

This won him the GDR's National Prize for Literature (First Class). He made himself available to be shown to visiting dignitaries and gave them a set speech denouncing West German rearmament. He signed protest telegrams. He wrote marching songs and other poems for the regime.

There were occasional rows, usually over money – for instance with the East German state film company over *Mother Courage*. The regime rejected *Kriegsfibel* at first as 'pacifist', but gave way when he threatened to bring the issue before the Communist-controlled World Peace Council. But as a rule it was Brecht who yielded. His 1939 play *The Trial of Lucullus*, originally written for radio as an anti-war diatribe, was set to music by Paul Dessau and a production planned to open on 17 March 1951 at the East Berlin State Opera. The regime became alarmed by the advance publicity. They decided it, too, was pacifist, and while it was too late to stop the production they reduced it to three performances and issued all the tickets to Party workers. But some were sold on the black market to West Berliners, who came and applauded wildly. The two remaining performances were cancelled. A week later the official Party paper, *Neues Deutschland*, published an attack under the heading: 'The Trial of Lucullus: the Failure of an Experiment at the German State Opera'. The fire concentrated on the music of Dessau, described as a follower of Stravinsky, 'a fanatical destroyer of the European musical tradition', but the text was also criticized for 'failing to correspond to reality'. Brecht as well as Dessau was summoned to a party meeting which lasted eight hours. At the end of it Brecht dutifully spoke up: 'Where else in the world can you find a government which shows such interest in artists and pays such attention to what they say?', and he made the alterations the party requested, changing the title to *The Condemnation of Lucullus*, while Dessau rewrote the music. But the new production on 12 October still did not satisfy. It was, said *Neues Deutschland*, 'a distinct improvement' but still lacked popular appeal and was 'dangerously close to symbolism'. Thus condemned, it disappeared from the East German stage, though Brecht got it put on in the West.[35]

The real test of Brecht's Faustian bargain came in June 1953, when the East German workers staged an uprising and Soviet tanks were brought in to suppress it. Brecht remained loyal, but at a price; indeed he cunningly used the tragedy to strengthen his own position and improve the terms of the bargain he had struck. When Stalin died in March 1953 Brecht was under growing pressure from the East German authorities to conform to Soviet arts policy, at that time boosting Stanislavsky's methods, which Brecht hated. *Neues Deutschland* which reflected the views of the State Commission for the Fine Arts – where Brecht had enemies and which was running a campaign against his Ensemble – warned that Brecht's company was 'undeniably in opposition to everything Stanislavsky's name stood for'. Moreover, at this time the Ensemble was still sharing a theatre, and the Commission was blocking Brecht's attempts to take over the Theater am Schiffbauerdamm. Brecht's aim was to destroy the Commission and grab the theatre.

The rising seems to have come as a complete surprise to him, thus revealing how completely out of touch he was with the lives of ordinary people. He had ample foreign currency and constantly travelled abroad, he and his wife doing most of their shopping there; in East Germany itself he had access to the special shops open only to senior party officials and other privileged elites. But the masses, many of whom were close to starving, were completely at the mercy of arbitrary switches in government rationing policy, and nearly 60,000 had taken refuge in West Berlin alone. In April, prices were abruptly raised and ration cards withdrawn from whole categories of people – the self-employed and house owners, for instance (Brecht, who was both, was exempted by his privileged status and his Austrian citizenship). On 11 June the policy was abruptly reversed, when ration cards were restored, and the prices and wages policy moved decisively against the factory workers. On 12 June the construction workers, finding their wages cut in half, demanded a mass meeting. The protests began in earnest on 15 June and continued with increasing fury until the Soviet tanks moved in.

Though surprised by the rising, Brecht, who was at his country house, was swift to take advantage of it. He realized how important his support would be to the regime at this juncture. On 15 June he wrote to the party boss, Otto Grotewohl, insisting that the takeover of the theatre by the Ensemble be decided and publicized. The understanding was that in return he would back the party line, whatever it might be. There was some difficulty in deciding the line until two days later, when an unemployed West Berliner called Willi Gottling, who had taken a short cut through the Eastern sector to collect his dole money, was arrested, secretly tried, convicted of being a 'Western agitator', and shot. 'Fascist agitation' thus became the explanation for the riots, and so the party line, which Brecht promptly adopted. By the end of the same day he had dictated letters to the party leaders Ulbricht and Grotewohl and to the Soviet political adviser, Vladimir Semionov, who was in effect the Russian Governor-General. On 21 June *Neues Deutschland* announced: 'National Prize Laureate Bertolt Brecht has sent the General Secretary of the Central Committee of the Socialist Unity Party, Walter Ulbricht, a letter in which he declared: "I feel the need to express to you at this moment my attachment to the Socialist Unity Party, Yours, Bertolt Brecht".' Brecht claimed later that his letter in fact contained a good deal of criticism of the government, and that the sentence quoted was preceded by two others: 'History will pay its respects to the revolutionary impatience of the Socialist Unity Party of Germany. The great discussion with the masses about the tempo of socialist construction will cause the Socialist achievements to be sifted and secured.' Gody Suter, a Swiss correspondent, wrote: 'It was the only time that I saw him helpless, almost small: when he pulled out eagerly the tattered original of that letter from his pocket. He had obviously shown it to many people.'[36] However, Brecht made no effort to publish the full text of his letter, then or later; and he would have possessed the carbon, not the top copy. If he had published it, the regime might have produced the original. Brecht was quite capable of sending one letter, then complaining privately he had sent a quite different one. Even if his version were true, his complaints about Ulbricht's behaviour have not much substance. The bosses of the GDR had more important things to think about than the subtleties of Brecht's support – how to save their necks, for example. In any case, was not Brecht bought and paid for already? Why should they hesitate to cut his little thank-you note?

Neues Deutschland published a long letter from him two days later which made his position brutally clear. It did indeed refer to 'the dissatisfaction of an appreciable section of Berlin's workers with a series of economic measures that had miscarried'. But it went on: 'Organized fascist elements tried to abuse this dissatisfaction for their bloody purpose. For several hours Berlin was standing on the verge of a Third World War. It was only thanks to the swift and accurate intervention of Soviet troops that these attempts were frustrated. It was obvious that the intervention of the Soviet troops was in no way directed against the workers' demonstrations. It was perfectly evident that it was directed exclusively against the attempt to start a new holocaust.'[37] In a letter to his West German publisher he repeated this version: 'a fascistic and war-mongering rabble', composed of 'all kinds of déclassé young people' had poured into East Berlin and only the Soviet Army had prevented world war. That was the party line down to the last 't'. But there was never the slightest evidence of 'fascist agitators'. Nor did Brecht himself believe in them. His private diary shows that he knew the truth. But of course it was not published until long after his death.[38] Moreover, Brecht found the truth – that ordinary German workers rejected the regime – hateful. Like most members of a ruling class he did not meet workers except as servants or occasionally as artisans making repairs in his house. He recorded a conversation with a plumber doing a job at his country place. The plumber complained that an apprentice he sacked for stealing was now in the People's Police, which was full of ex-Nazis. The plumber wanted free elections. Brecht replied: 'In that case the Nazis will be elected.' That was not at all the logic of the plumber's argument but it represented the bent of Brecht's mind. He did not trust the German people and he preferred Soviet colonial rule to democracy.[39]

Brecht got his *quid pro quo* for supporting the regime, though Ulbricht took the best part of a year to deliver. In his efforts to smash the Fine Arts Commission Brecht found he needed the help of Wolfgang Harich, the young and brilliant Professor of Marxist Philosophy at Humboldt University, who provided him with the doctrinal arguments, couched in the right jargon, which he could not himself produce. Early in 1954 the Commission was finally dissolved, being replaced by a new Ministry of Culture, with Brecht's crony Johannes Becher in charge. In March the final payment in the bargain was made when Brecht was given formal possession of the theatre he coveted. He celebrated his victory by pinching Harich's pretty wife, Isot Kilian, whom he made his principal mistress, *pro tem*, and promoted her from bit-player to assistant at his new headquarters. To the shattered Harich he gave the cynical advice: 'Divorce her now. You can marry her again in two years' time' – by then, he implied, he would have finished with her.

By that time, as it happened, he was virtually finished himself. He became ill towards the end of 1954. It was some time before heart trouble was diagnosed – odd in view of his medical history. He did not trust Communist medicine but used a clinic in West Berlin. He arranged to go into another clinic in Munich in 1956, but never got there: a massive coronary thrombosis carried him off on 14 August. He had played one last trick on the long-suffering Weigel. He devised a will leaving some of his literary copyrights to four women: his old secretary-mistress Elizabeth Hauptmann, who got *The Threepenny Opera*, the most valuable of all, poor Ruth Berlau, Isot Kilian, and Käthe Rulicke, whom he had seduced at the end of 1954 and double-banked with Kilian. However, Kilian, who was deputed by Brecht to get the will properly certified, was too impatient to wait in the lawyer's office while it was witnessed. So it turned out to be invalid. Weigel, as the sole legal wife, got the lot, and allocated the other women their share according to her good pleasure. But other desires of Brecht were carried out. He expressed the wish to be buried in a grey steel coffin, to keep out worms, and to have a steel stiletto put through his heart as soon as he was dead. This was done and published: the news being the first indication to many who knew him that he had a heart at all.

I have striven, in this account, to find something to be said in Brecht's favour. But apart from the fact that he always worked very hard – and sent food parcels to people in Europe during and just after the war (but this may have been Weigel's doing) – there is nothing to be said for him. He is the only intellectual among those I have studied who appears to be without a sole redeeming feature.

Like most intellectuals he preferred ideas to people. There was no warmth in any of his relationships. He had no friends in the usual sense of the word. He enjoyed working with people, provided he was in charge. But, as Eric Bentley noted, working with him was a series of committee or board meetings. He was not, Bentley said, interested in people as individuals. This was probably why he could not create characters, only types. He used them as agents for his purposes. This applied equally to his women, whom he saw not so much as individuals but as bedmates, secretaries, cooks. But what, in the end, were his purposes? It is not at all clear whether Brecht had any real, settled beliefs. His French translator, Pierre Abraham, said that Brecht told him, shortly before his death, that he intended to republish his didactic plays with a new preface, saying they were not meant to be taken seriously but as 'limbering-up exercises for those athletes of the spirit that all good dialecticians must be'.[40] These works were certainly presented seriously at the time, and if they were mere 'exercises', which of Brecht's works was not? In the winter of 1922–23, Arnolt Bronnen had a conversation with Brecht about the needs of the people. Bronnen was a major influence on Brecht. He had 'hardened' or 'lefted' his name by changing it from Arnold to Arnolt, and Brecht copied him. Not only did he drop his other two Christian names, Eugen and Friedrich, as 'too royalist', he also hardened Bertold into Bertolt. But when, on this occasion, Bronnen urged the need to change the world so that no one should ever go hungry, Brecht became angry. According to Bronnen, he said: 'What business is it of yours if people are starving? One must get on, make a name for oneself, get a theatre to put on one's own plays!' Bronnen added: 'He was not interested in anything else.'[41] Brecht loved to be ambivalent, ambiguous, mysterious. He veiled his mind cunningly, just as he clothed his body in worker's suits. But it may be that on this occasion, just for once, he said what he really thought.

8

Bertrand Russell:
A Case of Logical Fiddlesticks

*N*o intellectual in history offered advice to humanity over so
long a period as Bertrand Russell, third Earl Russell (1872–
1970). He was born in the year General Ulysses S. Grant was
reelected to the US Presidency and he died on the eve of Watergate.
He was a few months younger than Marcel Proust and Stephen Crane,
a few weeks older than Calvin Coolidge and Max Beerbohm; yet he
lived long enough to salute the revolting students of 1968 and enjoy
the work of Stoppard and Pinter. All this time he put forth a steady
stream of counsel, exhortation, information and warnings on an astonishing variety of subjects. One bibliography (almost certainly incomplete)
lists sixty-eight books. The first, *German Social Democracy*, was published
in 1896, when Queen Victoria still had five years to live; his posthumous
Essays in Analysis (1973) came out the year Nixon resigned. In between
he published works on geometry, philosophy, mathematics, justice,
social reconstruction, political ideas, mysticism, logic, Bolshevism,
China, the brain, industry, the ABC of atoms (this was in 1923; thirty-six
years later came a book on nuclear warfare), science, relativity, education,
scepticism, marriage, happiness, morals, idleness, religion, international
affairs, history, power, truth, knowledge, authority, citizenship, ethics,
biography, atheism, wisdom, the future, disarmament, peace, war
crimes and other topics.[1] To these should be added a huge output of
newspaper and magazine articles embracing every conceivable theme,
not excluding The Use of Lipstick, The Manners of Tourists, Choosing
Cigars and Wife-Beating.

Why did Russell feel qualified to offer so much advice, and why did people listen to him? The answer to the first question is not immediately obvious. Probably the biggest single reason he wrote so much was that he found writing was so easy, and in his case so well paid. His friend Miles Malleson wrote of him in the 1920s: 'Every morning Bertie would go for an hour's walk by himself, composing and thinking out his work for that day. He would then come back and write for the rest of the morning, smoothly, easily and without a single correction.'[2] The financial results of this agreeable activity were recorded in a little notebook, in which he listed the fees he had received for everything he had published or broadcast in the whole of his life. He kept it in an inner pocket and, in his rare moments of idleness or despondency, would bring it out for perusal, which he called 'a most rewarding occupation'.[3]

Certainly, Russell was not a man who ever acquired extensive experience of the lives most people lead or who took much interest in the views and feelings of the multitude. He was an orphan, both of whose parents died by the time he was four, and his childhood was spent in the household of his grandfather, the first Earl Russell, who as Lord John Russell had steered the Great Reform Bill (1832) through the old, unreformed House of Commons. Russell's background was that of the Whig aristocracy who, while sealing themselves hermetically from contact with the populace or even the gentry, had an arbitrary taste for radical ideas. The old earl, as a former Prime Minister, enjoyed a grace-and-favour residence, Pembroke Lodge in Richmond Park, which Queen Victoria accorded him, and Russell grew up there. I always assumed that his inimitable accent, of great clarity and antiquity, came straight from his grandfather, though it was often erroneously classed as 'Bloomsbury'. The chief influence on his childhood, however, was his grandmother, a high-principled, fiercely religious lady of marked puritanical views. Russell's parents had been atheists and ultra-radicals and had left instructions for Bertrand to be brought up under the aegis of John Stuart Mill. His grandmother would have none of this and kept Russell at home, in an atmosphere of Bibles and Blue Books, taught by a succession of governesses and tutors (one of whom, however, turned out to be an atheist). It all made little difference since Russell would have gone his own way whatever happened. By the age of fifteen, he was writing in his journal, using Greek characters to conceal his thoughts from prying eyes: 'I have ... come to look into the very foundations of the religion in which I was brought up.'[4] He became an unbeliever about this time and remained one for the rest of his days. The notion that most people recognize and need some kind of supreme being never had the smallest appeal to him. He believed that the answers to all the riddles of the universe would be found by the human mind, or not at all.

No man ever had a stronger confidence in the power of intellect, though he tended to see it almost as an abstract, disembodied force. It was his love of the abstract intellect and his suspicion of the bodily motions, derived very likely from his grandmother's puritanical teachings, which made Russell a mathematician. The science of numbers, than which nothing is more remote from people, was the first and greatest passion of his life. With the help of an army of crammers he got a scholarship to Trinity College, Cambridge, and in 1893 was listed as Seventh Wrangler in the Mathematical Tripos. There followed a Trinity Fellowship and, in due course, the draft of the great work he wrote with Alfred North Whitehead, *Principia Mathematica*, completed on the last day of the old century. He wrote: 'I like mathematics because it is *not* human.' In his essay, 'The Study of Mathematics', he rejoiced: 'Mathematics possesses not only truth, but supreme beauty – a beauty cold and austere, like that of sculpture, without appeal to any part of our weaker nature, sublimely pure and capable of a stern perfection such as only the greatest art can show.'[5]

Russell never believed that the populace could or should be encouraged to penetrate the frontiers of knowledge. His professional work in mathematics was carried out in a highly technical manner, making not the smallest concession to the non-specialist. Philosophical speculation, he argued, should be conducted in a special language and he fought not only to retain but to strengthen this hieratic code. He was a high priest of the intellect, forbidding outsiders to penetrate the arcana. He disagreed strongly with those of his philosophical colleagues, like G.E. Moore, who wanted to debate problems in an ordinary, common-sense language, insisting: 'common sense embodies the metaphysics of savages.' However, while the intellectual high priests had a duty, in his view, to keep the Eleusinian Mysteries to their caste, they had likewise a duty, on the basis of their store of knowledge, to regale the populace with some digestible fruits of their wisdom. He thus drew a distinction between professional philosophy and popular ethics, and practised both. Between 1895 and 1917, again in 1919–21 and in 1944–49, he was a Fellow of Trinity, and he also spent many years lecturing and teaching at a variety of American universities. But an even larger part of his life he spent in telling the public what they ought to think and do, and this intellectual evangelism completely dominated the second half of his long life. Like Dr Albert Einstein in the 1920s and 1930s, Russell became, for masses of people all over the world, the quintessence, the archetype of the abstract philosopher, the embodiment of the talking head. What was philosophy? Well: it was the sort of things Bertrand Russell said.

Russell was a gifted expositor. An early work of his had explained the work of Leibnitz, whom he always revered.[6] His brilliant survey, *A History of Western Philosophy* (1946), is the ablest thing of its kind ever written and was deservedly a best-seller all over the world. His fellow academics criticized, affected to deplore and doubtless envied his popular work. Ludwig Wittgenstein found his book *The Conquest of Happiness* (1930) 'quite unbearable'.[7] When his last major philosophical work, *Human Knowledge*, was published in 1949, the academic reviewers refused to take it seriously. One of them called it 'the patter of a conjuror'.[8] But the public likes a philosopher who goes out into the world. Moreover, there was a feeling that Russell, right or wrong, had the courage of his convictions and was prepared to suffer for them. Just as Einstein had gone into exile to escape Nazi tyranny, Russell was repeatedly at loggerheads with various authorities, and took his punishment manfully.

Thus, in 1916, he wrote an anonymous leaflet for the No Conscription Fellowship, protesting about a conscientious objector who was sent to jail despite the 'conscience clause' in the conscription law. The distributors were arrested, convicted and sent to prison. Russell wrote a letter to *The Times* saying he was the author. He was put on trial at the Mansion House in front of the Lord Mayor of London, convicted and fined £100. He refused to pay, so his furniture in Trinity was distrained and sold. The council of Trinity, the elite governing body of senior Fellows, then removed his Fellowship. They took the matter very seriously and most of them seem to have acted after great thought and on the highest principle.[9] But to the public it looked like a double punishment for the same offence.

On 11 February 1918 Russell was tried and convicted a second time. This was for writing in a radical paper called the *Tribunal* an article, 'The German Peace Offer', which stated: 'The American garrison, which will by that time be occupying England and France, whether or not they will prove efficient against the Germans, will no doubt be capable of intimidating strikers, an occupation to which the American army is accustomed at home.' For this rash, untruthful and indeed absurd statement he was charged, under the Defence of the Realm Act, with 'having in a printed publication made certain statements likely to prejudice His Majesty's relations with the United States of America', convicted at Bow Street and given six months.[10] When he was released the Foreign Office refused (for a time at least) to issue him with a passport, the Permanent Under-Secretary, Sir Arthur Nicolson, minuting on the file that he was 'one of the most mischievous cranks in the country'.[11]

Russell was again in trouble with the law in 1939–40, when he was appointed to a chair at the City University of New York. He was by this time notorious for his irreligious and supposedly immoral views.

In addition to countless anti-Christian articles, he perfected a parlour performance, 'The Atheist's Creed', which he recited in the nasal tones of a clergyman chanting: 'We do not believe in *God*. But we believe in the supremacy of *hu-man-ity*. We do not believe in life after *death*. But we believe in *immortality – through – good – deeds.*'[12] He delighted to recite this to the children of his progressive friends. When his New York appointment was announced, the local Anglican and Catholic clergy protested loudly. As the university was a municipal institution, citizens could litigate against its appointments, and a lady was induced to do so. She sued the City of New York, which by this time was as anxious to lose the case as she to win it. Her lawyer pronounced Russell's works 'lecherous, libidinous, lustful, venerous, erotomaniac, aphrodisiac, irreverent, narrow-minded, untruthful and bereft of moral fibre'. The judge, an Irish-American, added to the vituperation and ruled Russell unfit to hold the post as 'an alien atheist and exponent of free love'. The Mayor, Fiorello La Guardia, refused to appeal against the verdict and the Registrar of New York County said publicly that Russell should be 'tarred and feathered and driven out of the country'.[13]

Russell's final brush with authority came in 1961 when he was eighty-eight and made strenuous attempts to get himself arrested for acts of civil disobedience in protest against nuclear weapons. He took part in an illegal 'sit-down' outside the Defence Ministry in London on 18 February, and remained seated on the pavement for several hours. But nothing was done and he was obliged to go home. On 6 August, however, he was summoned to appear at Bow Street on 12 September for inciting the public to break the law, and in due course was convicted and sentenced to a month's jail, commuted to a week (which he spent in the hospital wing of the prison). When the sentence was announced a man shouted, 'Shame, shame, an old man of eighty-eight!', but the stipendiary magistrate merely remarked, 'You are old enough to know better.'[14]

Whether these episodes actually advanced any of Russell's views with the masses is doubtful. But they testified to his sincerity and his willingness to bring philosophy out of its ivory academic tower and into the marketplace. People thought of him, rather vaguely and inaccurately to be sure, as a modern Socrates taking poison or Diogenes emerging from his rain barrel. In fact the notion of Russell carrying philosophy into the world is quite misleading; rather he tried, unsuccessfully, to squeeze the world into philosophy and found it would not fit. Einstein's case was quite different, for he was a physicist concerned with the behaviour of the universe as it is and determined to apply to his description of this behaviour the most meticulous standards of empirical proof. By correcting Newtonian physics he changed the whole manner in which we see the universe and his work has countless and continuing applications – indeed his contribution to atomic theory was the first great milestone on the road to man-made nuclear energy.

By contrast, no one was more detached from physical reality than Russell. He could not work the simplest mechanical device or perform any of the routine tasks which even the most pampered man does without thinking. He loved tea but could not make it. When his third wife, Peter, had to go away and wrote down on the kitchen slate: 'Lift up bolster of the Esse [cooker]. Move kettle onto hot-plate. Wait for it to boil. Pour water from kettle onto teapot,' he failed dismally to carry out this operation.[15] In old age, he began to go deaf and was fitted with a hearing-aid; but he could never make it work without help. The human, as well as the physical world, constantly baffled him. He wrote that the coming of the First World War forced him 'to revise my views on human nature ... I had supposed until that time that it was quite common for parents to love their children, but the war persuaded me that it is a rare exception. I had supposed that most people liked money better than almost anything else, but I discovered that they liked destruction even better. I had supposed that intellectuals frequently loved truth, but I found here again that not 10 per cent of them prefer truth to popularity.'[16] This angry passage betrays such deep ignorance of how ordinary people's emotions function in wartime, or indeed at any other time, as almost to defy comment. There are many other statements in his volumes of autobiography which produce a sense of wonder in the normal reader that so clever a man could be so blind to human nature.

The curious thing is that Russell was quite capable of detecting – and deploring – in others the same dangerous combination of theoretical knowledge and practical ignorance of how people felt and what they wanted. In 1920 he visited Bolshevist Russia and on 19 May had an interview with Lenin. He found him 'an embodied theory'. 'I got the impression,' he wrote, 'that he despises the populace and is an intellectual aristocrat.' Russell saw perfectly well how such a combination disqualified a man from ruling wisely; indeed, he added, 'if I had met [Lenin] without knowing who he was I should not have guessed that he was a great man but should have thought him an opinionated professor.'[17] He could not or would not see that his description of Lenin applied in some degree to himself. He too was an intellectual aristocrat who despised, and sometimes pitied, the people.

Moreover, Russell was not merely ignorant of how most people actually behave; he had a profound lack of self-awareness too. He could not see his own traits mirrored in Lenin. Even more seriously he did not perceive that he himself was exposed to the forces of unreason and emotion that he deplored in common people. It was Russell's general position that the ills of the world could be largely solved by logic, reason and moderation. If men and women followed their reason rather than their emotions, argued logically instead of intuitively, and exercised moderation instead of indulging in extremes, then war would become impossible, human relationships would be harmonious and the condition of mankind could be steadily improved. It was Russell's view, as a mathematician, that pure mathematics had no concept which could not be defined in terms of logic and no problem which could not be solved by the application of reasoning. He was not so foolish as to suppose that human problems could be solved like mathematical equations but he nonetheless believed that given time, patience, method and moderation, reason could supply the answers to most of our difficulties, public and private. He was convinced it was possible to approach them in a spirit of philosophical detachment. Above all, he thought that, given the right framework of reason and logic, the great majority of human beings were capable of behaving decently.

The trouble was that Russell repeatedly demonstrated, in the circumstances of his own life, that all of these propositions rested on shaky foundations. At every great juncture, his views and actions were as liable to be determined by his emotions as by his reason. At moments of crisis logic was thrown to the winds. Nor could he be trusted to behave decently where his interests were threatened. There were other weaknesses too. When preaching his humanist idealism, Russell set truth above any other consideration. But in a corner, he was liable – indeed likely – to try to lie his way out of it. When his sense of justice was outraged and his emotions aroused, his respect for accuracy collapsed. Not least, he found it hard to achieve the consistency which the pursuit of reason and logic ought theoretically to impose on their devotees.

Let us follow Russell's opinions as they developed on the great themes of war and peace, which engaged his energies perhaps more than anything else. Russell regarded war as the supreme paradigm of irrational conduct. He lived through two world wars and countless minor ones and hated them all. His detestation of war was absolutely genuine. In 1894 he had married Alys Whitall, sister of Logan Pearsall Smith. She was a Quaker and her gentle, religious pacifism reinforced his robust and logical (as he saw it) variety. When war broke out in 1914 he declared himself totally opposed to it and did everything in his power, on both sides of the Atlantic, to bring about peace, jeopardizing his liberty and his career. But the remarks which led to his imprisonment were not those of an eirenic or a reasonable or a moderate man. His major philosophical statement in defence of pacifism, 'The Ethics of War' (1915), arguing that war is hardly ever justified, is logical enough.[18] But his pacifism, then and later, found expression in highly emotional, not to say combative, ways. For instance, when King George V took a wartime pledge to abstain from alcohol in 1915, Russell promptly abandoned the teetotalism he had embraced at Alys's desire: the King's motive, Russell wrote, 'was to facilitate the killing of Germans, and it therefore seemed as if there must be some connection between pacifism and alcohol'.[19] In the United States, he saw American power as a means to enforce peace and excitedly implored President Wilson, whom he then saw as a world saviour, to 'undertake the championship of mankind' against the belligerents.[20] He wrote a letter to Wilson in a messianic spirit: 'I am compelled by a profound conviction to speak for all the nations in the name of Europe. In the name of Europe I appeal to you to bring us peace.'

Russell may have hated war but there were times when he loved force. There was something aggressive, even bellicose, about his pacifism. After the initial declaration of war, he wrote, 'For several weeks I felt that if I should happen to meet Asquith or Grey I should be unable to refrain from murder.'[21] In fact, some time later he did come across Asquith. Russell emerged from swimming at Garsington Manor, stark naked, to find the Prime Minister sitting on the bank. But his anger had cooled by now and instead of murdering him, he embarked on a discussion of Plato, Asquith being a fine classical scholar. The great editor under whom I served, Kingsley Martin, who knew Russell well, often used to say that all the most pugnacious people he had come across were pacifists, and instanced Russell. Russell's pupil T. S. Eliot said the same: '[Russell] considered any excuse good enough for homicide.' It was not that Russell had any taste for fisticuffs. But he was in some ways an absolutist who believed in total solutions. He returned more than once to the notion of an era of perpetual peace being imposed on the world by an initial act of forceful statesmanship.

The first time this idea occurred to him was towards the end of the First World War when he argued that America should use its superior power to insist on disarmament: 'The mixture of races and the comparative absence of a national tradition make America peculiarly suited to the fulfilment of this task.'[22] Then, when America secured a monopoly of nuclear weapons, in 1945–49, the suggestion returned with tremendous force. Since Russell later tried to deny, obfuscate or explain away his views during this period, it is important to set them out in some detail and in chronological order. As his biographer Ronald Clark has established, he advocated a preventative war against Russia not once but many times and over several years.[23] Unlike most members of the left, Russell had never been taken in by the Soviet regime. He had always rejected Marxism completely. The book in which he described his 1920 visit to Russia, *The Practice and Theory of Bolshevism* (1920), was highly critical of Lenin and what he was doing. He regarded Stalin as a monster and accepted as true the fragmentary accounts of the forced collectivization, the great famine, the purges and the camps which reached the West. In all these ways he was quite untypical of the progressive intelligentsia. Nor did he share the complacency with which, in 1944–45, they accepted the extension of Soviet rule to most of Eastern Europe. To Russell this was a catastrophe for Western civilization. 'I hate the Soviet government too much for sanity,' he wrote on 15 January 1945. He believed that Soviet expansion would continue unless halted by the threat or use of force. In a letter dated 1 September 1945 he asserted: 'I think Stalin has inherited Hitler's ambition to world dictatorship.'[24] Hence, when the first nuclear weapons were exploded by the US over Japan, he immediately resurrected his view that America should impose peace and disarmament on the world, using the new weapons to coerce a recalcitrant Russia. To him it was a heaven-sent opportunity which might never recur. He first set out his strategy in articles in the Labour journal *Forward*, published in Glasgow 18 August 1945, and the *Manchester Guardian*, 2 October. There was a further article on the same theme in *Cavalcade*, 20 October. This was entitled 'Humanity's Last Chance' and included the significant remark 'A *casus belli* would not be difficult to find.'

Russell reiterated these or similar views over a period of five years. He set them out in *Polemic*, July–August 1946, in a talk to the Royal Empire Society on 3 December 1947 printed in the *United Empire*, January–February 1948 and *New Commonwealth*, January 1948, in a lecture at the Imperial Defence College, 9 December 1947, repeated on various occasions, at a student conference at Westminster School, November 1948, printed in the *Nineteenth Century and After*, January 1949, and again in ar article in *World Horizon* in March 1950. He did not mince his words. The Royal Empire Society talk proposed an alliance – adumbrating NATO – which would then dictate terms to Russia: 'I am inclined to think that Russia would acquiesce; if not, provided this is done soon, the world might survive the resulting war and emerge with a single government such as the world needs.' 'If Russia overruns Western Europe,' he wrote to an American disarmament expert, Dr Walter Marseille, in May 1948, 'the destruction will be such as no subsequent reconquest can undo. Practically the whole educated population will be sent to labour camps in north-east Siberia or on the shores of the White Sea, where most will die of hardship and the survivors will be turned into animals. Atomic bombs, if used, will at first have to be dropped on Western

Europe, since Russia will be out of reach. The Russians, even without atomic bombs, will be able to destroy all the big towns in England ... I have no doubt that America would win in the end, but unless Western Europe can be preserved from invasion, it will be lost to civilization for centuries. Even at such a price, I think war would be worth while. Communism must be wiped out, and world government must be established.'[25] Russell constantly stressed the need for speed: 'Sooner or later, the Russians will have atom bombs, and when they have them it will be a much tougher proposition. Everything must be done in a hurry, with the utmost celerity.'[26] Even when Russia exploded an A-bomb, he still pressed his argument, urging that the West must develop the hydrogen bomb. 'I do not think that, in the present temper of the world, an agreement to limit atomic warfare would do anything but harm, because each side would think that the other was evading it'. He then put the 'Better Dead than Red' argument in its most uncompromising form: 'The next war, if it comes, will be the greatest disaster that will have befallen the human race up to that moment. I can think of only one greater disaster: the extension of the Kremlin's power over the whole world.'[27]

Russell's advocacy of preventative war was widely known and much discussed in these years. At the International Congress of Philosophy at Amsterdam in 1948 he was furiously attacked for it by the Soviet delegate, Arnost Kolman, and replied with equal asperity: 'Go back and tell your masters in the Kremlin that they must send more competent servants to carry out their programme of propaganda and deceit.'[28] As late as 27 September 1953 he wrote in the *New York Times Magazine*: 'Terrible as a new world war would be, I still for my part would prefer it to a world communist empire.'

It must have been at about this time, however, that Russell's views began to change abruptly and fundamentally. The very next month, October 1953, he denied in the *Nation* that he had ever 'supported a preventative war against Russia'. The entire story, he wrote, was 'a communist invention'.[29] For some time, a friend recorded, whenever his post-war views were presented to him, he would insist: 'Never. That's just the invention of a communist journalist.'[30] In March 1959, in an interview on BBC television with John Freeman, in one of his famous *Face to Face* programmes, Russell changed his tack. Disarmament experts in America had sent him chapter and verse of his earlier statements and he could no longer deny they had been made. So he said to Freeman, who questioned him about the preventative war line: 'It's entirely true, and I don't repent of it. It's entirely consistent with what I think now.'[31] He followed this with a letter to the BBC weekly, the *Listener*, saying:

'I had, in fact completely forgotten that I had ever thought a policy of threat involving possible war desirable. In 1958 Mr Alfred Kohlberg and Mr Walter W. Marseille brought to my notice things which I said in 1947, and I read these with amazement. I have no excuses to offer.'[32] In the third volume of his autobiography (1968) he ventured a further explanation: '... at the time I gave this advice, I gave it so casually, without any real hope it would be followed, that I soon forgot I had given it.' He added: 'I had mentioned it in a private letter and again in a speech that I did not know to be the subject of dissection by the press.'[33] But as the investigation by Ronald Clark showed, Russell had argued the case for preventative war repeatedly, in numerous articles and speeches, and over a period of several years. It is hard to believe he could have forgotten so completely this tenacious and protracted stance.

When Russell told John Freeman that the views on nuclear weapons he held in the late 1950s were consistent with his post-war support for preventative war, he strained credulity in another way. Indeed, most people would say he was talking nonsense. But there was a certain consistency of quite another kind, the consistency of extremism. Both the preventative war case and the Better Dead than Red case, as presented by Russell, were examples of reasonable lines of argument pushed to the point of extremity by a ruthless and inhuman use of logic. There, indeed, was Russell's weakness. He attached a false value to the dictates of logic, in telling humanity how to conduct its affairs, allowing it to override the intuitive urgings of common sense.

Hence when, in the mid-1950s, Russell decided that nuclear weapons were intrinsically evil and should not be used in any circumstances, he tore off, following the howling banshees of logic, in quite another – but equally extreme – direction. He first declared his opposition to nuclear weapons in a 1954 broadcast about the Bikini Atoll tests, 'Man's Peril'; then came various international conferences and manifestos, with Russell's line hardening in favour of total abolition at all costs. On 23 November 1957 he published in the *New Statesman* 'An Open Letter to Eisenhower and Khrushchev', putting his case.[34] Next month, going through the paper's incoming letters box, I was astonished to find a huge translated screed, with a covering letter in Russian signed by Nikita Khrushchev. This was the Soviet leader's personal reply to Russell. It was of course largely propaganda, for the Soviets, with their huge superiority in conventional forces, had always been prepared to accept an agreed (though unsupervised) dismantling of nuclear weapons. But the letter, when published, created an immense sensation. In due course a more reluctant reply came from the American side, not indeed from the President himself but from his Secretary of State, John Foster Dulles.[35] Russell was delighted by so distinguished a response. His vanity, another weakness, was tickled and his judgment, never his strongest point, upset. The Khrushchev letter, which broadly sympathized with his position, not only drove him into a posture of extreme anti-Americanism but also stimulated him into making the abolition of nuclear weapons the centre of his life. Tolstoyan yearnings began to make their appearance.

The following year, 1958, Russell was made president of the new Campaign for Nuclear Disarmament, a moderate body designed by Canon John Collins of St Paul's, the novelist J.B.Priestley and others, to enlist the widest possible spread of opinion in Britain against the manufacture of nuclear weapons. It organized peaceful demonstrations, made a point of keeping strictly within the law, and in its early phase was impressive and highly effective. But, on Russell's part, signs of extremism were not slow to appear. Rupert Crawshay-Williams, whose intimate account of Russell is the best for these years, recorded in his diary, 24 July 1958, an illuminating outburst by Russell against John Strachey. Strachey was a former communist, later a right-wing Labour MP and War Minister in the post-war Attlee government. But in 1958 he had long been out of office and held no responsibilities, though he was known to believe in nuclear deterrence. When Russell heard Crawshay-Williams and his wife had been staying with Strachey, he asked about the latter's views on the H-bomb, and when given them assumed they were shared by the Williamses:

'You and John Strachey – you belong to the murderers' club,' he said, banging the arm of his chair. The murderers' club, he explained, consisted of people who did not really care what happened to the mass of the populace, since they as rulers felt that somehow they would survive in, and because of, privilege. 'They make sure of their own safety,' said Bertie, 'by building private bomb-proof shelters.'

When asked if he really thought Strachey had a private bomb-proof shelter, Russell roared: 'Yes, of course he has.' A fortnight later, they had a further discussion about the H-bomb, which 'began calmly'. Then suddenly, 'out of the blue, Bertie says in a voice of fury: "The next time you see your friend John Strachey, tell him I cannot understand why he wants Nasser [the then Egyptian dictator] to have the H-bomb" ... He was convinced that people like John are really endangering the world, and he felt justified in saying so.'[36]

This mounting anger, accompanied by a lack of concern for the objective facts, the attribution of the vilest motives to those holding different views, and signs of paranoia, found public expression in 1960, when Russell split CND by forming his own direct-action splinter, called the Committee of One Hundred, dedicated to civil disobedience. The original signatories of this group included leading intellectuals, artists and writers – Compton Mackenzie, John Braine, John Osborne, Arnold Wesker, Reg Butler, Augustus John, Herbert Read and Doris Lessing among others – many of whom were by no means extremists. But the group soon got out of control. History shows that all pacifist movements reach a point at which the more militant element becomes frustrated at the lack of progress and resorts to civil disobedience and acts of violence. This invariably marks the stage at which it ceases to retain a mass following. The Committee of One Hundred and the subsequent disintegration of CND was a classic example of this process. Russell's behaviour only accelerated what would probably have happened anyway. At the time it was attributed to the influence over him of his new secretary, Ralph Schoenman. I will examine his relationship with Schoenman shortly, but it is worth observing here that Russell's doings and sayings throughout the CND crisis were characteristic throughout. The meetings leading up to his resignation as president became increasingly unpleasant, with Russell attributing unworthy motives to Collins, accusing him of lying, and insisting that the private proceedings be tape-recorded.[37]

Indeed, once Russell was released from the restraining hand of Collins and his friends, extremism took over his mind completely and his statements became so absurd as to repel all but his most fanatical adherents. They contradicted what he knew to be, in calmer mood, the basic rules of persuasion. 'No opinion,' he wrote in 1958 in an essay on Voltaire, 'should be held with fervour. No one holds with fervour that seven times eight is fifty-six, because it can be known that this is the case. Fervour is only necessary in commending an opinion which is doubtful or demonstrably false.'[38] Many of Russell's sayings, from 1960 on, were not merely fervent but outrageous and were often made on the spur of the moment, when he had worked himself up into a state of righteous indignation with those who did not share his views. Thus, for a speech in Birmingham in April 1961, he had prepared notes which read: 'On a purely statistical basis, Macmillan and Kennedy are about fifty times as wicked as Hitler.' This was bad enough since (apart from anything else) it was comparing historical fact with futurist projection. But a recording shows that what Russell actually went on to say in his speech was: 'We used to think Hitler was wicked when he wanted to kill all the Jews. But Kennedy and Macmillan not only want to kill all the Jews but all the rest of us too. They're much more wicked than Hitler.' He added: 'I will not pretend to obey a government which is organizing the massacre of the whole of mankind . . . They are the wickedest people that ever lived in the history of man.'[39]

Granted Russell's premises, there was logic in his accusations. But even the logic was selectively applied. Sometimes Russell remembered that all powers possessing nuclear weapons were equally guilty of planning mass murder, and included the Russians in his polemics. Thus in a public letter of 1961, addressed 'from Brixton prison', he asserted: 'Kennedy and Khrushchev, Adenauer and De Gaulle, Macmillan and Gaitskell are pursuing a common aim: the ending of human life . . . To please these men, all the private affections, all the public hopes . . . [are] to be wiped out for ever.'[40] As a rule, however, he concentrated his fire on the West, particularly on Britain and above all on the United States.

This meant forgetting how much he hated not just the Soviet regime but Russia and the Russians themselves. In the immediate post-war period he had repeatedly said the Soviets were as bad as, or worse than, the Nazis. Crawshay-Williams recorded some of his fulminations: 'All Russians are eastern barbarians.' 'All Russians are imperialists.' He once 'managed even to say that *all* Russians would "crawl on their bellies to betray their friends".'[41] But in the late 1950s and thereafter, anti-Russian feeling was elbowed out of his mind as it was increasingly occupied by passionate anti-Americanism. This was deep-rooted and had surfaced before. It was propelled by old-fashioned British pride and patriotism, of an upper-class kind, contempt for upstarts and counter-jumpers, as well as liberal-progressive hatred for the world's largest capitalist state. His radical parents belonged to a generation which still associated America with democratic progress and had paid a long visit there in 1867 because, as he recorded, 'young people who hoped to reform the world went to America to discover how to do so.' He added: 'They could not foresee that the men and women whose democratic ardour they applauded and whose triumphant opposition to slavery they admired were the grandfathers and grandmothers of those who murdered Sacco and Vanzetti.'[42] He himself visited America many times and lived there for years, chiefly to make money: 'I am terribly hard up and am looking to America to reestablish my finances,' he wrote in 1913, a recurrent refrain. He was always critical of Americans – they were, he noted on his very first visit (1896), 'unspeakably lazy about everything except business'[43] – but his views on the impact of America on the world oscillated wildly. During the First World War, as we have seen, he regarded Wilson's America as a world saviour. Disappointed there, he switched to a strongly anti-American tack in the 1920s. He argued that socialism, which he currently favoured, would be impossible in Europe 'until America is either converted to socialism or at any rate willing to remain neutral'.[44] He accused America of 'the slow destruction of the civilization of China', predicted US democracy would collapse unless it embraced collectivism, called for 'a world-wide revolt' against American 'capitalist imperialism' and asserted that unless 'American belief in capitalism can be shaken' there would be 'a complete collapse of civilization'.[45]

Twenty years later, during and after the Second World War, he supported American military policy. But this was accompanied by a growing dislike for American politics. At the end of 1950, returning from a visit, he wrote to Crawshay-Williams: 'America was beastly – the Republicans are as wicked as they are stupid, which is saying a great deal. I told everybody I was finding it interesting to study the atmosphere of a police state ... I think the Third World War will begin next May.'[46] He bet Malcolm Muggeridge that Joseph McCarthy would be elected president (and had to pay up when the Senator died). When Russell began to campaign against the H-bomb, his anti-Americanism became totally irrational and remained so until his death. He developed a childish conspiracy-theory about Kennedy's assassination. Then, tiring of the H-bomb issue – like Tolstoy, Russell's attention-span was brief – he switched to Vietnam and organized a worldwide campaign of vilification of America's conduct there.

Russell, primed by his secretary Schoenman, fell an easy victim to the most extravagant inventions. Half a century before he had deplored the Allied use of atrocity stories about German behaviour in Belgium to whip up war-fever; he had been at pains, in his book *Justice in Wartime* (1916), to expose many of them as baseless. In the 1960s Russell used his prestige to circulate and give credence to tales from Vietnam which were even less plausible, and entirely for the purpose of whipping up hatred against America. This policy culminated in the 'War Crimes Tribunal' (1966–67) he organized and which eventually met to pronounce judgment against America in Stockholm. For this propaganda exercise he recruited such readily available intellectuals as Isaac Deutscher, Jean-Paul Sartre, Simone de Beauvoir, the Yugoslav author Vladimir Dedijer (who chaired it), a former president of Mexico and the Poet Laureate of the Philippines. But there was not even a pretence of justice or impartiality, since Russell himself said he was summoning it to try 'the war criminals Johnson, Rusk, McNamara, Lodge and their fellow criminals'.[47]

As a philosopher, Russell constantly insisted that words must be used with care and in their precise sense. As an adviser to humanity, he confessed, in his autobiography, to 'the practice of describing things which one finds unendurable in such a repulsive manner as to cause others to share one's fury'.[48] This is a curious admission from a man professionally devoted to the dispassionate analysis of problems, who tied his flag to the mast of reason. Moreover his attempts to infuriate only worked with those whose fury was not worth having or already available anyway. When Russell said (in 1951) that in America 'nobody ventures to make a political remark without first looking behind the door to make sure no one [is] listening,' no sensible person believed him.[49] When he announced, during the 1962 Cuban missile crisis, 'It seems likely that within a week you will all be dead to please American madmen,' he damaged himself, not President Kennedy.[50] When he said American soldiers in Vietnam were 'as bad as Nazis', his audience dwindled.[51]

It must be said, indeed, that Russell throughout his life was more impressive in a sustained argument than as a apothegmatist. His collected *obiter dicta* would read no better than Tolstoy's. 'A gentleman is a man whose grandfather had more than £1000 a year.' 'You will never get democratic government to work in Africa.' 'Children should be sent to boarding schools to get them away from mother love.' American mothers 'are guilty of instinctive incompetence. The fount of affection seems to have dried up.' 'The scientific attitude to life can scarcely be learned from women.'[52]

The last remark is a reminder that though Russell was almost exclusively associated in the last decades of his life with political pronouncements, he had at one time been even more notorious for his views on such inter-war topics as 'companionate marriage', free love, divorce reform and coeducation. In theory at any rate, he upheld the doctrine of women's rights as then expounded by its advocates. He demanded equality for women inside and outside marriage and portrayed them as victims of an antiquated system of morality which had no true ethical basis. Sexual freedom should be enjoyed and he castigated the 'doctrines of taboo and human sacrifice which pass traditionally as "virtue"'.[53] There were many echoes of Shelley in his views on women, social life, children and human relationships. He had, indeed, a particular devotion to Shelley, whose verse he claimed best expressed his attitude to life. He settled in the part of Wales where Shelley, in 1812–13, had tried to form a community, and his house, Plas Penrhyn, was the work of the same architect who built the house of Shelley's friend Maddox, across the Portmadoc estuary.

However, like Shelley, his actual behaviour towards women did not always accord with his theoretical principles. His first wife Alys, a gentle, loving, generous-minded American Quaker, was the victim of her husband's growing libertinism, as was Shelley's Harriet. Russell, as we have noted, was strictly brought up and remained strait-laced about sex until well into his twenties. Indeed in 1900, when his brother Frank, the second

Earl, left his first wife, got a Reno divorce and married again, Russell refused to recognize the new wife and suggested Frank should leave her behind when he came to dine. (Frank was later charged at the bar of the House of Lords with bigamy.) But as Russell grew older he became like Victor Hugo before him, more lecherous and less inclined to follow the rules of society, except when he found them convenient.

Alys was effectively dropped, after sixteen years, on 19 March 1911, when Russell visited the lively Bloomsbury hostess, Lady Ottoline Morrell, at her house at 44 Bedford Square. He found her husband Philip unexpectedly away, and made love to her. In his account, Russell said he did not have 'full relations' with Lady Ottoline that night but determined to 'leave Alys' and get Lady Ottoline to 'leave Philip'. What Morrell might feel or think 'was a matter of indifference to me'. He was assured the husband 'would murder us both' but he was 'willing to pay that price for one night'. Russell immediately broached the news to Alys, who 'flew into a rage, and said she would insist on a divorce, bringing in Ottoline's name'. After some argument, Russell said 'firmly' that if she did what she threatened, 'I should commit suicide in order to circumvent her.' Thereupon 'her rage became unbearable. After she had stormed for some hours, I gave a lesson in Locke's philosophy to her niece'.[54]

Russell's self-serving account does not accord with Alys's actual behaviour. She treated him throughout with great restraint, moderation and indeed affection, agreeing to go and live with her brother so he could carry on his affair with Lady Ottoline (her husband conniving, provided certain rules of public propriety were kept) and delaying the divorce till May 1920. She continued to love him. When Trinity College deprived him of his fellowship, she wrote: 'I have been saving up £100 to invest in Exchequer Bonds but I would rather give it to thee, if I may, as I am afraid all this persecution interferes very seriously with thy income.'[55] While he was in prison, she said: 'I thought of thee every day with greatest regret and dreamt of thee almost every night.'[56] Russell did not see her again until 1950.

The separation from Alys involved a good deal of lying, deception and hypocrisy. At one point Russell shaved off his moustache to conceal his identity during his clandestine meetings with Lady Ottoline. Russell's friends were shocked when they found out what was going on: he had always made such a point of truth and openness. The episode introduced a period of sexual confusion in his life. His relations with Lady Ottoline did not prove satisfactory. According to his account, 'I was suffering from pyorrhoea, although I did not know it, and this caused my breath to be offensive, which also I did not know. She could not bring herself to mention it.'[57] So their relations cooled. In 1913 he met 'the wife of a psychoanalyst' in the Alps and 'wished to make love to her, but thought I ought first to explain about Ottoline'. The woman was not so keen when she heard about the existing mistress, but 'decided, however, that for one day her objections could be ignored'. Russell 'never saw her again'.

Then in 1914 followed a discreditable episode with a young girl in Chicago. Helen Dudley was one of four sisters, the daughters of a leading gynaecologist, with whom Russell stayed while lecturing. According to Russell's account, 'I spent two nights under her parents' roof, and the second I spent with her. Her three sisters mounted guard to give warning if either of the parents approached.' Russell arranged that she should come to England that summer and live with him openly, pending a divorce. He wrote to Lady Ottoline telling her what had occurred. But she, in the meantime, hearing that he had been cured of his bad breath, told him she wished to resume their affair. In any case, by the time Helen Dudley arrived in London, in August 1914, war had been declared, Russell had decided to oppose it and 'I did not want to complicate my position with a private scandal, which would have made anything that I might say of no account.' So he told Helen their little plan was off, and though 'I had relations with her from time to time', the war 'killed my passion for her and I broke her heart'. He concludes: 'she fell a victim to a rare disease, which first paralysed her, and then made her insane.' So much for Helen.

Meanwhile Russell had indeed complicated his position by acquiring yet another mistress in the shape of Lady Constance Malleson, a society woman who acted under the name of Colette O'Neil. They met in 1916. The first time they confessed their love they 'did not go to bed' since 'there was too much to say'. Both were pacifists, and during their first copulation 'we heard suddenly a shout of bestial triumph in the street. I leapt out of bed and saw a Zeppelin falling in flames. The thought of brave men dying in agony was what caused the triumph in the street. Colette's love was in that moment a refuge to me, not from cruelty itself, which was inescapable, but from the agonizing pain of realizing that that is what men are.'[58]

Russell, as it happened, soon got over his agonizing pain and within a few years was being cruel to Lady Constance. She was content to share Russell with Lady Ottoline, and the two women visited him in alternate weeks during his sojourn in jail. According to Lady Constance's understanding, Lady Ottoline preferred to remain with her present husband, so she could have Russell when his divorce materialized. On that basis she provided the 'evidence' which allowed him to get a decree *nisi* in May 1920. However, Russell had now fallen for yet another and much younger woman, an emancipated feminist called Dora Black, and got her pregnant. She had no desire to marry since she disapproved of the institution, but Russell, not wishing to 'complicate his position' yet further, insisted on it, and after the degree absolute came through they went through a ceremony 'with six weeks to spare' before the child was born. So Lady Constance was discarded and Dora was forced into what she called 'the shame and disgrace of marrying'.[59]

Russell, now a man of fifty, was fascinated by Dora's 'elfin charm' and delighted to go 'bathing by moonlight or running with bare feet on the dewy grass'. She, for her part, was intrigued when he related that a militarist had scrawled on his house 'That F----- Peace Crank Lives Here' and that 'every word' was correct.[60] Russell was not, physically, to everyone's taste. He had by now developed a penetrating cackling laugh, which T. S. Eliot (a pupil of his at Cambridge) described as like 'the yaffle of a woodpecker'; George Santayana thought it was more like a hyena. He wore dark, old-fashioned three-piece suits, which he seldom changed (he rarely possessed more than one at a time), spats and high, stiff collars, rather like his contemporary Coolidge. At the time of his second marriage, Beatrice Webb recorded in her diary that he was a 'rather frowsty, unhealthy and cynical personage, prematurely old'. But Dora liked his 'thick and rather beautiful grey hair ... lifting in the wind, the large, sharp nose and odd, tiny chin, the long upper lip'. She noted that his 'broad but small feet turned outwards' and that he looked 'exactly like the Mad Hatter'.[61] She wanted – fatal desire – to 'protect him from his own unworldliness'.

They had two children, John and Kate, and in 1927 set up a progressive school, Beacon Hill near Petersfield. He told the *New York Times* that, ideally, 'cooperative groups of about ten families' should 'pool' their children and 'take it in turns to look after them'; every day there would be 'two hours of lessons' with a 'proper balance', and the rest of the time would be spent 'running wild'.[62] Beacon Hill was an attempt to give substance to this theory. But the school proved expensive and forced Russell to write potboilers to pay its bills. Moreover, like Tolstoy, he soon tired of its routine and left Dora, who for all her ultra-progressive views had a much stronger sense of responsibility, to run it.

They also quarrelled over sex. Mrs Webb had predicted that Russell's marriage to 'a girl of light character and materialist philosophy whom he does not and cannot reverence' was sure to fail. Russell, again like Tolstoy, insisted on a policy of 'openness', with which she agreed: 'Bertie and I ... left each other free as regards sexual adventures.' He did not object when she became secretary of the English branch of the World League for Sexual Reform, or attended (October 1926) the International

Sex Congress in Berlin, along with the pioneer of sex-change operations, Dr Magnus Hirschfeld, and the flamboyant gynaecologist Norman Haire. But when, quite openly, she had an affair with a journalist, Griffin Barry, and – following Russell's suggestion that eighteenth-century Whig ladies often had children by different fathers – had two babies by her lover, Russell grew uncomfortable. Many years later, in his autobiography, he admitted: 'In my second marriage I had tried to preserve this respect for my wife's liberty which I had thought that my creed enjoined. I found however that my capacity for forgiveness and what may be called Christian love was not equal to the demands I was making on it.' He added: 'Anyone else could have told me this in advance, but I was blinded by theory.'[63]

What Russell omitted to state was that there had been certain activities on his own part and, contrary to his policy of openness, they had been furtive. Indeed it is a significant fact that, in every case where intellectuals try to apply total disclosure to sex, it always leads in the end to a degree of guilty secrecy unusual even in normally adulterous families. Dora later related how she had been summoned home to their holiday house in Cornwall by a distraught cook, who refused to let the governess near their two acknowledged children as she had been 'sleeping with the Master'.[64] (The poor cook was sacked.) Dora also found out, many years afterwards, that in her absence Russell had also had his old love, Lady Constance, to stay for amorous purposes. At length, returning home with her new baby, she got an unpleasant surprise: 'Bertie administered the shock of telling me he had now transferred his affections to Peter Spence.' Margery ('Peter') Spence was an Oxford student who had come to look after John and Kate during the holidays. The Russells tried a foursome holiday in south-west France, each partner with his or her lover (1932). But the previous year Russell had become an earl on the death of his childless brother, and this made a difference. He became more lordly in his ways, Peter was anxious for a regular union, and so he took her to live with him in the family home. 'At first', said a shocked Dora, 'I could not believe that Bertie would do such a thing to me.' She added that it was 'inevitable' that 'such a man' should 'hurt many people on his way'; but his 'tragic flaw' was that he felt 'so little regret': 'Though he loved the multitudes and suffered with their suffering, he still remained aloof from them because the aristocrat in him lacked the common touch.'[65]

Dora also discovered, the hard way, that when it came to discarding one wife and taking on board another, Russell was by no means 'unworldly'. Like other men of his class and wealth, he promptly hired a high-powered team of lawyers and gave them carte blanche to get what he wanted. The divorce was extraordinarily complicated and bitter and took three years, partly because at an early stage the couple had signed a Deed of Separation, admitting adultery on both sides and agreeing that neither would invoke matrimonial offences committed before 31 December 1932 in any later litigation. But in fact this merely made the proceedings more difficult and messy, and Russell's lawyers more aggressive. Each parent was anxious for custody of the two children both acknowledged, and in the end Russell fought successfully to have them made Wards of Chancery, like Shelley's poor offspring. To do this his lawyers produced an affidavit by a chauffeur, dismissed from the school by Dora and now employed by Russell, to the effect that she was frequently drunk, had broken whisky bottles in her room, and had slept there with a male parent and a visitor.[66] Russell did not escape unscathed either. The President of the Divorce Court, in finally granting the decree in 1935, remarked that her adultery was 'preceded by at least two cases of infidelity on the part of her husband, and that he had been guilty of numerous acts of adultery in circumstances which are usually held to aggravate the offence ... infidelity of the respondent with persons in the household or engaged in the business in which they were mutually occupied'.[67] Reading the various accounts of this long, embittered wrangle, it is impossible not to feel sympathy for Dora, who throughout had been faithful to her principles, as opposed to Russell, who dropped his the moment they became personally inconvenient and then invoked the full force of the law. She had never wanted to be married in the first place, and it 'was March 1935 before I was finally free of my legal marriage. I was in my late thirties. The divorce had taken three years of my life and inflicted tragedies from which I was never fully to recover.'[68]

Russell's marriage to his third wife, Peter Spence, lasted the best part of fifteen years. He observed laconically: 'When in 1949 my wife decided she wanted no more of me, our marriage came to an end.'[69] Behind this misleading statement lay a long tale of petty adulteries on his part. Russell was never a positive Lothario, searching the highways for female prey. But he had no scruples about seducing any woman who fell in his way. Indeed he became quite an expert in the dodges the practised adulterer had to master in the pre-permissive age. Thus on one occasion we find him writing to Lady Ottoline: '... the safest plan will be for you to come to the station and wait in the First Class Waiting Room on the departure platform, and then go with me in a taxi to some hotel and walk in with me. That involves less risk than in any other plan, and does not look odd to the hotel authorities.'[70] Thirty years later he was giving unsolicited advice in such matters to Sidney Hook: 'Hook, if you ever take a girl to a hotel and the reception clerk seems suspicious, when he gives you the price of the room have her complain loudly, "It's *much* too expensive!" He's sure to assume she is your wife.'[71] Russell usually, however, preferred his women on the premises: it made things easier. In 1915 he offered his hard-up former pupil T.S. Eliot and his wife Vivien shelter in his London flat in Bury Street. The poet described Russell as Mr Apollinax, 'an irresponsible foetus', and said he 'heard the beat of centaur's hoofs over the hard turf' as his 'dry and passionate talk devoured the afternoon'. But Eliot was also a trusting soul who often left his wife alone with the centaur and his passionate talk. Russell gave his other mistresses conflicting versions of what happened. To Lady Ottoline he said that his flirtations with Vivien were platonic; to Lady Constance he confessed he had made love to her but found the experience 'hellish and loathsome'.[72] Very likely the truth was quite different to either of these accounts, and it is possible that Russell's behaviour contributed to Vivien Eliot's mental instability.

Russell's victims were often humbler creatures: chambermaids, governesses, any young and pretty female whisking around the house. Professor Hook, in his portrait of Russell, claims this was the essential reason his third marriage broke up. Hook said he learned 'on good authority' that Russell, 'despite his advanced age, was pursuing anything in skirts that crossed his path, and that he was carrying on flagrantly even with the servant girls, not behind [Peter's] back but before her eyes and those of his house guests'. She left him and returned, but Russell refused to take a pledge of marital fidelity and eventually she decided she was no longer prepared to be humiliated.[73] A divorce followed in 1952, when Russell was eighty. He then married a teacher from Bryn Mawr, Edith Finch, whom he had known for many years, and who looked after him for the rest of his life. When accused of being anti-American, he would reply smartly: 'Half my wives were American.'[74]

In theory Russell kept up with the twentieth-century movement to emancipate women; in practice he remained rooted in the nineteenth century, a Victorian – he was, after all, almost thirty when the old Queen died – and tended to see women as appendages to men. '. . . in spite of his championship of women's suffrage,' Dora wrote, 'Bertie did not 'really believe in the equality of women with men . . . he believed the male intellect to be superior to that of the female. He once told me that he usually found it necessary to talk down to women.'[75] He seems to have felt, in his heart, that the prime function of wives was to produce children for their husbands. He had two sons and a daughter, and at times tried to devote himself to them. But, like his hero Shelley, he combined fierce but sporadic possessiveness with a more general indifference. He became, Dora complained, 'remote from understanding of their problems and entirely absorbed in his role in world politics'; he himself was obliged to confess he had 'failed as a parent'.[76] As with so many famous intellectuals, people – and that included children and wives – tended to become servants of his ideas, and therefore in practice of his ego. Russell was in some ways a decent, kind-hearted and civilized man, capable of unselfish gestures of great generosity. He lacked the adamantine self-absorption of a Marx, a Tolstoy or an Ibsen. But the exploitative streak was there, especially in his relations with women.

Nor were women the only people he exploited, as the interesting case of Ralph Schoenman suggests. Schoenman was an American, a philosophy graduate of Princeton and the London School of Economics, who joined CND in 1958 and, two years later, aged twenty-four, wrote to Russell about his plans to organize a civil disobedience wing of the movement. The old man was intrigued, encouraged Schoenman to come and see him, and found him delightful. Schoenman's extreme ideas exactly coincided with his own. Their relationship strongly resembled that between old Tolstoy and Chertkov. Schoenman became Russell's secretary and organizer, in effect the prime minister at what, by 1960, had become the court of a prophet-king. Indeed there were two courts. One was in London, the centre of Russell's public activities. The other was his house at Plas Penrhyn on the Portmeirion peninsula in North Wales. Portmeirion, a fantasy Italian village, had been built by the rich left-wing architect Clough Williams-Ellis, who owned most of the surrounding land. His wife Amabel, sister of John Strachey, had been a prominent apologist for Stalin, and author of a propaganda book about the building of the White Sea Canal (by slave labour as we now know), one of the most repellent documents to emerge from the dark years of the 1930s. Many well-to-do progressives, such as Russell's Boswell Crawshay-Williams, Arthur Koestler, Humphrey Slater, the military scientist P.M.S. (later Lord) Blackett and the economic historian M.M.Postan, settled in this beautiful neighbourhood, to enjoy life and plan the socialist millennium. Russell was their monarch, and to his court came, in addition to the local middle-class intelligentsia, a host of pilgrims from all over the world, seeking wisdom and approval, as their predecessors had once sought it from Tolstoy at Yasnaya Polyana.

Russell enjoyed his much-publicized forays to London, to make speeches, demonstrate, get himself arrested and generally harass the Establishment. But he preferred life in Wales, and it was therefore highly convenient for him to have Schoenman, an unpaid but devoted, indeed fanatical, lieutenant, run things for him in London. So Schoenman played vizier to Russell's sultan and his reign lasted six years. He was with

Russell when he was arrested in September 1961, and also went to jail; when he was released, in November, the Home Office proposed to deport him, as an undesirable alien. Large numbers of prominent progressives signed a petition that he be allowed to stay, and the government relented. Later they bitterly regretted their intercession when Schoenman appeared to establish complete mastery over Russell's mind, as Chertkov had over Tolstoy's. It was sometimes difficult for old friends to speak to Russell on the phone: Schoenman answered the calls and merely undertook to relay messages. He was also accused of being the real author of the many letters Russell wrote to *The Times* or the statements sent in his name to press agencies, commenting on world events. Schoenman himself encouraged these beliefs. He claimed that 'every major political initiative that has borne the name of Bertrand Russell since 1960 has been my work in thought and deed'; it was, he said, at least 'a partial truth' that the old man had been 'taken over by a sinister young revolutionary'.[77]

Schoenman certainly had a great deal to do with the Committee of One Hundred, the Vietnam War Crimes Tribunal and the setting up of the Bertrand Russell Peace Foundation. During the 1960s Russell's London base became a kind of mini-Foreign Office, of a comic-subversive kind, dispatching endless letters and cables to prime ministers and heads of state – to Mao Tse-tung and Chou En-lai in China, Khrushchev in Russia, Nasser in Egypt, Sukarno in Indonesia, Haile Selassie in Ethiopia, Makarios in Cyprus, and numerous others. As these missives became longer, more frequent and wilder, fewer and fewer bothered to reply. There were also the public comments on home events, as they occurred: 'The Profumo Affair is grave not because of the fact the Cabinet consists of voyeurs, homosexuals or streetwalkers. It is grave because those in power have totally destroyed the integrity of the judiciary, faked evidence, intimidated witnesses, colluded with the police in the destruction of evidence and have even allowed the police to murder a man.' In time the newspapers ceased to print such nonsense.

Old friends who had lost touch with Russell assumed Schoenman was the author of all these communiqués. No doubt he did write many of them. But there was nothing new in that. Russell was quite capable of letting someone else write an article under his name if he was not much interested in the subject. In 1941 when Sidney Hook complained about a piece in *Glamour* entitled 'What To Do if You Fall in Love with a Married Man, by Bertrand Russell', Russell admitted he had got $50 for it: his wife had written the article, he had merely signed his name.[78] There is no evidence that Schoenman's efforts seriously misrepresented Russell's views, which were just as violent as those of his amanuensis.

The archives show that Schoenman altered and strengthened, in his own hand, certain phrases in Russell's texts; but this may well have been at Russell's own dictation (the statement on the Cuban missile crisis is a likely case of this). When Russell's emotions took control he was always liable to depart from a moderate prepared text. If many of the statements put out under his name seem childish today, it must be remembered that the 1960s was a childish decade and Russell one of its representative spirits. He was often guilty, especially in his last years, of nursery tantrums. Thus he arranged a special ceremony in public to witness the tearing up of his Labour Party card, and when at a reception Harold Wilson, then Prime Minister, came up to him with outstretched hand, saying 'Lord Russell', the old earl kept his hands ostentatiously in his pockets. What is quite clear, as his biographer Ronald Clark rightly insists, is that, contrary to what some then thought, Russell never became senile.[79] He allowed Schoenman a loose rein but in the last resort he remained firmly in control.

Indeed when he decided Schoenman no longer served his purposes, he acted quite ruthlessly. He did not object to Schoenman's extremism but he disliked any stealing of his limelight. Schoenman made various trips abroad as 'Earl Russell's Personal Representative', and these led to trouble. In China he infuriated Chou En-lai by exhorting the populace to disobey its government, and Chou complained to Russell. There was some much-publicized Schoenman misbehaviour in July 1965 at the World Congress for Peace in Helsinki. Russell received an indignant cable from the organizers: 'Speech of your personal representative caused uproar. Strongly rejected by audience. Tremendous provocation of Peace Congress. Foundation discredited. Essential you dissociate yourself from Schoenman and his speech. Friendly greetings.'[80] There then followed, in 1966–67, long public and behind-the-scenes wrangles over the Vietnam War Crimes Tribunal. In 1969 Russell, now ninety-seven, decided he had already derived any benefit likely to accrue from Schoenman's services and abruptly dispensed with them. On 9 July he further removed Schoenman from his will as executor and trustee and broke off relations completely in the middle of the same month. Two months later he removed him from the board of the Bertrand Russell Peace Foundation. In November he dictated to his fourth wife, Edith, a 7000-word statement about his whole relationship with Schoenman; it was typed by Edith and Russell initialled each sheet, with a signed accompanying letter done on a different typewriter. The tone was whiggish, condescending and dismissive; ending: 'Ralph must be well established in megalomania. The truth is, I suppose, that I have never taken Ralph as seriously as he likes to think I did. I was fond of him in the earlier years. But I never looked upon him as a man of parts and weight and much individual importance.'[81] It had some of the characteristics of Russell disposing of a wife who no longer appealed to him.

One of the reasons Russell retained Schoenman so long was that he was good at hauling in funds in a way Russell himself would have found distasteful. Russell was always keen on money: getting it, spending it and, to be fair, giving it away too. During the First World War, not wishing to hold £3000 of shares in an engineering firm, now making war weapons, which he had inherited, he handed them over to the poverty-stricken T. S. Eliot; 'years later,' he recalled, 'when the war was finished and [Eliot] was no longer poor, he gave them back to me.'[82] Russell often gave lavish presents, particularly to women. He was also capable of meanness and avarice. Hook claims his chief sins were vanity and greed: he says he would often, in the United States, write rubbishy articles, or introductions to books he thought little of, for quite small sums of money. Defending himself, Russell blamed first the school, which cost him £2000 a year, then his wives. He claimed his third wife was extravagant, and after they were divorced he asserted that, of the £11,000 he received from winning the Nobel Prize in 1950, £10,000 went to her. He had, he argued, to make a lot of cash and watch his money because he was paying two sets of alimony at once. But he also enjoyed the idea of a large income – hence those keen perusals of his little notebook. Crawshay-Williams noted in his diary: 'He enjoys being encouraged by us to dwell on the sheer quantity of money he is now making.' He particularly relished being awarded in 1960 the Danish Sonning Prize, worth £5000, tax-free. 'And no supertax,' he exulted, 'pure gain!' He told Crawshay-Williams he would spend only two days in Denmark: 'we're just going over to pick up the money and come straight back again.'[83]

Schoenman proved an excellent finance minister. He put slips into Russell's letters which read: 'If you believe that Bertrand Russell's work for peace is valuable, perhaps you would care to help to support it financially ... This note is inserted quite unknown to Lord Russell by his secretary.'[84] For those who wrote asking for Russell's autograph he charged £3 (later reduced to £2). Journalists were asked to pay £150 for the privilege of an interview. Russell certainly knew about these exactions, since he received a number of protests about Schoenman's American-style fund-raising. But he allowed it to continue, and he seems to have given his blessing to two of Schoenman's biggest schemes. Against the advice of Russell's old-fashioned publisher, Sir Stanley Unwin, Schoenman held an auction for the US rights of Russell's autobiography – almost an unknown trading device in those days – and pushed the bidding up to the then enormous sum of $200,000. He also took advantage of the fact that Russell, like Brecht, had accumulated a vast personal archive. Russell, like his contemporary Churchill, was among the first to perceive the financial value of letters from the famous, and kept all those he received (plus copies of his outgoing letters). By the 1960s it consisted of 250,000 documents and was termed 'the most important single archive of its kind in Britain'. Schoenman, a master of publicity, had the archive transported to London in two armoured cars, and after a good deal of razzmatazz, disposed of it to McMaster University, Hamilton, Ontario, for $250,000.[85] Schoenman's master-stroke was to set up the Peace Foundation, for which he obtained tax-free charitable status on the analogy of the Atlantic Peace Foundation. 'Rather against my will,' Russell noted complacently, 'my colleagues urged that the foundation should bear my name.'[86] In his last years, then, he was able to dispense substantial sums to all his favourite causes, sensible and foolish, enjoy a large income, and pay as little tax as was legally possible. When Schoenman had created this ingenious set-up he was, none too ceremoniously, shown the door. As for the charge that Russell, like his friend Williams-Ellis, was both a rich man and a socialist – why didn't either of them give their money away? – Russell had a stock answer: 'I'm afraid you've got it wrong. Clough Williams-Ellis and I are socialists. We don't pretend to be *Christians.*'

The ability to get the best of both worlds, the world of progressive self-righteousness and the world of privilege, is a theme which runs through the lives of many leading intellectuals, and none more so than Bertrand Russell's. If he did not often actively solicit, he never refused the good things his descent, fame, connections and title brought him. Thus, when the Bow Street magistrate sentenced him to six months in the second division (hard labour) in 1918, this was varied, on appeal, to the first division, the chairman announcing: 'It would be a great loss to the country if Mr Russell, a man of great distinction, were confined in such a manner that his abilities did not have full scope.'[87] Russell's own account, in his autobiography, suggests the leniency was due to a fellow philosopher, then Foreign Secretary: 'By the intervention of Arthur Balfour I was placed in the first division, so that while in prison I was able to read and write as much as I liked, provided I did no pacifist propaganda. I found prison in many ways quite agreeable.'[88] While in Brixton he wrote his *Introduction to Mathematical Philosophy* and began his *Analysis of Mind*. He was also able to get and read the latest books, including Lytton Strachey's subversive best-seller, *Eminent Victorians*, which caused him to laugh 'so loud that the officer came to my cell, saying I must remember that prison is a place of punishment'. Other, less well-connected fellow pacifists, such as E. D. Morel, had their health broken in the second division.

Russell delighted in small privileges, too, as when Schoenman arranged for him to receive an extra quota of thrillers from the public library: Russell devoured a vast number of detective stories, like many other Cambridge intellectuals of his generation (his old colleague J.E. McTaggart needed thirty volumes a week). He raised no protest – who would? – when, even during the worst post-war shortages, a famous Scotch distillery sent him a case of whisky every month marked 'The Earl Russell'.[89] Russell, not always deliberately, made it difficult to forget his social origins. He described his first wife as 'not what my grandmother would call a lady'. He called his twenty-first birthday the day when 'I came of age'. He often enjoyed being rude to people he termed middle-class, such as architects. If seriously annoyed, he would send for the police, as when, in imitation of his own activities, an actress and her agent staged a 'sit-down' in his London drawing room. He very much wanted the Order of Merit, thinking it scandalous that inferior men like Eddington and Whitehead had got it before him, and was suitably gratified when George VI finally bestowed it. The belief on the left that he never used his title was a myth. Unlike his third wife, who seems to have taken pleasure in it, Russell used it pragmatically, whenever he thought it would secure him an advantage. He was always an earl when necessary. When not, he was hail-fellow-well-met – up to a point. No one was allowed to take liberties.

As for logic, that too was only invoked when required. During the Soviet invasion of Czechoslovakia, Russell was persuaded to sign a letter of protest, along with a number of other writers. I had the job of negotiating its appearance in *The Times*. With the signatures in the customary alphabetic order, the heading on the letter would have been 'From Mr Kingsley Amis and others'. I decided, and the *Times* Letters Editor agreed, that it might have more effect in the Communist world if it read 'From Earl Russell OM and others'. So this was done. But Russell noticed this small deception and was angry. He telephoned to protest and eventually reached me at the printers, where I was putting the *New Statesman* to press. He said I had deliberately done it to give the false impression he himself had organized the letter. I denied this, and said the sole object was to give the letter maximum impact. 'After all,' I said, 'if you agreed to sign the letter at all, you cannot complain when your name is put first – it isn't logical.' 'Logical fiddlesticks!' said Russell sharply, and slammed down the receiver.

9

Jean-Paul Sartre:
'A Little Ball of Fur and Ink'

EAN-PAUL SARTRE, like Bertrand Russell, was a professional philosopher who also sought to preach to a mass audience. But there was an important difference in their approach. Russell saw philosophy as a hieratic science in which the populace could not participate. The most a worldly philosopher like himself could do was to distill small quantities of wisdom and distribute it, in a greatly diluted form, through newspaper articles, popular books and broadcasts. Sartre, by contrast, working in a country where philosophy is taught in the high schools and bandied about in the cafés, believed that by plays and novels he could bring about mass participation in his system. For a time at least it looked as though he had succeeded. Certainly no philosopher this century has had so direct an impact on the minds and attitudes of so many human beings, especially young people, all over the world. Existentialism was the popular philosophy of the late 1940s and 1950s. His plays were hits. His books sold in enormous quantities, some of them over two million copies in France alone.[1] He offered a way of life. He presided over a secular church, if a nebulous one. Yet in the end, what did it all amount to?

Like most leading intellectuals, Sartre was a supreme egoist. Nor is this surprising, given the circumstances of his childhood. He was the classic case of a spoiled only child. His family was of the provincial upper middle class, his father a naval officer, his mother a well-to-do Schweitzer from Alsace. The father was, by all accounts, an insignificant fellow, much bullied by *his* father; a clever man, though, a *Polytechnicien*, who grew ferocious moustaches to compensate for his small height (5 feet

2 inches). At all events he died when Sartre was only fifteen months old and became 'only a photo in my mother's bedroom'. The mother, Anne-Marie, married again to an industrialist, Joseph Mancy, boss of the Delaunay-Belleville plant in La Rochelle. Sartre, born 21 June 1905, inherited his father's height (5 feet 2½ inches), brains and books, but in his autobiography, *Les Mots*, went out of his way to dismiss him from his life. 'If he had lived,' he wrote, 'my father would have laid down on me and crushed me. Fortunately he died young.' 'No one in my family,' he added, 'has been able to arouse my curiosity about him.' As for the books, 'Like all his contemporaries he read rubbish ... I sold [them]: the dead man meant so little to me.'[2]

The grandfather, who crushed his own sons, doted on Jean-Paul and gave him the run of his large library. The mother was a doormat, the little boy her most precious possession. She kept him in frocks and long hair even longer than the little Hemingway, until he was nearly eight, when the grandfather decreed a massacre of the curls. Sartre called his childhood 'paradise'; his mother was 'This virgin, who lived with us, watched and dominated by everyone, was there to wait on me ... My mother was mine and no one challenged my quiet possession. I knew nothing of violence or hatred and was spared the harsh apprenticeship of jealousy.' There was no question of 'rebelling' since 'no one else's whim ever claimed to be my law.' He put salt into the jam once, aged four; otherwise, no crimes, no punishments. His mother called him Poulou. He was told he was beautiful 'and I believed it'. He said 'precocious things' and they were 'remembered and repeated to me'. So 'I learned to make up others.' He knew, he said, 'effortlessly, how to say things in advance of my age'.[3] There are times, indeed, when Sartre's account recalls Rousseau: 'Good was born in the depths of my heart and truth in the youthful darkness of my understanding.' 'I had no rights because I was overwhelmed with love; I had no duties because I did everything through love.' His grandfather 'believed in progress – and so did I: progress, that long and arduous road which led to myself'. He described himself as 'a cultural possession ... I was impregnated with culture and I returned it to the family like a radiance.' He recalls an exchange when he asked permission to read Flaubert's *Madame Bovary* (then still considered shocking). Mother: 'But if my little darling reads books like that at his age, what will he do when he grows up?' Sartre: 'I shall live them!' This witty riposte was repeated with delight in the family circle and beyond.[4]

As Sartre had little respect for the truth it is difficult to say how much credence should be placed on his description of his childhood and youth. His mother, when she read *Les Mots*, was upset: '*Poulou n'a rien compris à son enfance'* ('Poulou understood nothing about his childhood') was her comment.[5] What shocked her were his heartless comments about members of the family. There is no doubt that he was spoiled. But when he was four a catastrophe occurred: following a bout of influenza, a stye developed in his right eye, and he was never able to use it again. His eyes were always to cause him trouble. He invariably wore thick glasses, and in his sixties he went progressively blind. When Sartre finally got to school he found his mother had lied to him about his looks and that he was ugly. Though short, he was well-built: broad, barrel-chested, powerful. But his face was excessively plain and the faulty eye almost made him grotesque. Being ugly, he was beaten up. He retaliated with wit, scorn, jokes and became that bitter-sweet character, the school jester. Later he was to pursue women, as he put it, 'to get rid of the burden of my ugliness'.[6]

Sartre had one of the best educations available to a man of his generation: a good lycée in La Rochelle, two years as a boarder in the Lycée Henri Quatre in Paris, at the time probably the best high school in France; then the École Normale Superieure, where France's leading academics took their degrees. He had some very able contemporaries: Paul Nizan, Raymond Aron, Simone de Beauvoir. He boxed and wrestled. He played the piano, by no means badly, sang well in a powerful voice and contributed satirical sketches to the École's theatre reviews. He wrote poems, novels, plays, songs, short stories and philosophical essays. He was again the jester, but with a much wider range of tricks. He formed, and for many years maintained, the habit of reading about three hundred books a year.[7] The range was very wide; American novels were his passion. He also acquired his first mistress, Simone Jollivet; like his father, he preferred taller women if available, and Simone was a lanky blonde, a good head taller. Sartre failed his first degree exam, then passed it brilliantly the next year, coming top; de Beauvoir, three years his junior, was second. It was now June 1929, and like most clever young men at that time, Sartre became a schoolmaster.

The 1930s were rather a lost decade for Sartre. The literary fame which he expected and passionately desired did not come to him. He spent most of it as a teacher in Le Havre, the epitome of provincial dowdiness. There were trips to Berlin where, at Aron's suggestion, he studied Husserl, Heidegger and Phenomenology, then the most original philosophy in Central Europe. But mostly it was teaching drudgery. He hated the bourgeoisie. Indeed he was very class-conscious. But he was not a Marxist. In fact he never read Marx, except perhaps in extracts. He was certainly a rebel, but a rebel without a cause. He joined no party. He took no interest in the rise of Hitler. Spain left him unmoved. Whatever he later claimed, the record suggests he held no strong political views before the war. A photograph shows him decked out for an academic speech day in a black gown with ruffles and a yellow cloak with rows of ermine, both garments much too big. Normally he wore a sports jacket with an open-necked shirt, refusing to put on a tie; it was only in late middle age that he adopted an intellectual's uniform – white polo-neck pullover, weird half-leather jacket. He drank a lot. On his second speech day he was the central actor in a grotesque scene, adumbrating Kingsley Amis's *Lucky Jim*, when, drunk and incoherent, he was unable to make his contribution and had to be marched off the stage.[8] He identified then and throughout his life with youth, especially student youth. He let his pupils do more or less what they wanted. His message was: the individual is entirely self-responsible; he has a right to criticize everything and everybody. The boys could take off their jackets and smoke in class. They need not take notes or present essays. He never marked the roster or inflicted punishment or gave them marks. He wrote a lot but his early fiction could not find a publisher. He had the chagrin of seeing his friends, Nizan and Aron, getting published, acquiring a measure of fame. In 1936 he at last brought out a book, on his German studies, *Récherches philosophiques*. It attracted little attention. But he was beginning to see what he wanted to do.

The essence of Sartre's work was the projection of philosophical activism through fiction and drama. This had become firm in his mind by the late 1930s. He argued that all the existing novelists – he was thinking of Dos Passos, Virginia Woolf, Faulkner, Joyce, Aldous Huxley, Gide and Thomas Mann – were reflecting ancient ideas mostly derived, directly or indirectly, from Descartes and Hume. It would be much more interesting, he wrote to Jean Paulhan, 'to make a novel of Heidegger's time, which is what I want to do'. His problem was that in the 1930s he was working quite separately on fiction and on philosophy: he began to excite people only when he brought them firmly together and forced them on the public's attention through the stage. But a philosophical novel of a kind was slowly emerging. He wanted to call it *Mélancholie*. His publishers changed it to *La Nausée*, a much more arresting title, and finally brought it out in 1938. Again, there was little response at first.

What made Sartre was the war. For France it was a disaster. For friends like Nizan it was death. For others it brought danger and disgrace. But Sartre had a good war. He was conscripted into the meteorological section at Army Group Artillery headquarters, where he tossed balloons of hot air into the atmosphere to test which way the wind was blowing. His comrades laughed at him. His corporal, a maths professor, remarked:

'From the start we knew he would be no use to us in a military sense.' It was the nadir of French military morale. Sartre was notorious for never taking a bath and being disgustingly dirty. What he did was write. Every day he produced five pages of a novel, eventually to become *Les Chemins de la Liberté*, four pages of his *War Diary*, and innumerable letters, all to women. When the Germans invaded the front collapsed and Sartre was taken prisoner, still scribbling (21 June 1940). In the PoW camp near Trèves he was in effect politicized by the German guards who despised their French prisoners, especially when they were dirty, and kicked Sartre repeatedly on his broad bottom. As at school, he survived by jesting and writing camp entertainments. He also continued to work hard at his own novels and plays, until his release, classified 'partially blind', in March 1941.

Sartre made a beeline for Paris. He got a job teaching philosophy at the famous Lycée Condorcet, where most of the staff were in exile, underground or in the camps. Despite his methods, perhaps because of them, the school inspectors reported his teaching 'excellent'. He found wartime Paris exhilarating. He later wrote: 'Will people understand me if I say that the horror was intolerable but it suited us well... We have never been as free as we were under the German occupation.'[9] But that depended on who you were. Sartre was lucky. Having taken no part in pre-war politics, not even the 1936 Popular Front, he did not figure on any Nazi records or lists. So far as they were concerned he was 'clean'. Indeed among the cognoscenti he was looked on with favour. Paris was crowded with Francophile German intellectuals, in uniform, such as Gerhardt Heller, Karl Epting, Karl-Heinz Bremer. They influenced not only the censorship but such newspapers and magazines as were allowed, and not least their theatre and book reviews.[10] To them, Sartre's novels and plays, with their philosophical background from Central Europe and especially their stress on Heidegger, who was approved of by Nazi academic intellectuals, were highly acceptable. Sartre never actively collaborated with the regime. The nearest he came to it was to write for a collaborationist weekly, *Comoedia*, agreeing at one stage to contribute a regular column. But he had no difficulty in getting his work published and his plays presented. As André Malraux put it, 'I was facing the Gestapo while Sartre in Paris had his plays produced with the authorization of the German censors.'[11]

In a vague way Sartre yearned to contribute to the Resistance. Fortunately for him his efforts came to nothing. There is a curious irony here, the kind of irony one gets accustomed to, writing about intellectuals. Sartre's personal philosophy, what was soon to be called existentialism, was already shaping in his mind. In essence it was a philosophy of action, arguing that man's character and significance are determined by his actions, not his views, by his deeds, not words. The Nazi occupation aroused all Sartre's anti-authoritarian instincts. He wanted to fight it. If he had followed his philosophical maxims, ne would have done so by blowing up troop trains or shooting members of the SS. But that is not in fact what he did. He talked. He wrote. He was Resistance-minded in theory, mind and spirit, but not in fact. He helped to form a clandestine group, Socialism and Freedom, which held meetings and debated. He seems to have believed that, if only all the intellectuals could get together and blow trumpets, the walls of the Nazi Jericho would tumble. But Gide and Malraux, whom he approached, turned him down. Some members of the group, such as his philosopher-collea-gue Maurice Merleau-Ponty, were beginning to put their faith in Mar-xism. Sartre, in so far as he was anything, followed Proudhon: it was in this spirit that he wrote his first political manifesto of one hundred pages, dealing with post-war France.[12] So there were plenty of words but no deeds. One member, Jean Pouillon, put it thus: 'We were not an organized Resistance group, just a bunch of friends who had decided to be anti-Nazis together and to communicate our convictions to others.' Others, non-members, were more critical. George Chazelas, who opted for the Communist Party, said: 'They struck me from the very beginning as fairly childish: they were never aware, for instance, of the extent that their prattle jeopardized the work of others.' Raoul Lévy, another active Resistance man, called their work 'mere chitchat around a cup of tea' and Sartre himself 'a political illiterate'.[13] In the end the group died of inanition.

Sartre, then, did nothing of consequence for the Resistance. He did not lift a finger, or write a word, to save the Jews. He concentrated relentlessly on promoting his own career. He wrote furiously, plays, philosophy and novels, mainly in cafés. His association with St-Germain-des-Prés, soon to become world-famous, was in origin quite fortuitous. His major philosophy text, *L'Être et le Néant* (*Being and Nothingness*), which sets out the principles of Sartrean activism most comprehensively, was composed mainly in the winter of 1942–43, which was very cold. Mon-sieur Boubal, proprietor of the Café Flore on the Boulevard St Germain, was unusually resourceful at obtaining coal for heating and tobacco for smoking. So Sartre wrote there, every day, sitting in an ugly, ill-fitting but warm artificial fur coat, coloured bright orange, which he had some-how obtained. He would drink down a glass of milky tea, set out his inkpot and pen, then scribble relentlessly for four hours, scarcely lifting his eyes from the paper, 'a little ball of fur and ink'.[14] Simone de Beauvoir, who described him thus, noted that he was enlivening the tract, which was eventually 722 pages, with 'spicy passages'. One 'concerns holes in general and the other focuses on the anus and love-making Italian style'.[15] It was published in June 1943. Its success was slow in coming (some of the most important reviews were not published till 1945) but sure and cumulative.[16] It was through the theatre, however, that Sartre established himself as a major figure. His play *Les Mouches* opened the same month *L'Être* came out and at first sold comparatively few tickets. But it attracted attention and consolidated Sartre's rising reputation. He was soon in demand for screenplays for Pathé, writing three of them (including the brilliant *Les Jeux sont faits*) and making, for the first time, a good deal of money. He was involved in the creation of a new and influential review, *Les Lettres françaises* (1943) and the following spring was coopted onto the jury of the Prix de la Pléiade, along with André Malraux and Paul Élouard, a sure sign that he had arrived as a literary power-broker. It was at this point, on 27 May 1944, that his play *Huis clos* (*No Exit*) opened at the Vieux Colombier. This brilliant work, in which three people meet in a drawing room which turns out to be an ante-chamber to hell, operated at two levels. At one level it was a comment on character, with the message 'Hell is other people.' At another it was a popular presention of *L'Être et le Néant*, a radicalized version of Heidegger, given a flashy Gallic gloss and a contemporary relevance and presenting a message of activism and concealed defiance. It was the kind of thing at which the French have always been outstandingly gifted – taking a German idea and making it fashionable with superb timing. The play was a huge success both with the critics and the public, and has been well described as 'the cultural event which inaugurated the golden age of St-Germain-des-Prés'.[17]

Huis clos made Sartre famous, and it is another instance of the unrivalled power of the theatre to project ideas. But, oddly enough, it was through the old-fashioned forum of the public lecture that Sartre became world-famous, indeed notorious, a *monstre sacré*. Within a year of the play's opening France was at peace. Everyone, especially youth, was catching up greedily on the lost cultural years and searching for the post-war elixir of truth. The Communists and the new-born Catholic Social Democrats (MRP) were fighting a fierce battle for paramountcy on the campus. Sartre used his new philosophy to offer an alternative: not a church or a party but a challenging doctrine of individualism in which each human being is seen as absolute master of his soul if he chooses to follow the path of action and courage. It was a message of liberty after the totalitarian nightmare. Sartre had already established his gifts and drawing power as a lecturer by a successful series on 'The Social Techniques of the Novel' which he had given in the rue St Jacques in Autumn 1944. Then he had merely hinted at some of his notions. A year later, with France free and agog for intellectual stimulation, he announced a public lecture in the Salle des Centraux, 8 rue Jean Goujon, for 29 October 1945. The word 'existentialism' was not his. It seems to have been invented by the press. The previous August, when asked to define the term, Sartre had replied: 'Existentialism? I don't know what it is. My philosophy is a philosophy of existence.' Now he decided to embrace what the media had coined, and entitled his lecture: 'Existentialism is a Humanism'.

Nothing is so powerful, Victor Hugo had laid down, as an idea whose time has come. Sartre's time had come in two distinct ways. He was preaching freedom to people who were hungry and waiting for it. But it was not an easy freedom. 'Existentialism,' said Sartre, 'defines man by his actions... It tells him that hope lies only in action, and that the only thing that allows man to live is action.' So, 'Man commits himself to his life, and thereby draws his image, beyond which there is nothing.' The new European of 1945, Sartre said, was the new, existentialist individual – 'alone, without excuses. This is what I mean when I say we are condemned to be free.'[18] So Sartre's new, existentialist freedom was immensely attractive to a disillusioned generation: lonely, austere, noble, slightly aggressive, not to say violent, and anti-elitist, popular – no one was excluded. Anyone, but especially the young, could be an existentialist.

Secondly, Sartre was presiding over one of those great, periodic revolutions in intellectual fashion. Between the wars, sickened by the doctrinaire excesses of the long battle over Dreyfus and the Flanders carnage, the French intelligentsia had cultured the virtues of detachment. The tone had been set by Julien Benda, whose immensely successful book *La Trahison des clercs* (1927) had exhorted intellectuals to avoid 'commitment' to creed and party and cause, to concentrate on abstract principles and keep out of the political arena. One of the many who had heeded Benda had been precisely Sartre himself. Up to 1941 nobody could have been less committed. But now, just as he had tested the atmosphere with his hot-air balloons, he sniffed a different breeze. He and his friends had put together a new review, *Les Temps modernes*, with Sartre as editor-in-chief. The first issue, containing his editorial manifesto, had appeared in September. It was an imperious demand that writers become committed again:

> The writer has a place in his age. Each word has an echo. So does each silence. I hold Flaubert and [Edmond] Goncourt responsible for the repression that followed the Commune because they did not write

a single line to prevent it. You may say: it was none of their business. But then, was the Calas trial Voltaire's business? Was the condemnation of Dreyfus Zola's business?[19]

This was the background to the lecture. There was an extraordinary cultural tension in Paris that autumn. Three days before Sartre spoke, there had been a scene at the opening of two new ballets, *Les Forains* and *Le Rendez-vous*, at the Théâtre des Champs-Elysées when Picasso's drop-curtain had been hissed by the packed society audience. Sartre's lecture had not been widely advertised: a few insertions in the small-ads of *Libération*, *Le Figaro*, *Le Monde* and *Combat*. But the word-of-mouth build-up was evidently tremendous. When Sartre arrived near the hall at 8.30 the mob in the street outside was so big he feared it was an organized CP demonstration. It was in fact people frantically trying to get in, and as the hall was already packed, only celebrities were allowed to pass through. His friends had to force an entrance for Sartre himself. Inside, women fainted, chairs were smashed. The proceedings began an hour late. What Sartre had to say was in all essentials a technical academic philosophy lecture. But in the circumstances it became the first great post-war media event. By a remarkable coincidence, Julien Benda also gave a public lecture that evening, to a virtually empty hall.

Sartre's press coverage was astounding.[20] Many newspapers produced thousands of words of Sartre's text, despite the paper shortage. Both what he had to say, and the way he said it, were passionately denounced. The Catholic daily *La Croix* called existentialism 'a graver danger than eighteenth-century rationalism or nineteenth-century positivism', and joined hands with the communist *L'Humanité* in calling Sartre an enemy of society. In due course Sartre's entire works were placed upon the Vatican Index of Prohibited Books, and Stalin's cultural commissar, Alexander Fadayev, called him 'a jackal with a typewriter, a hyena with a fountain-pen'. Sartre likewise became the object of fierce professional jealousy. The Frankfurt School, which hated Brecht, hated Sartre still more. Max Horkheimer called him 'a crook and a racketeer of the philosophic world'. All these attacks merely accelerated Sartre's juggernaut. He was by now, like so many leading intellectuals before him, an expert in the art of self-promotion. What he would not do himself his followers did for him. *Samedi Soir* commented sourly: 'We have not seen such a promotional triumph since the days of Barnum.'[21] But the more the Sartre phenomenon was moralized over, the more it flourished. The November issue of *Les Temps modernes* pointed out that France was a beaten and demoralized country. All it had left was its literature and the fashion industry, and existentialism was designed to give the French a bit of dignity and to preserve their individuality in an age of degrada-tion. To follow Sartre became, in a weird way, a patriotic act. A hastily expanded book version of his lecture sold half a million copies in a month.

Moreover, existentialism was not just a philosophy to be read, it was a craze to be enjoyed. An *Existentialist Catechism* insisted: 'Existentialism, like faith, cannot be explained: it can only be lived,' and told readers where to live it.[22] For St-Germain-des-Près to become the centre of intel-lectual fashion was not new. Sartre was in fact treading in the footsteps of Voltaire, Diderot and Rousseau, who had patronized the old Café Procope, further down the boulevard. It had again been lively under the Second Empire, in the age of Gautier, George Sand, Balzac and Zola; that was when the Café Flore had first opened, with Huysmans and Apollinaire among its patrons.[23] But in pre-war Paris the intellectual focus had been Montparnasse, whose tone had been politically uncom-mitted, slightly homosexual, cosmopolitan, its cafés adorned by slim, bisexual girls. The shift to St Germain, which was social and sexual as well as intellectual, was therefore dramatic, for Sartre's St Germain was leftish, committed, strongly heterosexual, ultra-French.

Sartre was a convivial soul, loving whisky, jazz, girls and cabaret. When not seen at the Flore or at the Deux Magots, a block away, or eating at the Brasserie Lipp across the road, he was in the new cellar nightclubs or *caves* which now abruptly opened in the bowels of the Quartier Latin. At the Rose Rouge there was the singer Juliette Greco, for whom Sartre wrote a delightful song; the writer and composer Boris Vian played the trombone there and contributed to *Les Temps modernes*. There was the Tabou in the rue Dauphine and Bar Verte on the rue Jacob. Not far away, at 42 rue Bonaparte, lived Sartre himself, in a flat which overlooked the church of St Germain itself and the Deux Magots. (His mother lived there too and continued to look after his laundry.) The movement even had its daily house organ, the newspaper *Combat*, edited by Albert Camus, whose best-selling novels were widely hailed as existentialist. Simone de Beauvoir later recalled: '*Combat* reported favourably everything that came from our pens or our mouths.' Sartre worked all day, scribbling hard: he wrote millions of words at this time, lectures, plays, novels, essays, introductions, articles, broadcasts, scripts, reports, philosophical diatribes.[24] He was described, by Jacques Audiberti, as 'a truck parking everywhere with great commotion, in the library, in the theatre, in the movies'. But at night he played, and by the end of the evening he was usually drunk and often aggressive. Once he gave Camus a black eye.[25] People came to goggle at him. He was king of the *quartier*, of the *enragés* (the angry ones), of those who were *branché* (in the know), of the *rats des caves* (the cellar rats); in the words of his chief publicist, Jean Paulhan, he was 'the spiritual leader of thousands of young people'.

But if Sartre was king, who was queen? And if he was the young people's spiritual leader, where was he leading them? These are two separate, though linked, questions, which need to be examined in turn. By the winter of 1945–46, when he became a European celebrity, he had been associated with Simone de Beauvoir for nearly two decades. De Beauvoir was a Montparnasse girl actually born in an apartment over the famous Café de la Rotonde. She had a difficult childhood, coming from a family ruined by a disgraceful bankruptcy in which her grandfather was jailed; her mother's dowry was never paid and her father was a worthless boulevardier who could not get a proper job.[26] She wrote bitterly of her parents: 'My father was as convinced of the guilt of Dreyfus as my mother was of the existence of God.'[27] She took refuge in schoolwork, becoming a bluestocking, though a remarkably elegant one. At Paris University she was an outstanding philosophy student and was taken up by Sartre and his circle: 'From now on,' he told her, 'I'm going to take you under my wing.' That remained in a sense true, though for her their relationship was a mixed blessing. She was an inch taller than Sartre, three years younger and, in a strictly academic sense, abler. One of her contemporaries, Maurice de Gandillac, described her work as 'rigorous, demanding, precise, very technical'; despite her youth she almost beat Sartre for first place in the philosophy degree, and the examiners, Georges Davy and Jean Wahl, thought her the better philosopher.[28] She, like Sartre, was also a compulsive writer and in many respects a finer one. She could not write plays but her autobiographical works, though equally unreliable as to facts, are more interesting than his, and her major novel, *Les Mandarins*, which describes the French post-war literary world and won her the Prix Goncourt, is far better than any of Sartre's. In addition, she had none of Sartre's personal weaknesses, except lying.

Yet this brilliant and strong-minded woman became Sartre's slave from almost their first meeting and remained such for all her adult life until he died. She served him as mistress, surrogate wife, cook and manager, female bodyguard and nurse, without at any time acquiring legal or financial status in his life. In all essentials, Sartre treated her no better than Rousseau did his Thérèse; worse, because he was notoriously unfaithful. In the annals of literature, there are few worse cases of a man exploiting a woman. This was all the more extraordinary because de Beauvoir was a lifelong feminist. In 1949 she produced the first modern manifesto of feminism, *La Deuxième sexe*, which sold widely all over the world.[29] Its opening words, *'On ne nait pas femme, on la devient'* ('One is not born a woman, one becomes one') are a conscious echo of the opening of Rousseau's *Social Contract*. De Beauvoir, in fact, was the progenitor of the feminist movement and ought, by rights, to be its patron saint. But in her own life she betrayed everything it stood for.

Quite how Sartre established and maintained such a dominance over de Beauvoir is not clear. She could not make herself write honestly about their relationship. He never troubled to write anything at all about it. When they first met he was much better read than she was and able to distill his reading into conversational monologues she found irresistible. His control over her was plainly of an intellectual kind. It cannot have been sexual. She was his mistress for much of the 1930s but at some stage ceased to be so; from the 1940s their sexual relations seem to have been largely non-existent: she was there for him when no one better was available.

Sartre was the archetype of what in the 1960s became known as a male chauvinist. His aim was to recreate for himself in adult life the 'paradise' of his early childhood in which he was the centre of a perfumed bower of adoring womanhood. He thought about women in terms of victory and occupation. 'Every single one of my theories,' he says in *La Nausée*, 'was an act of conquest and possession. I thought that one day, with the help of them all, I'd conquer the world.' He wanted total freedom for himself, he wrote, and 'I dreamed above all of asserting this freedom against women.'[30] Unlike many practised seducers, Sartre did not dislike women. Indeed he preferred them to men, perhaps because they were less inclined to argue with him. He noted: 'I prefer to talk to a woman about the tiniest things, than about philosophy to Aron.'[31] He loved writing letters to women, sometimes a dozen a day. But he saw women not so much as persons but as scalps to add to his centaur's belt, and his attempts to defend and rationalize his policy of conquest in progressive terms merely add a layer of hypocrisy. Thus he said he wanted 'to conquer a woman almost like you'd conquer a wild animal' but 'this was only in order to shift her from her wild state to one of equality with man.' Or again, looking back on his early seductions, he reflected on 'the depth of imperialism there was in all that'.[32] But there is no evidence that such thoughts ever deflected him from a potential capture; they were for the record.

When Sartre first seduced de Beauvoir he outlined to her his sexual philosophy. He was frank about his desire to sleep with many women. He said his credo was 'Travel, polygamy, transparency.' At university, a friend had noted that her name was like the English word 'beaver', which in French is *castor*. To Sartre, she was always *Castor* or *vous*, never *tu*.[33] There are times when one feels he saw her as a superior trained animal. Of his policy of 'asserting' his 'freedom against women', he wrote: 'The Castor accepted this freedom and kept it.'[34] He told her there were two kinds of sexuality: 'necessary love' and 'contingent love'. The latter was not important. Those on whom it was bestowed were 'peripherals', holding his regard on no more than 'a two-year lease'. The love he had for her was of the permanent, 'necessary' kind; she was a 'central', not a 'peripheral'. Of course she was entirely free to pursue the same policy. She could have her peripherals so long as Sartre remained her central, necessary love. But both must display 'transparency'. This was just another word for the favourite intellectual game of sexual 'openness', which we came across in the cases of Tolstoy and Russell. Each, said Sartre, was to tell the other what he or she was up to.

The policy of transparency, as might have been expected, merely led in the end to additional and more squalid layers of concealment. De Beauvoir tried to practise it but the indifference with which Sartre greeted news of her affairs, most of which seem to have been tentative or half-hearted, clearly gave her pain. He merely laughed at her description of being seduced by Arthur Koestler, which figures in *Les Mandarins*. Moreover, those dragged into the transparency policy did not always like it. Her own great peripheral, in some ways the love of her life, was the American novelist Nelson Algren. When he was seventy-two and their affair just a memory, he gave an interview in which he revealed his fury at her disclosures. Putting him in *Les Mandarins* was bad enough, he said, but at least he was there disguised under another name. But in her second volume of autobiography, *The Prime of Life*, she had not only named him but quoted from his letters, to which he had felt reluctantly obliged to consent: 'Hell, love letters should be private,' he raged. 'I've been in whorehouses all over the world and the women there always close the door, whether it's in Korea or India. But this woman flung the door open and called in the public and the press.'[35] Algren apparently grew so indignant at the thought of de Beauvoir's behaviour that he had a massive heart attack after the interviewer had left, and died that night.

Sartre also practised transparency, but only up to a point. In conversation and letters he kept her informed about his new girls. Thus: 'this is the first time I've slept with a brunette... full of smells, oddly hairy, with some black fur in the small of her back and a white body... A tongue like a kazoo, endlessly uncurling, reaching all the way down to my tonsils.'[36] No woman, however 'central', can have wished to read such things about one of her rivals. When Sartre was in Berlin in 1933, and de Beauvoir briefly joined him there, the first thing he told her was that he had acquired a new mistress, Marie Ville. With Sartre, as with Shelley, there was a childish longing for the old love to approve of the new. However, Sartre never told all. When de Beauvoir, who was teaching at Rouen for most of the 1930s, stayed with him in Berlin or anywhere else, he gave her a wedding ring to wear. But that was the nearest she got to marriage. They had their private language. They signed themselves into hotels as Monsieur et Madame Organatique or Mr and Mrs Morgan Hattick, the yankee millionaires. But there is no evidence he ever wanted to marry her or gave her the choice of a more formal union. Quite unknown to her, he did on several occasions propose marriage to a peripheral.

That the life they led went against the grain for her is clear. She was never able to bring herself to accept Sartre's mistresses with equanimity. She resented Marie Ville. She resented still more the next one, Olga Kosakiewicz. Olga was one of two sisters (the other, Wanda, also became a mistress of Sartre) and, to envenom matters, one of de Beauvoir's pupils. De Beauvoir disliked the affair with Olga so much that she put her into her novel, *L'Invitée*, and murdered her in it.[37] She admitted in her autobiography, 'I was vexed with Sartre for having created this situation and with Olga for having taken advantage of it.' She fought back: 'I had no intention of yielding to her the sovereign position that I had always occupied, in the very centre of the universe.'[38] But any woman who feels obliged to refer to her lover as 'the very centre of the universe' is not in a strong position to frustrate his divagations. What de Beauvoir did was to attempt to control them by a form of participation. The three, Sartre, de Beauvoir and the girl – usually a student, either his or hers – formed a triangle, with de Beauvoir in a supervisory position. The term 'adoption' was frequently used. By the early 1940s, Sartre seems to have become dangerously well-known for seducing his own female students. In a hostile criticism of *Huis clos*, Robert Francis wrote: 'We all know Monsieur Sartre. He is an odd philosophy teacher who has specialised in the study of his students' underwear.'[39] But as de Beauvoir taught many more suitable girls, it was her students who provided most of Sartre's victims; indeed de Beauvoir seems to have been close, at time, to the role of a procuress. She also, in her confused desire not to be excluded from love, formed her own close relationships with the girls. One such was Nathalie Sorokine, the daughter of Russian exiles, and de Beauvoir's best pupil at the Lycée Molière in Passy where she taught during the war. In 1943 Nathalie's parents laid formal charges against de Beauvoir for abducting a minor, a serious criminal offence which carried a jail sentence. Mutual friends intervened and the criminal charge was eventually dropped. But de Beauvoir was barred from the university and had her licence to teach anywhere in France revoked for the rest of her life.[40]

During the war de Beauvoir came closest to being Sartre's real wife: cooking, sewing, washing for him, looking after his money. But with the end of the war he suddenly found himself rich and surrounded by women, who were after his intellectual glamour as much as his money. The year 1946 was his best for sexual conquests and it marked the virtual end of his sexual relationship with de Beauvoir. 'At a relatively early stage,' as John Weightman has put it, 'she tacitly accepted the role of senior, sexually-retired, pseudo-wife on the fringe of his fluctuating seraglio.'[41] She grumbled about 'all that money he spent on them'.[42] She noted with concern that, as Sartre grew older, his girls became younger – seventeen- or eighteen-year-olds, whom he spoke of 'adopting' in a legal sense, meaning they would inherit his copyrights. She could give them advice and warnings, as Helene Weigel did to Brecht's girls, though she did not possess the German woman's legal status. She was constantly lied to. In 1946 and 1948, while Sartre was on trips to the Americas, she was given a detailed account of his torrid affair with a certain Dolores; but Sartre, while telling her he was tiring of the girl's 'exhausting passion' for him, was actually proposing marriage to her. Then there was Michelle, the honey-blonde wife of Boris Vian, Olga's pretty sister Wanda, Evelyne Rey, an exotic blonde actress for whom Sartre wrote a part in his last play, *Les Séquestrés d'Altona*, Arlette, who was only seventeen when Sartre picked her up – she was the one de Beauvoir hated most – and Hélène Lassithiotakis, a Greek youngster. At one time in the late 1950s he was running four mistresses at once, Michelle, Arlette, Evelyn and Wanda, as well as de Beauvoir, and deceiving all of them in one way or another. He dedicated his *Critique de la raison dialectique* (1960) publicly to de Beauvoir, but got his publisher Gallimard to print privately two copies with the words 'To Wanda'; when *Les Séquestrés* was produced Wanda and Evelyne were each told he had dedicated it to her.

One reason de Beauvoir disliked these young women was that she believed they encouraged Sartre to lead a life of excess – not just sexual excesses, but drink and drugs too. Between 1945 and 1955 Sartre got through a phenomenal amount of writing and other work, and to do this he steadily increased his intake of both alcohol and barbiturates. While in Moscow in 1954 he collapsed from over-drinking and had to be rushed into a Soviet clinic. But, once recovered, he continued to write thirty to forty pages a day, often taking an entire tube of Corydrane pills (a drug withdrawn as dangerous in 1971) to keep going. The book on dialectical reason, indeed, appears to have been written under the influence of both drink and drugs. His biographer Annie Cohen-Solal says that he often drank a quart of wine over two-hour lunches at Lipp, the Coupole, Balzar or other favourite haunts, and she calculates that his daily intake of stimulants at this time included two packets of cigarettes, several pipes of black tobacco, a quart of alcohol (chiefly wine, vodka, whisky and beer), 200 milligrams of amphetamines, fifteen grams of aspirin, several grams of barbiturates, plus coffee and tea.[43] In fact de Beauvoir did not do the young mistresses justice. They all tried to reform Sartre, and Arlette, the youngest, tried hardest, even extracting a written promise from him that he would never again touch Corydrane, tobacco or alcohol – a promise he promptly broke.[44]

Thus surrounded by adoring, though often fractious, women, Sartre had little time for men in his life. He had a succession of male secretaries, some like Jean Cau of great ability. He was always surrounded by a crowd of young male intellectuals. But all these were dependent on him for wages, charity or patronage. What he could never stomach for long were male intellectual equals, of his own age and seniority, who were liable at any moment to deflate his own often loose and windy arguments. Nizan was killed before a break could come, but he quarrelled with all the rest: Raymond Aron (1947), Arthur Koestler (1948), Merleau-Ponty (1951), Camus (1952), to mention only the more prominent.

The quarrel with Camus was as bitter as Rousseau's rows with Diderot, Voltaire and Hume, or Tolstoy's with Turgenev – and, unlike the last case, there was no reconciliation. Sartre seems to have been jealous of Camus' good looks, which made him immensely attractive to women, and his sheer power and originality as a novelist: *La Peste*, published in June 1947, had a mesmeric effect on the young and rapidly sold 350,000 copies. This was made the object of some ideological criticism in *Les Temps modernes* but the friendship continued in an uneasy fashion. As Sartre moved towards the left, however, Camus became more of an independent. In a sense he occupied the same position as George Orwell in Britain: he set himself against all authoritarian systems and came to see Stalin as an evil man on the same plane as Hitler. Like Orwell and unlike Sartre he consistently held that people were more important than ideas. De Beauvoir reports that in 1946 he confided in her: 'What we have in common, you and I, is that individuals count most of all for us. We prefer the concrete to the abstract, people to doctrines. We place friendship above politics.'[45]

In her heart of hearts de Beauvoir may have agreed with him, but when the final break came, over Camus' book *L'Homme révolté* in 1951–52, she of course sided with Sartre's camp. He and his acolytes at *Les Temps modernes* saw the book as an assault on Stalinism and decided to go for it in two stages. For the first, Sartre put up the young Francis Jeanson, then only twenty-nine, remarking at the editorial meeting which decided it, 'He will be the harshest but at least he'll be polite.' Then, when Camus replied, Sartre himself wrote a long and extraordinarily unpleasant attack addressed to Camus personally: 'A violent and ceremonial dictatorship has taken possession of you, supported by an abstract bureaucracy, and pretends to rule according to moral law'; he was suffering from 'wounded vanity' and indulging in a 'petty author's quarrel'; 'Your combination of dreary conceit and vulnerability always discouraged people from telling you unvarnished truths.'[46] By now Sartre had all the organized far left behind him and his attack did Camus damage; it may also have hurt – Camus *was* a vulnerable man – and at times he was depressed by his break with Sartre. At other times he just laughed and saw Sartre as a figure of fun, 'a man whose mother has to pay his income tax'.

Sartre's inability to maintain a friendship with any man of his own intellectual stature helps to explain the inconsistency, incoherence and at times sheer frivolity of his political views. The truth is he was not by nature a political animal. He really held no views of consequence before he was forty. Once he had parted with men like Koestler and Aron, both of whom had matured by the late 1940s into political heavyweights, he was capable of supporting anyone or anything. In 1946–47, very conscious of his immense prestige among the young, he dithered about which, if any, party to back. It seems to have been a belief of his that an intellectual had a kind of moral duty to back 'the workers'. The trouble with Sartre is that he did not know, and made no effort to meet, any workers, apart from his brilliant secretary Jean Cau who, being of proletarian origin and retaining a strong Aude accent, counted as one. Must one not, then, back the party most workers support? In France in the 1940s that meant the Communists. But Sartre was not a Marxist; indeed Marxism was almost the exact opposite of the strongly individualistic philosophy he preached. All the same, even in the late 1940s he could not bring himself to condemn the Communist Party or Stalinism – one reason why he quarrelled with Aron and Koestler. His former pupil Jean Kanapa, now a leading communist intellectual, wrote disgustedly: 'He is a dangerous animal who likes flirting with Marxism – because he has not read Marx, though he knows more or less what Marxism is.'[47]

Sartre's only positive move was to help organize an anti-Cold War movement of the non-communist left, called the Rassemblement Démocratique Révolutionaire, in February 1948. It aimed to recruit world intellectuals – he called it 'The International of the Mind' – and its theme was Continental unity. 'European youth, unite!' proclaimed Sartre in a speech in June 1948. 'Shape your own destiny!. . . By creating Europe, this new generation will create democracy.'[48] In fact if Sartre had really wanted to play the European card and make history, he might have given support to Jean Monnet, who was then laying the foundations of the movement which would create the European Community ten years later. But that would have meant a great deal of attention to economic and administrative detail, something Sartre found impossible. As it was, his fellow organizer of the RDR, David Rousset, found him quite useless: 'despite his lucidity, he lived in a world which was totally isolated from reality.' He was, said Rousset, 'very much involved in the play and movement of ideas' but took little interest in actual events: 'Sartre lived in a bubble.' When the party's first national Congress took place, in June 1949, Sartre was nowhere to be found: he was in Mexico with Dolores, trying to persuade her to marry him. The RDR simply dissolved, and Sartre transferred his fluctuating attention to Gary Davis's absurd World Citizens' Movement. François Mauriac, the great novelist and sardonic Catholic independent, gave Sartre some sensible public advice about this time, echoing the sneering words of Rousseau's dissatisfied girlfriend: 'Our philosopher must listen to reason – give up politics, Zanetto, *e studia la mathematica*!'[49]

Instead, Sartre took up the case of the homosexual thief, Jean Genet, a cunning fraud who appealed strongly to the credulous side of Sartre's nature – the side which wanted some substitute for religious faith. He wrote an enormous and absurd book about Genet, nearly 700 pages long, which was really a celebration of antinomianism, anarchy and sexual incoherence. This was the point, in the opinion of his more sensible friends, when Sartre ceased to be a serious, systematic thinker, and became an intellectual sensationalist.[50] It is curious that de Beauvoir, a more rational creature, who in some ways looked and dressed and thought like an old-fashioned schoolmarm, was able to do so little to save him from such follies. But she was anxious to retain his love and her position in his court – as John Weightman put it, Madame de Maintenon to his Louis XIV – and worried too about his drinking and pill-taking. To retain his confidence she felt she had to go along with him. Thus she served as his echo rather than his mentor, and that became the pattern of their relationship: she reinforced his misjudgments and endorsed his silliness. She was no more of a political animal than he was and in time she came to talk equal nonsense about world events.

In 1952 Sartre resolved his dilemma about the Communist Party and decided to back it. This was an emotional not a rational judgment, reached via involvement in two Communist Party agitprop campaigns: 'L'Affaire Henri Martin' (Martin was a naval rating who went to prison for refusing to participate in the Indo-China War), and the brutal suppression of riots organized by the Communist Party against the American NATO commander, General Matthew Ridgeway.[51] As many foresaw at the time, the Communist Party campaign to get Martin released actually led the authorities to keep him in jail longer than they had originally intended; the Communist Party did not mind this – his incarceration was serving their purpose – but Sartre should have had more sense. The level of his political perception is revealed by his accusing the Prime Minister Antoine Pinay, an old-fashioned parliamentary conservative, of setting up a dictatorship.[52] Sartre never showed any real knowledge of or interest in – let alone enthusiasm for – parliamentary democracy. Having the vote in a multi-party society was not at all what he meant by freedom. What did he mean then? That was more difficult to answer.

Sartre's aligning himself with the Communists in 1952 made no logical sense at all. That was just the time when other left-wing intellectuals were leaving the Communist Party in droves, as Stalin's dreadful crimes were documented and acknowledged throughout the West. So Sartre now found himself standing on his head. He observed an uneasy silence about Stalin's camps, and his defence of his silence was a total contradiction of his manifesto on commitment in *Les Temps modernes*. 'As we were not members of the Party or avowed sympathizers,' he argued feebly, 'it was not our duty to write about Soviet labour camps; we were free to remain aloof from quarrels over the nature of this system, provided no events of sociological significance occurred.'[53] He likewise forced himself to keep silent about the appalling trials in Prague of Slansky and other Czech Jewish communists. Worse, he allowed himself to be made a performing bear at the absurd conference which the Communist World Peace Movement held in Vienna in December 1952. This meant truckling to Fadayev, who had called him a hyena and a jackal, telling the delegates that the three most important events in his life were the Popular Front of 1936, the Liberation and 'this congress' – a blatant lie – and, not least, cancelling the performance in Vienna of his old, anti-communist play, *Les Mains sales*, at the behest of the Communist Party bosses.[54]

Some of the things Sartre did and said during the four years when he consistently backed the Communist Party line almost defy belief. He, like Bertrand Russell, reminds one of the disagreeable truth of Descartes' dictum: 'There is nothing so absurd or incredible that it has not been asserted by one philosopher or another.' In July 1954, after a visit to Russia, he gave a two-hour interview to a reporter from the fellow-travelling *Libération*. It ranks as the most grovelling account of the Soviet state by a major Western intellectual since the notorious expedition by George Bernard Shaw in the early 1930s.[55] He said that Soviet citizens did not travel, not because they were prevented but because they had no desire to leave their marvellous country. 'The Soviet citizens,' he insisted, 'criticize their government much more and more effectively than we do.' Indeed, he maintained, 'There is total freedom of criticism in the USSR.' Many years later he admitted his mendacity:

> After my first visit to the USSR in 1954, I lied. Actually, lie might be too strong a word: I wrote an article . . . where I said a number of friendly things about the USSR which I did not believe. I did it partly because I considered that it is not polite to denigrate your hosts as soon as you return home, and partly because I didn't really know where I stood in relation to both the USSR and my own ideas.[56]

This was a curious admission from 'the spiritual leader of thousands of young people'; moreover it was just as deceptive as his original false-hoods, since Sartre was consciously and deliberately aligning himself with Communist Party aims at that time. In fact it is more charitable to draw a veil over some of the things he said and did in 1952–56.

By the latter date Sartre's public reputation, both in France and in the wider world, was very low, and he could not avoid perceiving it. He fell upon the Soviet Hungarian invasion with relief as a reason, or at any rate an excuse, for breaking with Moscow and the Communist Party. Equally, he took up the burgeoning Algerian war – especially after de Gaulle's return to power supplied a convenient hate-figure from 1958 – as a reputable good cause to win back his prestige among the independent left and especially the young. To some extent this manoeuvre was genuine. To a limited degree it succeeded. Sartre had a 'good' Algerian War, as he had had a 'good' Second World War. Unlike Russell he did not actually succeed in getting himself arrested, though he tried hard. In September 1960 he persuaded some 121 intellectuals to sign a statement asserting 'the right to disobedience [of public servants, army, etc.] in the Algerian War'. A Fourth Republic government would almost certainly have jailed him but the fifth was a more sophisticated affair, dominated by two men of outstanding intellect and culture, de Gaulle himself and André Malraux. Malraux said: 'Better to let Sartre shout "Long live the [terrorists]" in the Place de la Concorde, than arrest him and embarrass ourselves.' De Gaulle told the Cabinet, citing the cases of François Villon, Voltaire and Romain Rolland, that it was better to leave intellectuals untouched: 'These people caused a lot of trouble in their day but it is essential that we continue to respect freedom of thought and expression in so far as this is compatible with the laws of the state and national unity.'[57]

Much of Sartre's time in the 1960s was spent travelling in China and the Third World, a term invented by the geographer Alfred Sauvy in 1952 but which Sartre popularized. He and de Beauvoir became familiar figures, photographed chatting with various Afro-Asian dictators – he in his First World suits and shirts, she in her schoolmarm cardigans enlivened by 'ethnic' skirts and scarves. What Sartre said about the regimes which invited him made not much more sense than his accolades for Stalin's Russia, but it was more acceptable. Of Castro: 'The country which has emerged out of the Cuban revolution is a direct democracy.' Of Tito's Yugoslavia: 'It is the realization of my philosophy.' Of Nasser's Egypt: 'Until now I have refused to speak of socialism in connection with the Egyptian regime. Now I know I have been wrong.' He was particularly warm in praise of Mao's China. He noisily condemned American 'war crimes' in Vietnam and compared America to the Nazis (but then he had compared de Gaulle to the Nazis, forgetting the General was fighting them when he himself was having his plays staged in occupied Paris). Both he and de Beauvoir were always anti-American: in 1947, following a visit, de Beauvoir had written an absurd piece in *Les Temps modernes*, full of hilarious misspellings ('Greeniwich Village', 'Max Tawin' [Mark Twain], 'James Algee') and dotty assertions, e.g. that only the rich are allowed inside the shops on Fifth Avenue; virtually every statement in it is false, and it became the butt of a brilliant polemic by Mary McCarthy.[58] Now in the 1960s Sartre played a leading part in Bertrand Russell's discredited 'War Crimes Tribunal' in Stockholm. None of these somewhat vacuous activities had much effect on the world and merely blunted the impact of anything serious which Sartre had to say.

Nevertheless, there was a more sinister side to the advice Sartre proffered to his admirers in the Third World. Though not a man of action himself – it was one of Camus's more hurtful gibes that Sartre 'tried to make history from his armchair' – he was always encouraging action in others, and action usually meant violence. He became a patron of Frantz Fanon, the African ideologue who might be called the founder of modern black African racism, and wrote a preface to his Bible of violence, *Les Damnés de la terre* (1961), which is even more bloodthirsty than the text itself. For a black man, Sartre wrote, 'to shoot down a European is to kill two birds with one stone, to destroy an oppressor and the man he oppresses at the same time.' This was an updating of existentialism: self-liberation through murder. It was Sartre who invented the verbal technique (culled from German philosophy) of identifying the existing order as 'violent' (e.g. 'institutionalized violence'), thus justifying killing to overthrow it. He asserted: 'For me the essential problem is to reject the theory according to which the left ought not to answer violence with violence.'[59] Note: not 'a' problem but 'the essential' problem. Since Sartre's writings were very widely disseminated, especially among the young, he thus became the academic godfather to many terrorist movements which began to oppress society from the late 1960s onwards. What he did not foresee, and what a wiser man would have foreseen, was that most of the violence to which he gave philosophical encouragement would be inflicted by blacks not on whites but on other blacks. By helping Fanon to inflame Africa, he contributed to the civil wars and mass murders which have engulfed most of that continent from the mid-1960s onwards to this day. His influence on South-East Asia, where the Vietnam War was drawing to a close, was even more baneful. The hideous crimes committed in Cambodia from April 1975 onwards, which involved the deaths of between a fifth and a third of the population, were organized by a group of Francophone middle-class intellectuals known as the Angka Leu ('the Higher Organization'). Of its eight leaders, five were teachers, one a university professor, one a civil servant and one an economist. All had studied in France in the 1950s, where they had not only belonged to the Communist Party but had absorbed Sartre's doctrines of philosophical activism and 'necessary violence'. These mass murderers were his ideological children.

Sartre's own actions, in the last fifteen years of his life, did not add up to much. Rather like Russell, he strove desperately to keep in the vanguard. In 1968 he took the side of the students, as he had done from his first days as a teacher. Very few people emerged with any credit from the events of May 1968 – Raymond Aron was an outstanding exception in France[60] – so Sartre's undignified performance does not perhaps deserve particular censure. In an interview on Radio Luxembourg he saluted the student barricades: 'Violence is the only thing remaining to the students who have not yet entered into their fathers' system . . . For the moment the only anti-Establishment force in our flabby Western countries is represented by the students . . . it is up to the students to decide what form their fight should assume. We can't even presume to advise them on this matter.'[61] This was an odd statement from a man who had spent thirty years advising young people what to do. There were more fatuities: 'What is interesting about your action,' he told the students, 'is that it puts the imagination in power.' Simone de Beauvoir was equally elated. Of all the 'audacious' slogans the students had painted on the Sorbonne walls, she enthused, the one that 'touched' her most was 'It is forbidden to forbid.' Sartre humbled himself to interview the ephemeral student leader, Daniel Cohn-Bendit, writing it up in two articles in *Nouvel-Observateur*. The students were '100 per cent right', he felt, since the regime they were destroying was 'the politics of cowardice... a call to murder'. Much of one article was devoted to attacking his former friend Aron, who almost alone in that time of folly was keeping his head.[62]

But Sartre's heart was not in these antics. It was his young courtiers who pushed him into taking an active role. When he appeared on 20 May in the amphitheatre of the Sorbonne to address the students, he seemed an old man, confused by the bright lights and smoke and being called 'Jean-Paul', something his acolytes had never dared to do. His remarks did not make much sense, ending: 'I'm going to leave you now. I'm tired. If I don't go now I'll end by saying a lot of idiotic things.' At his last appearance before the students, 10 February 1969, he was disconcerted to be handed, just before he began to orate, a rude note from the student leadership which read: 'Sartre, be clear, be brief. We have a lot of regulations we need to discuss and adopt.' That was not advice he had ever been accustomed to receive, or was capable of following.[63]

By this time however he had acquired a fresh interest. Like Tolstoy and Russell, Sartre's attention-span was short. His interest in student revolution lasted less than a year. It was succeeded by an equally brief, but more bizarre, attempt to identify himself with 'the workers', those mysterious but idealized beings about whom he wrote so much but who had eluded him throughout his life. In spring 1970 a belated attempt was made by the far left in France to Europeanize Mao's violent Cultural Revolution. The movement was called Proletarian Left and Sartre agreed to join it; in theory he became editor-in-chief of its journal, *La Cause du peuple*, largely to prevent the police from confiscating it. Its aims were violent enough even for Sartre's taste – it called for factory managers to be imprisoned and parliamentary deputies to be lynched – but it was crudely romantic, childish and strongly anti-intellectual. Sartre really had no place in it and he seems to have felt this himself, muttering: 'if I went on mingling with activists I'd have to be pushed around in a wheelchair and I'd be in everyone's way.' But he was hustled along by some of his younger followers and in the end he could not resist the temptations of political show-biz. So Paris was treated to the spectacle of the sixty-seven-year-old Sartre, whom even de Gaulle (to Sartre's annoyance) addressed as 'Cher Maître', selling crudely written newspapers in the street and pressing leaflets on bored bypassers. A photographer caught him thus occupied in the Champs-Elysées, on 26 June 1970, dressed in his new proletarian rig of white sweater, anorak and baggy trousers. He even contrived to get himself arrested, but was released in less than an hour. In October he was at it again, standing on an oil barrel outside the Renault factory in Billaincourt, haranguing the car workers. A report in *L'Aurore* sneered: 'The workers were not having it. Sartre's congregation consisted entirely of the few Maoists he had brought with him.'[64] Eighteen months later he was back at another Renault factory, this time being smuggled inside to give verbal support to a hunger strike; but the security guards found him and threw him out. Sartre's efforts do not appear to have aroused even a flicker of interest among the actual car workers; all his associates were middle-class intellectuals, as they always had been.

But for the man who failed in action, who had indeed never been an activist in any real sense, there were always 'the words'. It was appropriate that he called his slice of autobiography by this title. He gave as his motto *Nulla dies sine linea*, 'Not a day without writing'. That was one pledge he kept. He wrote even more easily than Russell and could produce up to 10,000 words a day. A lot of it was of poor quality; or, rather, pretentious, high-sounding but lacking in muscular content, inflated. I discovered this myself in Paris in the early 1950s, when I occasionally translated his polemics: they often seemed to read well in French but collapsed once expressed in concrete Anglo-Saxon terms. Sartre did not set much store by quality. Writing to de Beauvoir in 1940 and reflecting on the vast amount of words he put down on paper, he admitted: 'I have always considered quantity a virtue.'[65] It is odd that in his last decades he became increasingly obsessed by Flaubert, a writer of exceptional fastidiousness, especially where words were concerned, who revised his works with maniacal persistence. The book he eventually produced on Flaubert ran to three volumes and 2802 pages, many of them almost unreadable. Sartre produced many books, some of them enormous, and many more which were not finished – though often material was recycled in other works. There was a giant projected tome on the French Revolution, a second on Tintoretto. Another huge enterprise was his autobiography, rivalling Chateaubriand's *Mémoires d'outre-tombe* in length, of which *Les Mots* is in effect an extract.

Sartre confessed that words were his whole life: 'I have invested everything in literature... I realize that literature is a substitute for religion.' He admitted that words were to him more than their letters, their meanings: they were living things, rather as the Jewish students of the Zohar or the Kabbala felt the letters of the Torah had religious power: 'I felt the mysticism of words... little by little, atheism has devoured everything. I have disinvested and secularized writing... as an unbeliever I returned to words, needing to know what speech meant... I apply myself, but before me I sense the death of a dream, a joyous brutality, the perpetual temptation of terror.'[66] This was written in 1954, when Sartre still had millions of words to go. What does it mean? Very little, probably. Sartre always preferred to write nonsense rather than write nothing. He is a writer who actually confirms Dr Johnson's harsh observation: 'A Frenchman must be always talking, whether he knows anything of the matter or not.'[67] As he put it himself: '[Writing] is my habit and also my profession.' He took a pessimistic view of the effectiveness of what he wrote. 'For many years I treated my pen as my sword: now I realize how helpless we are. No matter: I am writing, I shall continue to write books.'

He also talked. At times he talked interminably. He sometimes talked when no one was listening. There is a brilliant vignette of Sartre in the autobiography of the film director John Huston. In 1958–59 they were working together on a screenplay about Freud. Sartre had come to stay at Huston's house in Ireland. He described Sartre as 'a little barrel of a man and as ugly as a human being can be. His face was both bloated and pitted, his teeth were yellowed and he was wall-eyed.' But his chief characteristic was his endless talk: 'There was no such thing as a conversation with him. He talked incessantly. You could not interrupt him. You'd wait for him to catch his breath, but he wouldn't. The words came out in an absolute torrent.' Huston was amazed to see that Sartre took notes of his own words while he talked. Sometimes Huston left the room, unable to bear the endless procession of words. But the distant drone of Sartre's voice followed him around the house. When Huston returned to the room, he found Sartre still talking.[68]

This verbal diarrhoea eventually destroyed his magic as a lecturer. When his disastrous book on dialectic appeared, Jean Wahl nonetheless invited him to give a lecture on it at the Collège de Philosophie. Sartre started at 6 pm, reading from a manuscript taken from a huge folder, 'in a mechanical, hurried tone of voice'. He never raised his eyes from the text. He appeared to be completely absorbed in his own writing. After an hour, the audience was restless. The hall was packed and some had to stand. After an hour and three quarters, the audience was exhausted and some were lying on the floor. Sartre appeared to have forgotten they were there. In the end Wahl had to signal to Sartre to stop. Sartre picked up his papers abruptly, and walked out without a word.[69]

But there was always the court to listen to him. Gradually, as Sartre got older, there were fewer courtiers. In the late 1940s and early 1950s he made prodigious sums of money. But he spent it just as quickly. He had always been careless about money. As a boy, whenever he wanted any, he simply took it from his mother's purse. As a schoolteacher he and de Beauvoir borrowed (and lent) freely: 'We borrowed from everybody,' she admitted.[70] He said: 'Money has a sort of perishability that I like. I love to see it slip through my fingers and vanish.'[71] This carelessness had its agreeable side. Unlike many intellectuals, and especially famous ones, Sartre was genuinely generous about money. It gave him pleasure to pick up the tab in a café or restaurant, often for people he scarcely knew. He gave to causes. He provided the RDR with over 300,000 francs (over $100,000 at the 1948 exchange rate). His secretary Jean Cau called him 'incredibly generous and trusting'.[72] His liberality and his (occasional) sense of fun were the best sides of his character. But his attitude to money was also irresponsible. He pretended to be professional about royalties and agents' fees – at his one meeting with Hemingway in 1949 the two writers discussed nothing but such topics, a conversation very much to Hemingway's taste[73] – but this was for show. Cau's successor, Claude Faux, testified: '[Sartre] obstinately refused to have anything to do with money. He saw it as a waste of time. And yet he was in constant need of it, to give it away, to help others.'[74] As a result he ran up huge debts with his publishers and faced horrifying income-tax demands for back payments. His mother secretly paid his taxes – hence Camus's jibe – but her resources were not limitless and by the end of the 1950s Sartre was in deep financial trouble, from which he never really extricated himself. Despite continued large earnings, he remained in debt and often short of cash. He once complained he could not afford a new pair of shoes. There were always a number of people on his payroll in one capacity or another, or receiving handouts. They constituted his outer court, the women forming the inner one. In the late 1960s the number sharply diminished as his financial position weakened, and the outer court contracted.

In the 1970s Sartre was an increasingly pathetic figure, prematurely aged, virtually blind, often drunk, worried about money, uncertain about his views. Into his life stepped a young Jew from Cairo, Benny Levy, who wrote under the name of Pierre Victor. His family had fled from Egypt at the time of the Suez Crisis in 1956–57, and he was stateless. Sartre helped him to get permission to stay in France and made him his secretary. Victor had a taste for mysteries, wearing dark glasses and sometimes a false beard. His views were eccentric, often extreme, forcefully held and earnestly pressed on his master. Sartre's name would appear over strange statements or pieces which the two men wrote together.[75] De Beauvoir feared that Victor would turn into another Ralph Schoenman. She became particularly bitter when he formed an alliance with Arlette. She began to hate and fear him, as Sonya Tolstoy had hated and feared Chertkov. But by this time Sartre was incapable of much public folly. His private life remained sexually varied and his time was shared out among his harem. His holidays were spent as follows:

three weeks with Arlette at the house they jointly owned in the South of France; two weeks with Wanda, usually in Italy; several weeks on a Greek island with Hélène; then a month with de Beauvoir, usually in Rome. In Paris he often moved between the various apartments of his women. His last years were brutally described by de Beauvoir in her little book, *Adieux: A Farewell to Sartre*: his incontinence, his drunkenness, made possible by girls slipping him bottles of whisky, the struggle for power over what was left of his mind. It must have been a relief to them all when he died, in Broussais Hospital, on 15 April 1980. In 1965 he had secretly adopted Arlette as his daughter. So she inherited everything, including his literary property, and presided over the posthumous publication of his manuscripts. For de Beauvoir it was the final betrayal: the 'centre' eclipsed by one of the 'peripheries'. She survived him five years, a Queen Mother of the French intellectual left. But there were no children, no heirs.

Indeed Sartre, like Russell, failed to achieve any kind of coherence and consistency in his views of public policy. No body of doctrine survived him. In the end, again like Russell, he stood for nothing more than a vague desire to belong to the left and the camp of youth. The intellectual decline of Sartre, who after all at one time did seem to be identified with a striking, if confused, philosophy of life, was particularly spectacular. But there is always a large section of the educated public which demands intellectual leaders, however unsatisfactory. Despite his enormities, Rousseau was widely honoured at and after his death. Sartre, another *monstre sacré*, was given a magnificent funeral by intellectual Paris. Over 50,000 people, most of them young, followed his body into Montparnasse Cemetery. To get a better view, some climbed into the trees. One of them came crashing down onto the coffin itself. To what cause had they come to do honour? What faith, what luminous truth about humanity, were they asserting by their mass presence? We may well ask.

10

Edmund Wilson: A Brand from the Burning

*T*HE case of Edmund Wilson (1895–1972) is illuminating because it enables us to draw a distinction between the traditional man of letters and the intellectual of the kind we have been examining. Wilson, in fact, could be described as a man who began his career as a man of letters, became an intellectual looking for millenarian solutions, and then – a sadder and wiser man – reverted to his youthful preoccupation with literature, his true metier. By the time he was born, the American man of letters was a solidly established institution. Indeed in Henry James it had found an outstanding exemplar. For James, letters were life. He rejected with disdain the notion of the secular intellectual that the world and humanity could be transformed by ideas conjured up out of nothing. For him, history, tradition, precedence and established forms constituted the inherited wisdom of civilization and the only reliable guides to human behaviour. James took a serious, if detached, interest in public affairs; and his gesture of taking out British citizenship in 1915, thus identifying himself with a cause he believed to be just, showed he thought it right for the artist to come forward on momentous issues. But literature always came first, and those who consecrated their lives to it – the priests who tended its altars – should never go whoring after the false gods of politics.

Wilson was at heart a man of similar inclinations, though much more ruggedly and incorrigibly an American. Unlike James, he saw Europe, especially England, as constitutionally corrupt, and America, with all its imperfections, as the embodiment of a noble ideal. That explains why, within his traditionalist carapace, an activist sometimes struggled to get out. All the same, by birth, background and – for a time at least – by inclination, he followed the Jacobean path. He came from an immense New England Presbyterian family and in childhood knew practically nobody outside it. His father was a lawyer, one-time Attorney-General of the state of New Jersey. He had the instincts of a judge and Wilson inherited them. He said his father dealt with people 'on their merits' but 'to some extent *de haut en bas*'; and, as Leon Edel, who edited Wilson's papers, has pointed out, the propensity to cross-question literary plaintiffs and to sit in Olympian judgment on them were Wilson's most marked characteristics as a critic.[1] But he also got from his father a passionate love of truth and an obstinate determination to find it. This in the end was his salvation.

Wilson's mother was a regular philistine. She loved gardening and followed college football. To the end of her life she attended the Princeton games. Her desire was for Wilson to be a distinguished athlete and she took no interest in his writings. This may have been just as well, avoiding the destructive tensions which grew up between Hemingway and his clever, literary mother. Wilson went to the Ivy League prep school, Hill School, then Princeton, from 1912–15, where he was well taught by Christian Gauss. He had a spell in an army camp and hated it, worked as a reporter on the *New York Evening Sun*, went out to France in a hospital unit and ended the war as a sergeant in Intelligence.

Wilson was always a man capable of hard, persistent and systematic reading. His notes show that between August 1917 and the Armistice fifteen months later he read over two hundred books: not only older writers like Zola, Renan, James and Edith Wharton, but a wide range of contemporaries from Kipling and Chesterton to Lytton Strachey, Compton Mackenzie, Rebecca West and James Joyce. No man ever read more thoroughly and thoughtfully than Wilson; in his judge-like way, he read as though the author was on trial for his life. As a writer, however, he was much less systematic. He seemed incapable of long-term forward planning. His books evolved and elongated themselves, his non-fiction works starting as mere essays, his novels as short stories. Initially, he had the attention-span of a journalist; then, as he got emotionally involved in a subject, his judicial passion to get at the truth would force him to burrow ever deeper. But it was some time before he found what he wanted to do. In the 1920s he worked on *Vanity Fair*, then the *New Republic*; tried drama criticism on the *Dial*, and went back to the *New Republic*; wrote verse, stories, a novel, *I Thought of Daisy*, and worked hard on a study of modern writers, *Axel's Castle*. He had the privileged life of a bachelor Ivy-Leaguer, briefly (1923–25) tried marriage with an actress, Mary Blair, went footloose again, then married a second time, to Margaret Canby, in 1929. By this point he was already a junior man of letters, with a wide range of literary interests and an enviable reputation for shrewd and objective judgment.

The prosperity of the 1920s was so spectacular, and seemed so durable, as to inhibit political radicalism. Even Lincoln Steffens, whose *Shame of the Cities* (1904) – his collected 'muckraking' articles – had been a milestone in the progressive era, suggested that US capitalism might be just as valid as Soviet collectivism – 'The race is saved one way or the other and, I think, both ways.'[2] The *Nation* began a three-month series on the permanence of prosperity by Stuart Chase, the opening episode of which was published on Wednesday 23 October 1929, the first big break in the market. But when the full magnitude of the crash and the subsequent depression became clear, intellectual opinion ricocheted off in the opposite direction. Writers were particularly hard hit by the slump. By 1933 sales of books were only 50 per cent of the 1929 figure; the old Boston firm of Little, Brown described 1932–33 as 'the worse so far' since they began publishing books in 1837. John Steinbeck complained he could not sell anything at all: 'When people are broke, the first things they give up are books.'[3] Not all writers turned left, but most of them did, joining in a broad, vague, loosely organized and often disputatious but unquestionably radical movement. Looking back on it, Lionel Trilling saw the emergence of this force in the early 1930s as a great turning-point in American history:

> It may be said to have created the American intellectual class as we
> now know it in its great size and influence. It fixed the character
> of this class as being, through all mutations of opinion,
> predominantly of the left. And quite apart from opinion, the political
> tendency of the thirties defined the style of the class – from that
> radicalism came the moral urgency, the sense of crisis, and the
> concern with personal salvation that mark the existence of American
> intellectuals.[4]

Trilling noted that the essence of intellectuals had been defined in W. B. Yeats's observation that one could not 'shirk' the 'spiritual intellect's great work', and that there was

> no work so great
> As that which cleans man's dirty slate.

The trouble, Trilling added, was that in the 1930s there were far too many people anxious to reverse James's attitude and 'to scrub the slate clean of the scrawls made on it by family, class, ethnic or cultural group, [and] the society in general'.[5]

Swept into this seething mob of intellectuals anxious for a *tabula rasa* on which to write anew the foundation documents of civilization was Edmund Wilson. In the winter of 1930–31, the shaken and demoralized *New Republic* was without a policy, and it was Wilson who now proposed it should adopt socialism. In 'An Appeal to Progressives' he argued that, up to the Wall Street crash, American liberals and progressives had been betting on capitalism to deliver the goods and create a reasonable life for all. But capitalism had broken down and he hoped that 'Americans would be willing now for the first time to put their idealism and their genius for organization behind a radical social experiment'. Russia would act as a challenger to the US since the Soviet state had 'almost all the qualities that Americans glorify – the extreme of efficiency and economy combined with the ideal of a Herculean feat to be accomplished by common action in an atmosphere of enthusiastic boasting – like a Liberty Loan Drive – the idea of putting over something big in five years'.[6]

Wilson's comparing Stalin's first Five-Year Plan with Liberty Loans showed how innocent, at this stage, the newly fledged radical intellectual was. But he began reading with his customary Stakhanovite energy the entire political works of Marx, Lenin and Trotsky. By the end of 1931 he was convinced that the changes must be enormous and that intellectuals had to find specific political and economic solutions and embody them in detailed programmes. In May 1932 he drafted, with John Dos Passos, Lewis Mumford and Sherwood Anderson, a manifesto, couched in the hieratic of political theology, proposing 'a socio-economic revolution'.[7] He followed this in the summer with a personal statement of his own beliefs beginning, 'I expect to vote for the Communist candidates in the elections next fall.' He never seems to have contemplated actually joining the Communist Party but he thought its leaders 'authentic American types' who, while insisting on 'that obedience to a central authority without which serious revolutionary work is impossible', had 'not lost their grasp of American conditions'. The CP was right to insist that 'the impoverished public has no choice but to take over the basic industries and run them for the common benefit.'[8]

Wilson was well aware that he and his friends might be seen as well-to-do outsiders playing with working-class politics. Indeed the perception was just. Apart from reading Marxism, his contribution to the cause was to give a cocktail party for the CP leader William Z. Foster, at which Foster answered questions from newly-radicalized writers. Wilson quoted with relish a vignette of Walter Lippmann in his big Washington house during a rainstorm, in full evening dress, 'holding out a small frying pan with which he was trying to contend with a veritable inundation caused by a leak in the ceiling' – the perfect image of the intellectual coping impotently with crisis.[9] But he gives, quite unconsciously, an equally revealing vignette of himself, thanking his faithful black servant Hatty who had 'marvellously enlarged and patched up' his old evening-dress trousers so he could go to a party at the Soviet consulate to celebrate their 'new constitution'.[10]

But Wilson, having a genuine passion for truth, and unlike virtually all the intellectuals described in this book, did make a serious, sincere and prolonged effort to brief himself on the social conditions about which he wished to pontificate. Once he had finished *Axel's Castle* in 1931, he plunged into on-the-spot reporting, writing articles from all over the US which were later collected in *The American Jitters* (1932). Wilson was a good listener, a sharp observer and a scrupulously accurate recorder. He examined the steel industry in Bethlehem, Pennsylvania, then went on to Detroit to look at the car industry. He reported on a textile strike in New England and mining in West Virginia and Kentucky. He went to Washington, through Kansas and the Midwest up to Colorado, then down to New Mexico and into California. His descriptions are notable for their lack of tendentiousness, their gift for arresting detail, their concern for the normal, the non-political and the bizarre, as well as the class war, above all for their interest in people as well as ideas – in short, the exact opposite of Engels's *Condition of the Working Class in England*. Henry Ford was 'a queer combination of imaginative grandeur with cheapness, of meanness with magnificent will, of a North-Western plainness and bleakness with a serviceable kind of distinction'. Wilson noted: 'Wide use of spats in Detroit.' He recorded anecdotes about quarrels, crimes and murders which had nothing to do with the crisis, described winter in Michigan, the fantastic architecture of California and the dude ranches of New Mexico. John Barrymore's wife was 'a soft little doughnut'. A Midwestern girl told him she was 'making the best of the last twenty-four hours of capitalism'. The old derricks near Laguna Beach were 'like druids of old with beards that hang on their bosoms'. At San Diego a distant lighthouse going on and off reminded him 'of a rhythmically expanding penis in a vagina'.[11]

In the terrible winter of 1932, when there were more than thirteen million unemployed, Wilson joined a great cluster of intellectuals who had come to observe the Kentucky coal strike and wrote a harrowing description of what he saw. The writers brought emergency supplies and were told by the County Attorney: 'you can distribute food as much as you want to, but as soon as you buck the law it will be my pleasure as well as my duty to prosecute you.' Wilson described the novelist Waldo Frank threatening a mayor with publicity. Frank: 'The pen, as

Shakespeare said, is mightier than the sword.' Mayor: 'I'm not scared of a Bolshevik pen any day.' The visiting intellectuals were searched for guns, some kicked out, others beaten. At CP headquarters he noted: 'Deformed people ... hunchback running the elevator, dwarf woman with glasses, woman with part of face discoloured as if by a burn but with a protruding growth of some kind from the discoloured part.' He showed a healthy scepticism about the value of such visits, writing to Dos Passos: 'The whole thing was very interesting for us – though I don't know that it did much for the miners.'[12]

The most remarkable aspect of Wilson's thirties' radicalism was the way in which his independence of mind and his real concern for truth prevented him from becoming, like Hemingway, a pliable instrument of the CP. As he told Dos Passos, writers ought to form their own independent group precisely 'so that the comrades can't play them for suckers'. He had already perceived that the radical middle-class intellectual tended to lack one essential human characteristic, the ability to identify with his own social group. In a note on 'The Communist Character' (1933) he put his finger on the weakness of the intellectual:

> he can only identify his interests with those of an outlawed minority
> ... his human solidarity lies only in his imagination of general human
> improvement – a motive force, however, the strength of which cannot
> be overestimated – what he loses in immediate human relationships
> is compensated by his ability to see beyond them and the persons
> with whom one has them: one's family and one's neighbours.[13]

To a man strongly interested in human life and character, as Wilson was, such compensation was not nearly enough. Yet he determined to explore communism not only in its theoretical origins – he was already working on what was to become a major account of Marxist history, *To the Finland Station* – but in its practical applications in the Soviet Union. In certain ways he made a bigger effort to get at the truth than any other intellectual of the 1930s. He learned to read and speak Russian. He mastered much of its literature in the original. In spring 1935 his application for a Guggenheim scholarship to study in Russia was answered with a $2000 grant. He went to Leningrad on a Russian ship and was soon talking to people. From Leningrad he travelled to Moscow, then down the Volga by boat to Odessa. The great purges were just beginning but visitors could still move about in some freedom. In Odessa, however, he had a dose of scarlatina, followed by an acute kidney attack. He spent many weeks in a battered, filthy but curiously easy-going quarantine hospital, a mixture of kindness and bedbugs, socialism and squalor. Many of the characters could have come straight from the pages of Pushkin; indeed the place had been built when Pushkin was still alive. It gave him an entry into Russian society he could not otherwise have found. As a result he left Russia with a growing dislike for Stalin and an uneasy scepticism about the whole system, but with a huge respect for the Russian people and an overwhelming admiration for their literature.

Clearly it was Wilson's irrepressible interest in people, his unwillingness to allow them to be effaced by ideas, which prevented him from sustaining the posture of the intellectual for long. By the end of the 1930s all the instincts and itches of the man of letters were returning. But the process of emancipating himself from the lure of Marxism and the left was not easy. *To the Finland Station* grew and grew. It was not finally published till 1940, and not until the second edition did Wilson denounce Stalinism as 'one of the most hideous tyrannies the world has ever known'. The book itself is a mixture, containing passages dating from the period when he found the impact of Marx intellectually overwhelming. Thus he links together Marx's three propaganda diatribes, *The Class Struggles in France* (1848–50), *The Eighteenth Brumaire of Louis Bonaparte* (1852) and *The Civil War in France* (1871) as 'one of the great cardinal productions of the modern art-science of history', when they are in fact an unscrupulous blend of falsehood, wishful thinking and invective, and historically quite worthless. He defends or dismisses Marx's anti-Semitism – 'If Marx is contemptuous of his race, it is primarily perhaps with the anger of Moses at finding the children of Israel dancing before the golden calf.' He describes Marx's attitude to money as springing from 'almost maniacal idealism', without mentioning his cheating tradespeople, longing for his relatives, including his mother, to die, borrowing without the slightest intention of repaying or speculating on the stock exchange (it is possible Wilson was unaware of this last activity). Wilson is not in the least distressed by the sufferings which Marx, in the cause of his 'art-science', inflicted on his family; he can imagine doing it himself, at any rate in theory.

But in practice? Wilson clearly lacked the disregard for truth and the preference for ideas over people which marks the true secular intellectual. But did he nonetheless possess the monumental egotism which is, as we have seen, equally characteristic of the group? When we look at this aspect of his character, and examine his personal behaviour, the evidence is inconclusive. Wilson had four wives. He parted from the first by mutual agreement since their respective careers proved incompatible; they remained on friendly terms. His second, attending a Santa Barbara party in September 1932 in high heels, tripped and fell down some steps and died of a fractured skull. He remained single during his most intense Marxist-Russian period, but in 1937 he met, and the following year married, Mary McCarthy, a brilliant young writer seventeen years his junior.

The third wife added a new dimension to Wilson's political existence. Mary McCarthy was an extraordinary mixture of origins and inclinations. She came from Seattle. On her mother's side she had both Jewish and New England Protestant blood. Her father's parents had been second-generation Irish farm settlers who had become rich in the grain-elevator trade. She was born 21 June 1912, three younger brothers followed, but all were then orphaned. Mary was brought up first by an oppressive Catholic uncle and aunt, then by her Protestant grandparents.[14] Her education was conducted at one extreme in a Catholic convent and at the other at Vassar, the distinguished woman's college.[15] As might have been predicted, she emerged a mixture of spoilt nun and bluestocking. Her real ambitions were theatrical and she took to writing as a *pis aller.* But she was very good at it, quickly establishing a reputation as an extraordinarily sharp reviewer, first of books, then of the theatre. She married but soon outdistanced an unsuccessful actor-writer called Harold Johnsrud, and when their marriage broke up three years later dissected it neatly in a superb story, 'Cruel and Barbarous Treatment'.[16] Her next adventure, in 1937, was to share an apartment with Philip Rahv, the Russian-born editor of *Partisan Review,* and this brought her into the heart of the New York radical scene.

The true if paradoxical point has been made that, in the 1930s, New York 'became the most interesting part of the Soviet Union . . . the one part of that country in which the struggle between Stalin and Trotsky could be openly expressed'.[17] The battle was fought to a great extent in and around *Partisan Review* itself. It had been founded in 1934 and initially dominated by the Communist Party. But its editor, Rahv, was in his own way an unbiddable spirit. His formal education had ended at sixteen and thereafter he was on his own, sleeping on New York park benches and reading in the Public Library. Early in the 1930s at the same time as Wilson, he became a Marxist convert, signalling his conversion in 'An Open Letter to Young Writers', which insisted: 'We must sever all ties with this lunatic civilization known as capitalism.'[18] In *Partisan Review* he struck with unerring accuracy the prevailing note of the epoch – the middle-class intellectual grubbing down to the worker-peasant level: 'I have thrown off,' he wrote, 'the priestly robes of hypocritical spirituality affected by bourgeois writers, in order to become an intellectual assistant of the proletariat.'[19] He was the great organizer of what he called 'The Literary Class War', the title of one of his articles.[20]

But he broke with the Communists in 1936 over the Moscow trials, which he was certain were a frame-up. Rahv was a skilful herdsman of literary cattle and extraordinarily sensitive to their collective moods. He suspended *Partisan Review* for a time to see which way literary opinion was moving, then resumed it as a quasi-Trotskyist organ, and found he had guessed right: most of the writers who mattered in that milieu were with him. They included Mary McCarthy, who became his mistress as well, a worthwhile bonus since she was a pretty and vivacious young woman.[21]

What attracted her to the Stalin–Trotsky war was not politics as such but the histrionic excitement it generated. 'There is now,' wrote James T. Farrell, the Chicago novelist, 'a line of blood drawn between the supporters of Stalin and those of Trotsky, and that line of blood appears like an impassable river.'[22] Earl Browder, the CP boss, said that Trotskyists caught distributing leaflets at CP meetings should be 'exterminated'. Mary McCarthy later pictured the *Partisan Review* offices as an isolated garrison in Union Square: 'The whole region was Communist territory; "they" were everywhere – in the streets, in the cafeterias; nearly every derelict building contained at least one of their front-groups or schools or publications.' When *Partisan Review* moved to Astor Place it shared a building with the CP's *New Masses*: 'meeting "them" in the elevator, riding down in silence, enduring their cold scrutiny, was a prospect often joked about but dreaded.'[23] She seems to have found this religious warfare, with its pungent atmosphere of *odium theologicum*, exciting. Indeed it is interesting the way her Catholic moral training survived as ideological priggishness, such as a refusal to talk to, lunch or associate with anyone who broke one of her moral-intellectual-political rules, often narrowly defined in doctrinal terms. Her actual knowledge or concern for politics as such was slight. She later admitted she drifted into her political postures, often from the desire to show off or have fun. She was too critical to be a comrade in the thirties' sense. She later compared Trotsky to Gandhi, suggesting she knew little about either. Even at the time she would cause uproar at left-wing parties by revealing royalist underpinnings when tipsy and bringing up the brutal murder of the Tsar's family.[24] In retrospect she strikes one as not a political animal at all: first totally ignorant of communism, then a communist, then almost by accident a Trotskyist, then an anti-communist; then nothing at all but mild, all-purpose left. But all the time she was ultra-critical, partly by nature, partly by training in Eng. Lit. Crit.; and, at bottom, not really interested in ideas but in people, and as such more of an intellectual's girl than an intellectual herself, by the definition we have been using here.

But did she prefer to be the girl of an intellectual or a man of letters? Rahv was unquestionably an intellectual but he was not an attractive man. While expert at guiding what has been well called 'The Herd of Independent Minds',[25] he was extraordinarily close at concealing his own inner feelings. He was, wrote William Styron, 'so secretive as to be almost unknowable'. Mary McCarthy herself noted: 'If no two people are alike, he was less like anyone else than anybody.'[26] He was a man, Norman Podhoretz later testified, 'With a great appetite for power'.[27] Moreover, this appetite expressed itself most commonly by exerting power over other people, as his new mistress quickly discovered.

So Mary McCarthy, a romantic soul who loved New York's partisan warfare but was not easily dominated for long, slipped from under Rahv's influence and found herself married to Wilson. In theory, this might have become a literary alliance, an intellectual union of the distinction and duration of Sartre's association with de Beauvoir. In practice, however, it would have required two very different people to succeed. To be sure, Wilson's attitude to women had something in common with Sartre's: that is, it was self-centred and exploitative. His illuminating record of a conversation with Cyril Connolly, made in 1956, on the subject of wives reveals that, in his view, the primary function of the wife was to serve the husband. He told Connolly to get rid of his present wife, Barbara Skelton: 'he ought to get a different kind of woman, who would take better care of him.' Connolly replied that he was indeed trying to take that advice and extricate himself: 'I'm still on the flypaper – I've got most of my leg loose but I haven't yet quite got off.' Both these men talked about wives as though they were some kind of upper servant.[28]

But Wilson, unlike Sartre, regarded women with suspicion and a certain amount of fear. Women, he noted as a young man, were 'the most dangerous representatives of those forces of conservatism' against which the literary hero's 'whole life was a protest'. He protected himself, as he thought, by pursuing a variation of the usual policy of 'openness' of which intellectuals are so fond: he jotted down, in his notebooks, long passages describing his women in their most intimate postures, and in particular their sexual relations with him. Wilson was a writer of fiction as well as a critic and when he formed his note-taking habits was much under the influence of the James Joyce of *Ulysses*. He seems to have thought that, by actually writing down what happened, he could exorcise some of the terrors of sex and the power of women over him. He wrote a great deal about Edna St Vincent Millay, the beautiful poetess, his first and perhaps strongest love, who mesmerized him. He described how he and the young man who shared his apartment, John Peale Bishop – also in love with her – came to an arrangement whereby, obliged to share her, Bishop fondled the upper half of her body and Wilson the lower; she called them 'the choir-boys of hell'.[29] He described (1920) buying his first condom: 'I went to a drugstore on Greenwich Avenue and watched nervously from outside to be sure there were no ladies there'. The shop assistant 'produced a condom of rubber, which he highly recommended, blowing it up like a balloon to show me how reliable it was'. But the thing burst 'and this turned out to be something of an omen'. He described getting venereal disease. He wrote that he was 'a victim of many of the hazards of sex . . . abortions, gonorrhoea, entanglements, a broken heart'.[30] He took a gruesome interest in the garments women had to discard to allow him entry: getting off 'one of those confounded girdles' was 'like eating shellfish'.[31]

Many of the most relentless passages concerned his second wife, Margaret, 'standing up with her clothes off in the sitting room at 12th Street, her round, soft, broad bosom (white skin)'. She had a 'short little figure when I'd embrace her without her shoes, standing up nude, fat hips and big soft breasts and big torso and tiny feet'. He noted also 'little strong paws of hands (with hard grip) . . . when lying on bed, little arms and legs, turtle paws, sticking out at each corner'. He described making love to her in her Beaux Arts Ball costume on an armchair, 'it had been a little hard managing it – had she put one leg over an arm?'; or 'the time she took off her dress and her underthings came with it . . . I'm one of those ready girls, she said.'[32]

Then there were the adulterous encounters. One woman 'rather shocked me by telling me she wanted me to beat her; one of her friends liked to switch his wife. I bought a hairbrush with wire bristles . . . and first scraped, then spanked her with it. I found this rather difficult, perhaps because of inhibitions. She said afterwards she had thoroughly enjoyed it.' Another 'thought that men's cocks were stiff all the time because whenever they got close enough to her for her to notice, they always were'. A whore picked up in Curzon Street 'worked energetically and authoritatively'. Many women – too many to be plausible, perhaps – express admiration: 'You were so big!' and so forth.[33]

His fourth wife, Elena, got the same treatment. Thus, during the 1956 election campaign: '. . . we sat on the couch and listened to [Adlai] Stevenson campaigning in Madison Square Garden, I began to feel her – she was half sitting up – and she opened her legs and loved it . . . when they cut it off we went onto something more active.' He continued: 'Nowadays I never seem to get enough.' In England, fed up with 'the monastic staleness' of All Souls, Oxford, he returned hurriedly to London where 'I leapt on Elena, who had gone to bed.'[34]

There was none of this quasi-pornographic material in the notebooks he kept during his third marriage to Mary McCarthy; or none has been published anyway. The union lasted from February 1938 to the end of the war but seems to have been a failure from the start. Sartre may have treated de Beauvoir like a slave, but he never told her what to write. Wilson, however, insisted that Mary McCarthy write fiction and treated her like a clever schoolgirl who needed academic supervision. She apparently married him at his insistence and as a spouse found him domineering: he delivered not so much opinions as judgments, which she termed 'the Authorized Version'. He drank a good deal and, when drunk, sometimes became violent if her fiery spirit rebelled. Tipsy, aggressive, red-haired men (Wilson had red hair, though his eyes were brown) later figured in her stories, as did women with black eyes and bruises inflicted by husbands.[35]

The marriage dragged on until 1946 but the critical break came in the summer of 1944, as was described by McCarthy herself in her testimony during the separation petition. There had been a party for eighteen people; everyone had gone home and she was washing the dishes:

> I asked him if he would empty the garbage. He said, 'Empty it yourself.' I started carrying out two large cans of garbage. As I went through the screen door, he made an ironical bow, repeating: 'Empty it yourself.' I slapped him – not terribly hard – went out and emptied the cans, then went upstairs. He called me and I came down. He got up from the sofa and took a terrible swing and hit me in the face and all over. He said, 'You think you're unhappy with me. Well, I'll give you something to be unhappy about.' I ran out of the house and jumped into my car.[36]

The row over the garbage was later described by her in *A Charmed Life* (1955), where Martha is terrified of red-haired Miles Murphy: 'Nobody, except Miles, had ever browbeat her successfully ... with Miles she had done steadily what she hated.' When Mary McCarthy wrote to Wilson saying that Miles was not him, he replied that he had not read the book but 'I assume it is just another of your malignant, red-haired Irishmen.'

The truth is, Mary McCarthy was too strong a character and had too distinctive a talent to be a satisfactory mate to such an Olympian and demanding figure. She may, initially, have prolonged his involvement with left-wing politics but ultimately, with her spirit of independence, she helped to disgust him with the whole idea of prescribed progressive attitudes. Her departure marked the point at which he finally ceased to be an intellectual and reassumed the far more congenial role of man of letters. In 1941 he had bought a large, old-fashioned house at Wellfleet,

Cape Cod, and later he inherited the family's stone mansion in upstate New York, thereafter moving majestically between the two, according to the seasons. His fourth wife, Elena, was born Hélène-Marthe Mumm, a half-German vintner's daughter from the Rheims champagne country. He noted, with complacent approval, that 'Her frank and uninhibited animal spirits contrasted with her formal and aristocratic manners'; he found her 'a great relief' and began 'to function normally again'. She ran his houses with a certain amount of old-style European discipline, introducing comfort and elegance into his life. He accepted this routine with satisfaction, working with his habitual relentless concentration all day, in pyjamas and dressing-gown, then emerging at 5 pm for what he called 'the social date', in a well-pressed suit, fresh shirt and tie.

On 19 January 1948 he made a note about this new life, as a member of the traditional gentry with a literary turn. He had gone for a walk with the dogs: 'they looked quite handsome against the sheet of snow.' The marsh was 'wide, blond and mellow under the greying Cape Cod sky'. He had had 'a good day's work' and had 'taken two straight drinks of good Scotch'. Now he 'stood in the house enjoying its brightness and pleasantness – bow window in the dining room, gleam of candle-sticks . . .'[37] Eight years later, he penned an essay, 'The Author at Sixty', which is, in its quiet way, a hymn to the importance of tradition and continuity. 'Life in the United States,' he noted, 'is much subject to disruptions and frustrations, catastrophic collapses and gradual peterings-out.' When young he had felt threatened by such a fate but now, 'in my sixty-first year, I find that one of the things which most gratify me is the sense of my continuity.' He was back in the country, 'surrounded by the books of my boyhood and furniture that belonged to my parents'. Was he then just 'in a pocket of the past?' Not at all: he was 'at the centre of things – since the centre can be only in one's head – and my feelings and thoughts may be shared by many'.[38] Here was an approach to life not too far distant from Henry James's.

Yet it is worth observing that Wilson retained, even in his reincarnation as a literary gentleman, some at least of the characteristics which had pushed him into radical intellectual life. He was a man who usually strove for truth with great earnestness. But there were areas of prejudice in his mind where he kept the truth at bay with great ferocity. His Anglophobia, an amalgam of anti-imperialism, hatred of the English class system and sheer insecurity, survived the decline of all his other radical impulses. In his post-war notes one has the impression that he is actually grinding his teeth while writing: Churchill is 'disgusting and intolerable'. He observes (quite seriously): '*The British* are said quietly but carefully to be getting *the hemp business* into their hands' – the sort of fact, or non-fact, a second-rate French consul might report. He took careful bearings on 'the Oxford brush-off', the 'competitive spitefulness of the English', 'the two ways they have of saying "Yes"' – the chilly and the insincere – the fact that 'they have a special word, "civil", for what is elsewhere mere ordinary politeness', their propensity to 'foment violence', and their 'international reputation as hypocrites'. He refers to 'perfide Albion', 'la morgue anglaise' and admits: 'I have become so *anti-British* over here that I have begun to feel sympathetic with Stalin because he is making things difficult for the English.'[39]

Of another visit to England in 1954 we have not only his own, envenomed account but a delightful sketch by Isaiah Berlin, who was his host at All Souls. 'My present policy in England,' he announced, was to be 'discreetly aggressive'. He was pleased to find that English intellectuals were even more 'provincial' and 'isolated' than before, that Oxford had become 'shabby and crumbling, scrofulous and leprous'. His All Souls room was a 'dismal little cell like a fourth-rate New York rooming-house' and the college servants were 'obviously disaffected'. Meeting at a party E.M.Forster, 'a tiny little man who might at first glance be some sort of clerk or a man in an optician's shop', he said aggressively that, while he shared Forster's enthusiasm for his three favourite books, *War and Peace*, *The Divine Comedy* and Gibbon's *Decline and Fall*, 'I thought that *Das Kapital* almost belonged to the same category.' This was an amazing remark for a man of letters as opposed to an intellectual to make, and Wilson noted that it 'disconcerted' Forster, as well it might. Forster countered by hastily bringing up the safe subject of Jane Austen, then edging away: 'Well, I mustn't keep you from other people,' a phrase whose dismissive irony Wilson fortunately failed to detect.[40] When Berlin asked him if he had 'disliked every literary person he had met in London', Wilson replied: 'No, I liked Evelyn Waugh and Cyril Connolly best.' 'Why?' 'Because I thought they were so nasty.'[41]

Wilson's personal hostility towards other writers was, indeed, another characteristic he shared with many intellectuals: not even a Marx could have recorded his impressions of them with more malice. D.H.Lawrence's head was 'disproportionately small. One saw that he belonged to an inferior caste – some bred-down unripening race of the collieries.' There is a horrific picture of Scott Fitzgerald, pathetically drunk, lying in a corner on the floor, of Robert Lowell mad and manic, of e.e.cummings with his 'feminine' voice, of W.H.Auden 'portly and man-of-the-world . . . he suddenly began telling us he was no good at flagellation.' Dorothy Parker wears too much cheap scent. Van Wyck Brooks 'doesn't understand great literature', Cyril Connolly 'never listens to anyone else's sallies or stories', T.S.Eliot had 'a scoundrel' somewhere 'inside him', the Sitwells were 'of no interest'.[42] There was a good deal of hatred within the Olympian judge.

There was also a lack of balance about ordinary affairs of the world which we come across time and again in the ranks of the intellectuals and which lingered on in Wilson long after he broke from them. It emerged suddenly and disastrously in Wilson's embittered battle with the officials of the American Internal Revenue Service, about which he wrote an indignant book. His problem was quite simple: between 1946 and 1955 he did not file any income-tax returns, a serious offence in the United States as in most other countries. Indeed in America it is normally so heavily punished, by fines and jail, that when Wilson first confessed his delinquency to a lawyer he 'told me at once that I was evidently in such a mess that the best thing I could do was to become a citizen of some other country'.[43] The reasons he gave for his failure to comply with the law strike one as feeble. For most of his adult life he had been a freelance. At the end of 1943 he got a regular job on the *New Yorker*, with the tax withheld from his earnings there. In 1946 he published *Memoirs of Hecate County*, his one big commercial success. Until then his top earnings had been $7500 as Associate Editor of the *New Republic*. However it was in this year he got married again, and had to pay the bills for two divorces. For these he used the *Hecate* windfall. He said he then intended to catch up with his income-tax obligations as the book was still selling well and money was coming in. But it suddenly ran into obscenity trouble and the money from it ceased. Hence: 'I thought that before filing for the years since 1945 it would be better to wait until I was earning more money.' That happened in 1955 when the *New Yorker* published his long and much admired study of the Dead Sea Scrolls, which was made into a successful book. It was then that he went to the tax lawyer, whose advice came as a shock: 'I had no idea at that time of how heavy our taxation had become or of the severity of the penalties for not filing tax returns.'[44]

This was an extraordinary admission. Here was a man who had written extensively on social, economic and political problems throughout the 1930s and who had offered vehement advice to the authorities involving heavy public expenditures and the nationalization of major industries. He had also published a large book, *To the Finland Station*, tracing with enthusiasm the development of ideas designed to revolutionize the position of ordinary people by seizing the assets of the bourgeoisie. How did he think the state paid its way during the high-spending New Deal, of which he strongly approved? Did he not feel it the personal responsibility of all to make such reforms work, especially those, like himself, who had expressed direct moral obligations towards the less-favoured?

What about the Marxist tag, which he endorsed: 'From each according to his ability, to each according to his need'? Or did he think that applied to others but not to himself? Was this a case, in short, of a radical who favoured humanity in general but did not think of human beings in particular? If so, he was in good, or rather bad, company, since Marx seems never to have paid one penny of income tax in his life. Wilson's attitude was in fact a striking example of the intellectual who, while telling the world how to run its business, in tones of considerable moral authority, thinks the practical consequences of such advice have nothing to do with those like himself – they are for 'ordinary people'.

It took two lawyers, plus accountants, five years to square Wilson's accounts with the IRS. Naturally the IRS gave him a hard time. They charged him $69,000, made up of 6 per cent interest over ten years, plus 90 per cent legal penalties – 50 per cent for fraud, 25 per cent for delinquency, 5 per cent for failure to file and 10 per cent for alleged under-declaration of income. But this was comparatively lenient treatment since Wilson could have gone to jail for a year for each failure to file. Moreover, since he pleaded poverty and had to meet $16,000 in legal fees, the IRS in the end let him off with a compromise settlement of $25,000. So he should have considered himself lucky. Instead Wilson wrote his diatribe, *The Cold War and the Income Tax: a Protest*. This was in every way an irrational response to his troubles. They had given him a frightening insight into the harshness of the modern state at its most belligerent – the tax-gathering role – but this should have come as no surprise to an imaginative man who had made it his business to study the state in theory and in practice. The person who is in the weakest moral position to attack the state is he who has largely ignored its potential for evil while strongly backing its expansion on humanitarian grounds and is only stirred to protest when he falls foul of it through his own negligence. That exactly describes Wilson's position. In his book he tried to evade his own inconsistencies by arguing that most of the income tax went on defence spending induced by Cold War paranoia. But then he had not paid his state income taxes either, and they did not go on defence. Nor did he meet the point that, by the time he settled, a rapidly rising proportion of federal income tax was going on welfare. Was it morally justifiable to evade that too? In short, the book shows Wilson at his worst and makes one grateful that, in general, he ceased being a political intellectual by the time he was forty.

As it was, by reverting to his true role as a man of letters, Wilson's maturity was remarkably productive. It included *The Scrolls of the Dead Sea* (1955), *Apologies to the Iroquois* (1959) on the Indian confederacy, and *Patriotic Gore* (1962) on the literature of the American Civil War. These books, and other works, were characterized by courage as well as industry – writing on the Scrolls involved learning Hebrew – and by an undeviating and relentless concern for truth. That in itself set him apart from most intellectuals. But still more so did the way in which Wilson's research and writings centred around a strong, warm, penetrating and civilized concern for people, both as groups and individuals, rather than abstract ideas. It was the same concern which gave colour and vivacity to his literary criticism, and made it so enjoyable. For Wilson at his best kept in the forefront of his mind the realization that books are not disembodied entities but come from the hearts and brains of living men and women, and that the key to their understanding lies in the interaction of theme and author. The cruelty of ideas lies in the assumption that human beings can be bent to fit them. The beneficence of great art consists in the way in which it builds up from the individual illumination to generality. Discussing Edna St Vincent Millay, about whom he wrote with moving brilliance, Wilson produced the perfect definition of how a poet should function:

> In giving supreme expression to profoundly felt personal experience,
> she was able to identify herself with more general human experience
> and stand forth as a spokesman for the human spirit, announcing
> its predicaments, its vicissitudes, but as a master of human
> expression, by the splendour of expression itself, putting herself
> beyond common embarrassments, common oppressions and
> panics.[45]

It was Wilson's humanism, which enabled him to understand such processes, which saved him from the millenarian fallacy.

11

The Troubled Conscience of
Victor Gollancz

*O*NE thing which emerges strongly from any case-by-case study of intellectuals is their scant regard for veracity. Anxious as they are to promote the redeeming, transcending Truth, the establishment of which they see as their mission on behalf of humanity, they have not much patience with the mundane, everyday truths represented by objective facts which get in the way of their arguments. These awkward, minor truths get brushed aside, doctored, reversed or are even deliberately suppressed. The outstanding example of this tendency is Marx. But all those we have looked at suffered from it to some extent, the only exception being Edmund Wilson, who perhaps was not a true intellectual at all. Now come two intellectuals in whose work and lives deception – including self-deception – played a central, indeed determining role.

The first, Victor Gollancz (1893–1967) was important not because he gave birth to a salient idea himself but because he was the agent by whom many ideas were impressed on society – impressed with great force and with palpable results. He was, perhaps, the outstanding intellectual publicist of our century. He was in no sense an evil man and even when he did wrong he was usually aware of it and his conscience pricked him. But his career shows strikingly the extent to which deception plays a part in the promotion of millenarian ideas. Even in his lifetime, people who had dealings with him were aware how cavalier he could be with the truth. But now, thanks to the honesty of his daughter, Livia Gollancz, who opened his papers for inspection, and the skilful fair-mindedness of a first-class biographer, Ruth Dudley

Edwards, the exact nature and extent of his deceits can be examined.[1]

Gollancz was fortunate in his birth and still more so in his marriage. He came from one highly gifted and civilized family and married into another. The Gollanczes were Orthodox Jews, originally from Poland; the grandfather was a *chazan* or cantor in the Hambro synagogue. Gollancz's father Alexander was a hardworking and successful jeweller and a man of piety and learning. His uncle Sir Hermann Gollancz was a rabbi and Semitic scholar who performed a huge range of public services; another uncle, Sir Israel Gollancz, a Shakespearean scholar, was Secretary to the British Academy and virtually created the English department at London University.[2] One of his aunts was a Cambridge scholar, another a brilliant pianist. His wife Ruth was also a well-educated woman who had been to St Paul's School for Girls and trained as an artist; her family, the Lowys, were likewise remarkable for combining scholarship, art and business success, the women being as vigorous as the men in the pursuit of culture (Graetz's famous *History of the Jews* was translated into English by Bella Lowy).

Throughout his life, then, Gollancz was surrounded by people steeped in all that is best in European civilization. From the earliest age he was given every opportunity to enjoy it himself. The only son, he was cosseted by adoring parents and obsequious sisters, and treated in effect as an only child. He had plenty of pocket-money with which to indulge a passion for opera. He acquired it very early – by the age of twenty-one he had already seen *Aida* forty-seven times – and touring Europe's opera houses remained his standard vacation to the end of his life.[3] He won a scholarship to St Paul's, received a superb classical education – twice a week he translated the *Times* first leader into both Greek and Latin – and went up to New College, Oxford, as an Open Scholar. In due course he won the Chancellor's Latin Essay Prize and took a First in Classics.

He was already a radical intellectual, who had derived fiery sustenance from Ibsen, Maeterlink, Wells, Shaw and Walt Whitman. He seems to have made up his mind at a very early age on most great issues and never saw any reason to change his views later. At school and university his contemporaries found him dogmatic and over-sure of himself, and he was popular at neither. He abandoned Orthodox Judaism early, saying he could not abide the forty-minute walk (transport being forbidden on the Sabbath) from his home in Maida Vale to the Bayswater synagogue; this was a characteristic exaggeration – it was only fifteen minutes. He trod the usual path via Reform Judaism to nothing at all, helped at Oxford by Gilbert Murray, a high minded atheist. But he later constructed for himself an idiosyncratic version of Platonic Christianity, centred on Jesus, 'the Supreme Particular'. This osmotic religion had the great advantage of providing religious sanction for whatever secular positions Gollancz happened to adopt. But he exercised the Jewish privilege of telling innocuous anti-Jewish jokes.

For a time poor eyesight kept him out of the First World War. Then followed a disastrous spell as a second lieutenant in the Northumberland Fusilliers, during which he broke the rules, made himself thoroughly unpopular and was threatened with a subaltern's court-martial. He escaped to teach Classics at Repton. Taking the Upper Sixth, all of whom expected to be soon at the front, and probably killed there, he proved himself a brilliant if subversive master. He was already half a pacifist (though an exceptionally aggressive one), a theoretical feminist, a socialist of sorts, an opponent of capital punishment, a penal reformer and – at that time – an agnostic. He was determined to proselytize on all these issues: 'I took my decision,' he wrote later. 'I would talk politics to these boys and to any others I could get hold of, day in and day out.'[4] This was to be the watchword of his life: he was a seer, a magus, who had got hold of a Truth, or The Truth, and was determined to pound it into the heads of others. The thought that the parents of the boys might not wish them to be subjected to what they would regard as subversive propaganda by a person given privileged access to them, and that there was something inherently dishonest in this abuse of his position, did not disturb him. In fact with his colleague D.C. Somervell he defended his approach, producing two pamphlets, *Political Education at a Public School*, a plea for 'the study of politics as the basis of public school education', and *The School and the World*. His headmaster, the crafty Geoffrey Fisher (later Archbishop of Canterbury), recognized Gollancz's outstanding ability, noted most of the staff could not stand him, warned him he was going too far, then – at the behest of the War Office, which had compiled a file on 'pacifist activities' at Repton – abruptly sacked him at Easter 1918.

Gollancz's career continued with a job at the Ministry of Food in charge of kosher rationing, a spell in Singapore, then work for the Radical Research Group and for the Rowntree Trust. He finally found his metier as a publisher at Benn Brothers. The firm put out a large number of magazines like *The Fruit Grower* and *Gas World*, which Gollancz found dull, as well as books, mostly works of reference. He persuaded Sir Ernest Benn to let him turn the book division into a separate company, with a commission and a shareholding, and within three years he had achieved an astonishing success. 'It reflects,' Benn wrote in his diary, the 'greatest credit on the genius of Victor Gollancz, who is alone responsible. Gollancz is a Jew, and a rare combination of education, artistic knowledge and business ability.'[5] Gollancz's secret was to produce groups of books which covered the whole price range, and which were collectively immune to seasonal and fashionable fluctuations, and push them selectively with shamelessly brash advertising. He put out works on new technical subjects, like automatic telephones, which those in the business had to have, but also permissive fiction. He started the hugely successful Benn's Sixpenny Library, an adumbration of Penguins, and at the other end of the scale expensive art books, such as *The Sleeping Princess*, using Bakst's designs. According to Douglas Jerrold, the brilliant assistant he recruited, the art books involved some cheating since the colour plates were fakes painted by miniaturists, then photographed.[6] By 1928 he was earning £5000 a year. But he wanted a half-share of the company, under a new title, Benn & Gollancz, and when Sir Ernest refused, Gollancz set up his own firm, taking some of Benn's best authors, such as Dorothy L. Sayers, with him.

The new firm had a curious company structure which bore all the marks of Gollancz's astonishing ability to persuade people into arrangements which favoured his interests at the expense of their own.[7] He put up considerably less than half the capital but he had himself made Governing Director with absolute voting control and 10 per cent of the net profits before dividends were paid. This was rather like the arrangements Cecil Rhodes devised for his diamond and gold ventures in South Africa, and maybe that was where Gollancz got the idea. It worked primarily because the firm made large profits almost from the start and the investors received quite enough to keep them satisfied. Gollancz succeeded because he sold vast numbers of books, especially fiction; and he did this by keeping the prices low, decking out the cheaply produced volumes in a new style of uniform yellow-and-red cover brilliantly designed by a typographer of genius, Stanley Morison, and then boosting the product with high-pressure publicity of a kind never before seen in British or even American publishing.

In addition to these sound, commercial reasons for the firm's prosperity, there was also constant corner-cutting, sharp practice and humbug. He had spies who reported on the internal affairs of other firms and especially on discontented authors. If Gollancz thought such a writer worth having, he would write a long, ingratiating letter of a kind he perfected. Some came to him without prompting because Gollancz, in his heyday, was better at launching a newcomer or turning a steady-seller into a best-seller than any other publisher on either side of the Atlantic. He perfected the art of hype before the word was even known in London. But, once in the Gollancz camp, authors found there were drawbacks. Gollancz genuinely believed that his publicity methods were far more important in selling books than the texts. So he had no scruple in forcing authors to take smaller advances and royalties in order to raise the advertising budget. He hated agents because they did not like this kind of thing. If at all possible, he would persuade his authors not to employ an agent at all. The kind of writer he loved was Daphne du Maurier, who was not interested in money. He would often make verbal agreements on 'a friendly basis'. He believed he had a perfect memory. What he had, rather, was an astonishing capacity to rewrite history in his head and then defend the new version with passionate conviction. There were thus rows and recriminations. When the novelist Louis Golding accused him of non-payment of a promised bonus over his best-seller *Magnolia Street*, Gollancz replied with a six-page letter, blazing with sincerity and injured rectitude, proving that his conduct was impeccable. To an agent who tried to challenge his memory he wrote: 'How dare you! I am incapable of error.'[8] These bold commercial tactics were backed up by formidable displays of rage and shouting. When roused, his voice could be heard all over the building. He liked to have a phone with a long cord so he could march up and down his office while bellowing into the receiver at agents or other enemies. His letters varied from almost hysterical rage to unctuous pleading – at which he was superb – sometimes within the compass of a single epistle. When furious, he would hold up their dispatch for a day to allow 'the sun to set on my wrath'; and in consequence many in his files were marked 'not sent'. Some authors cowered and submitted. Others sneaked off to calmer shores. But during the 1930s and 1940s at least the balance of arrivals was in the firm's favour.

There were other reasons why profits were high. Gollancz always paid low wages. When real need was pleaded, he would make an *ex gratia* payment or offer a loan, rather than raise a salary or an advance. In many ways he was like a character from Dickens. When being particularly mean, he would invoke his rubber-stamp board, which he claimed enforced parsimony on him, and state: 'My board, which is here as I dictate this letter, instructs me to add ...'[9] One reason he could keep wages low, even by the standards of the publishing trade, was that whenever possible he employed women rather than men. This could be justified, indeed made a virtue of, on feminist grounds, but the real reasons were twofold. First, women could be induced to accept much smaller salaries and harsher conditions of service. Second, they were more amenable to his highly personal way of running things. He would storm at them, reduce them to tears, embrace them – his habit of promiscuous kissing was unusual in the 1930s – call them by their first names, or pet names, and tell them how pretty they were. Some of the women staff enjoyed this highly charged emotional office atmosphere. They knew, too, that Gollancz was the one firm where they stood a good chance of promotion to senior executive posts, albeit ill-paid ones. He also gave them opportunities to tyrannize. A staff memorandum of April 1936 gives the flavour of Gollancz's office in his heyday:

> I have detected a certain absence for some time now of the old spirit which used to animate the staff . . . The absence of the old happiness causes me personally a great deal of unhappiness. I think we may get back to the old position by a little more leadership; and I have decided to make Miss Dibbs general leader and supervisor of all the female staff on the main floor . . . She will, in fact, be occupying the position which is occupied in a Russian factory by the leader of a factory Soviet.[10]

Some women flourished under this patriarchal regime. One, Sheila Lynd, was promoted to be his mistress, taken on holiday three times, and allowed to address him as 'Darling Boss'. The men led an uneasy existence. It was not that Gollancz was incapable of discovering male talent. On the contrary, he was very good at it. But he did not like men and men did not like him. He could not work with them for long. He discovered Douglas Jerrold, one of the best publishers of his generation, but reneged on his promise to bring him into the new firm. He discovered Norman Collins, another outstanding media entrepreneur, but eventually picked a quarrel with him and drove him out, replacing him with a servile woman. His relations with Stanley Morison, one of the architects of the firm's success, ended in a shouting match and Morison's departure. There were some epic rows with male authors. In the post-war he brought in his nephew, Hilary Rubinstein, another exceptionally able executive, with the clear understanding that in due course he would inherit Elijah's mantle; but, after many years of exploitation, Rubinstein was driven out.

It is one of the themes of this book that the private lives and the public postures of leading intellectuals cannot be separated: one helps to explain the other. Private vices and weaknesses are almost invariably reflected in conduct on the world stage. Gollancz was an outstanding example of this principle. He was a monster of self-deception and, deceiving himself, he went on to deceive others on a heroic scale. He believed himself a man of great, instinctive benevolence, a true friend of humanity. He was in fact incorrigibly selfish and self-centred. This was most notably illustrated in his conduct to women. He professed devotion to the interests of women, especially his own. In fact he loved them only in so far as they served him. Like Sartre, he wanted to be the baby-adult in the *berceau*, surrounded by devoted, scented femininity. Because his mother's existence revolved around his father – not himself – he dismissed her from his life. She figures scarcely at all in his autobiography, and he admitted, in a letter written in 1953: 'I do not love her.' All his life he surrounded himself with women, but he had to be their paramount interest. He found the idea of male competition intolerable. In youth he had his adoring sisters. In maturity he had his adoring wife (from another family of sisters) who in due course presented him with a series of adoring daughters. So he was the one male in a family of six. Ruth had brains and ability, but Gollancz had to be her career. She would not yield to him on one point, his desire that she cease attending synagogue. But in all other respects she was his slave. She not only ran his houses in London and the country but also drove him when necessary, cut his hair, ran his personal finances (which, strangely enough, he could not handle) and gave him pocket-money; and, in conjunction with his valet, supervised all his intimate concerns. He was child-like and helpless in many ways, quite deliberately so perhaps, and loved to call her 'Mummy'. When they went abroad, the children and their nannies were put in a different and cheaper hotel, so that Ruth could devote herself entirely to him. She put up with his many infidelities and his disagreeable habit of pawing women which led J.B.Priestley to remark that any adultery was pure compared to Gollancz's flirtations. He would clearly have liked her to supervise his mistresses – in the manner of Brecht's Helene Weigel or Sartre's de Beauvoir – as this would have formally signified her forgiveness and thus absolved him from guilt. But she could not steel herself to comply. From all his women, family and employees alike, he demanded unswerving loyalty, even in matters of opinion. He refused to give one woman a job solely because she would not endorse his view that capital punishment should be abolished.

He needed unqualified female devotion at least in part to still irrational fears. His mother had believed that, when his father left for work in the morning, he would never return, and she would perform elaborate anxiety rituals. Gollancz inherited this fear which he focused on Ruth. The curious working habits he developed as a boy led to chronic insomnia and this in turn heightened his many terrors. Though his capacity for humbug was prodigious he could never quite still his lurking conscience. It constantly ambushed him in the form of guilt. His hypochondria, which became more intense and varied as he grew older, often expressed this guilt. He believed his frequent adulteries must inevitably end in venereal disease, about which he knew very little. In fact his biographer thinks he suffered from 'hysterical VD'. In the middle of the war he had a breakdown marked by agonized itching and pain in the skin, fear and a sense of terrible degradation. Lord Horder thought he suffered from hypersensitivity of the nerve-endings. But the most remarkable symptom was his belief he would lose the use of his penis. As he put it in one of his autobiographical volumes, 'the instant I sat down . . . my member would disappear. I would feel it retiring into my body.' Like Rousseau, he was obsessed by his penis, though with less apparent reason. He would constantly take it out to inspect it, to discover whether it showed signs of VD or indeed whether it was still there at all. In his office he would perform this ritual several times a day, near a frosted window he believed to be entirely opaque. The staff in the theatre opposite pointed out that this was not so and that his habits were disturbing.[11]

Gollancz's self-deceptions inflicted suffering on himself as well as others. But clearly a man whose grasp of objective reality was so weak in some ways was not naturally suited to give political advice to humanity. He was a socialist of one kind or another all his life, which he believed was devoted to helping 'the workers'. He was convinced he knew what 'the workers' thought and wanted. But there is no evidence that he ever knew a single working-class man, unless one counts the British Communist Party boss, Harry Pollitt, who had once been a boiler-maker. Gollancz had ten servants at his London house in Ladbroke Grove and three gardeners at Brimpton, his country place in Berkshire. But he could rarely bring himself to communicate with any of them except by letter. He hotly denied, however, that he was out of touch with the proletariat. When one of his authors, Tom Harrisson, who ran the 'Mass Observation' surveys, accused him of withholding sums he needed to pay his staff, he received a characteristically indignant reply: 'If by the time you reach my age you have worked as hard for the working class as I have done you won't be doing badly. And let me tell you that not when I was your age but very much later . . . I had a damned sight less to live on than you have got.'[12] It was Gollancz's belief that he led a quasi-monastic existence. In fact from the mid-1930s he always enjoyed a chauffeur-driven car, big cigars, vintage champagne and a daily lunch table at the Savoy. He always stayed in the best hotels. There is no evidence he ever denied himself anything he wanted.

It is a curious fact that Gollancz's participation in the active anti-capitalist cause dates from 1928-30, just at the time when he was becoming a highly successful capitalist himself. He argued that it encouraged man's natural propensity to greed and so to violence. By September 1939 we find him writing to the playwright Benn Levy that Marx's *Capital* was 'in my view the fourth most enthralling volume in the world's literature'; it combined 'the attractions of an A-plus detective story and a gospel' (can he actually have read it?).[13] This was the prelude to a long love-affair with the Soviet Union. He swallowed whole the Webbs' fantastic account of how the Soviet system functioned.[14] He described it as 'amazingly fascinating'; the chapters designed to eliminate 'misconceptions' about the democratic nature of the regime were 'much the most important in the book'.[15] In due course – at the height of the great purges, as it happens – he nominated Stalin 'Man of the Year'.

Gollancz began his own political activities by asking Ramsay Macdonald, the Labour leader, for a seat in Parliament; nothing came of it, then or later. Instead he concentrated on didactic publishing. By the early 1930s he was putting out a growing proportion of left-wing political books, at low prices and in enormous quantities. They included G.D.H. Cole's brilliant best-seller *The Intelligent Man's Guide Through World Chaos* and *What Marx Really Meant*, and John Strachey's *The Coming Struggle for Power*, which probably had more influence, on both sides of the Atlantic, than any other political book at this time.[16] It was at this point that Gollancz ceased to be a commercial publisher as such and became a political propagandist; at this point, too, that the systematic deception began. A sign of his new policy was a letter to the Reverend Percy Dearmer, Canon of Westminster, commissioned to edit *Christianity and the Crisis*. The book, he laid down, had to be and look 'official', containing contributions from 'a considerable number of high dignitaries of the Church'. But, he wrote, 'I am perhaps a rather peculiar kind of publisher in that, on topics which I believe to be of vital importance, I am anxious to publish nothing with which I am not in agreement.' Hence the book *must* start out from the position that 'Christianity is not solely a religion of personal salvation but must essentially concern itself with politics' and it must then 'go "all out" for immediate and practical socialism and internationalism'.[17]

Despite these clear elements of deception and direction, the Canon complied and the book duly appeared in 1933. Instructions were given to other authors in the same spirit. Leonard Woolf, who was editing *The Intelligent Man's Way to Prevent War*, was told by Gollancz that the climactic chapter was the last, 'International Socialism the Key to Peace', and that the others must 'lead up tendentiously to this final section'; however, to conceal this purpose it was 'desirable' that earlier chapters should not be written 'by people definitely associated in the public mind with socialism'.[18] As the 1930s progressed, the element of deceit became larger and more blatant. In an internal letter to an editor criticizing a book on trade unions by the communist John Mahan, Gollancz complained: 'As it goes on the thing becomes very much of a left-wing exposition; and *particularly on this subject* this is to be avoided.' What he wanted, he continued, was not a 'left-wing exposition but an apparently impartial exposition from a left-wing pen'. 'All sorts of devices will occur to you,' he wrote meaningfully, and concluded: '... both points of view can be represented in such a way that, while there is a grand atmosphere of impartiality which no one can attack, the readers inevitably draw the right conclusion.'

In Gollancz books there began to be, indeed, all sorts of 'devices' to deceive readers. For instance, whenever possible, 'left wing' was always substituted for 'Communist Party'. There was also outright suppression, reflected in many of Gollancz's letters and often accompanied by a self-pitying harping on his agonies of conscience. Thus in a letter to Webb Miller about a book on Spain he ordered the suppression of two chapters he knew to be true, beginning 'I feel distressed, and almost ashamed, to write this letter.' He knew that Miller's account was not 'exaggerated in any way', but it was 'absolutely inevitable' that 'a great number of passages will be picked out from those chapters and widely quoted for propaganda purposes as a proof of "communist barbarism".' He felt he could not 'publish anything which, by giving occasion for propaganda on the other side', will 'weaken [Communist] support'. Miller might think, he added, that 'this is playing with truth. It isn't really: one must consider ultimate results.' Then his final plea: 'Forgive me, please' – rather as he wanted Ruth's formal absolution for his delinquency in keeping a mistress.[19]

Some of Gollancz's instructions to authors and editors, though plainly enjoining dishonesty, were extraordinarily muddled – no doubt because of the writhing agony of his conscience – and it was not at all clear what particular dishonesty they were being told to commit. Thus, to an author of history textbooks, he wrote: 'I want the thing done with the utmost degree of impartiality – but I also want my impartial author to be of radical mind.' He added that 'the author's radicalism' would give him 'the guarantee that if any tendency does, in spite of all efforts, get through', it shall not be 'a tendency in the wrong direction'. What in effect Gollancz was saying, as his letters at this time constantly suggest, was that he wanted slanted books but books which did not appear slanted.

These letters which have survived in the Gollancz files are peculiarly fascinating because they constitute one of the few occasions when direct evidence can be produced of an intellectual poisoning the wells of truth, knowing he was doing wrong and justifying his actions by claiming a higher cause than truth itself. Gollancz was soon practising dishonesty on a large scale. After Hitler's accession to power in January 1933 he decided to cut out of his list any book which did not make money or serve a propaganda purpose. He also launched huge ventures designed primarily to promote socialism and the image of the Soviet Union. The first was the New Soviet Library, a series of propaganda books by Soviet authors arranged directly through the Soviet Embassy and government. But unforeseen difficulties occurred in getting hold of the texts, since the gestation of the series coincided with the great purges. Several of the proposed authors abruptly disappeared into the Gulag archipelago or were hauled in front of firing squads. Some of the texts were sent to Gollancz with the name of the author blank, to be filled in later when the executed writer had been officially replaced. A further, and grue-some, setback was that Andrei Vishinsky, the Soviet public prosecutor, who played the same part in Stalin's regime as Roland Freisler, chairman of the People's Court, in Hitler's, was down to contribute the volume on *Soviet Justice*; but he was too busy getting death sentences passed on former comrades to write it. When the text finally arrived, it was too hastily and badly written to be published. Gollancz's readers were kept blissfully unaware of these problems.

In any case, by the time the series was out Gollancz was involved in a much bigger venture, the Left Book Club, originally set up to counter the unwillingness of booksellers to stock ultra-left propaganda. The LBC was launched with a huge advertising campaign in February–March 1936, coinciding with the Comintern's adoption of a 'Popular Front' policy throughout Europe: suddenly the democratic socialist parties, like Labour, ceased to be 'social fascists' and became 'companions in the struggle'. Members of the LBC agreed to buy at half a crown a month, for a minimum of six months, books chosen by a committee of three, Gollancz himself, John Strachey and Professor Harold Laski of the London School of Economics. They also got free the monthly *Left Book News*, and the right to participate in a huge range of activities – summer schools, rallies, film shows, discussion groups, plays, joint foreign holidays, lunches and Russian language classes, as well as use of the Club Centre.[20] The 1930s was the great age for participatory groups. One of the reasons why Hitler was so successful in Germany was that he created so many of them, for all ages and interests. The CP belatedly followed him and the Left Book Club showed just how effective the technique could be. Gollancz's original hope was that he would get 2500 subscribers by May 1936; in fact he got 9000 and the figure eventually rose to 57,000. The impact of the Club was even wider than these figures suggest; of all the institutions of the thirties' media, it was the one which most success-fully set the agenda and directed the trend of discussion. However, it was based on a series of lies. The first lie, contained in its brochure, was that the selection committee 'together adequately represent most shades of opinion in the active and serious "left" movement'. In fact for all practical purposes the LBC was run in the interests of the Communist Party. John Strachey was completely controlled by the CP at this period.[21] Laski was a Labour Party member and had just been elected to its National Executive; but he had been converted to Marxism in 1931 and usually followed the CP line until 1939.[22] Gollancz was also a dependable fellow-traveller until the end of 1938. He did everything the CP asked of him. For the *Daily Worker*, the CP's organ, he wrote a fulsome article, 'Why I Read the *Daily Worker*', which was used in its publicity material. He singled out its devotion to truth, accuracy and trust in its readers' intelligence – all of which he knew to be baseless – and noted: 'it is characteristic of men and women, as opposed to ladies and gentlemen. For my own part, who meet a lot of ladies and gentlemen and find a lot of them exceedingly tiresome, I find this quality extraordinarily refreshing.'[23] He also visited Russia (1937) and declared: 'For the first time I have been completely happy ... while here one can forget the evil in the rest of the world.'[24]

However, Gollancz's greatest practical service to the CP was to staff the LBC with its people. Sheila Lynd, Emile Burns and John Lewis, who edited all the manuscripts, and Betty Reid, who organized the LBC groups, were all at this time CP members or party-controlled. All policy decisions, even of quite a minor nature, were discussed with CP officials; often Gollancz dealt direct with Pollitt himself, the CP's General Secretary. None of this was known to the public. The LBC deliberately referred to CP members as 'socialists', to conceal their affiliation. Of the first fifteen books selected, all but three were by CP members or crypto-Communists; this worried Gollancz, not the fact itself but the impression it might create that the Club was not independent. Its putative independence was, indeed, its biggest single asset in CP eyes. As the leading CP ideologue, R. Palme Dutt, rejoiced in a letter to Strachey, the fact that the public believed it to be 'an independent commercial enterprise' and not 'the propaganda of a particular political organization' constituted its value to the Party.

The second lie was Gollancz's repeated assertion that the whole LBC organization, with the groups, rallies and events, was 'essentially democratic'. That had no more validity than Miss Dibbs and her 'office Soviet'. Behind a pretence of oligarchy, it was in fact a personal despotism of Gollancz himself, for the simple reason he controlled its finances. Indeed, he kept no separate accounts for the LBC and its income and expenditure were absorbed in Victor Gollancz Limited. The consequence is that there is no means of knowing whether Gollancz made or lost by the venture. When critics asserted he had made a fortune out of it, he sued them for libel. He told authors in private letters that its losses were appalling but added: '*this is absolutely confidential*: from many points of view it is less dangerous that we should be considered to be making huge profits than we should be known to be making a loss.'[25] But this may have been simply to justify paying authors miniscule royalties or none at all. One has to assume that the LBC benefited the firm, if only by sharing overheads and boosting its other books. In any case, as Gollancz handled the receipts and paid the wages and bills, he took all the ultimate decisions. Any idea that Club members had a say in anything was fantasy. Seeking a man to edit the *LBC News*, he laid down that he must 'combine initiative with absolutely immediate and unquestioning obedience to my instructions however foolish they may seem to him'.[26]

The third lie was uttered by John Strachey: 'We do not dream of refusing to select a book simply because we do not agree with its conclusions.' Apart from one or two token Labour volumes – Clement Attlee, the Labour Party leader, was invited to contribute *The Labour Party in Perspective* – there is overwhelming evidence that adherence to the CP line was usually the chief criterion of selection. A particularly flagrant case was August Thalheimer's *Introduction to Dialectical Materialism* which Gollancz, believing it orthodox, had agreed to publish in May 1937. But in the meantime the author had become involved in some obscure dispute with Moscow, and Pollitt asked Gollancz to suppress it. The book had already been announced and Gollancz protested that the Club's enemies would seize on the cancellation as 'proof positive that the LBC was simply a part of the CP'. Pollitt replied with his pseudo-proletarian, Old Soldier act: 'Don't publish it! Not when I've got to cope with the Old Bugger, the Long Bugger and that bloody red arse of a dean!' (By these he meant Stalin, Palme Dutt and the Very Reverend Hewlett Johnson, Dean of Canterbury.) Gollancz complied and the book was suppressed, but he later wrote a whining letter of complaint to Pollitt: 'I hated and loathed doing this: I am made in such a way that this kind of falsehood destroys something inside me.' Another book the party wanted suppressed was *Why Capitalism Means War*, by the highly respected veteran socialist H.N. Brailsford, because it critized the Moscow trials. When in September 1937 the manuscript was shown to Burns, he advised that even with massive cuts and changes the book was unacceptable to the Party. On this occasion Gollancz too was all for suppression. He wrote to the author: 'I cannot act against my conscience in the matter.' To publish a book criticizing the trials would be like 'committing the sin against the Holy Ghost'. But Laski, who was unhappy about the trials himself, and an old friend of Brailsford, said the book must go out and threatened resignation, which would have destroyed the

LBC's Popular Front facade. So Gollancz reluctantly did as Laski asked, but brought the book out in August with a total absence of publicity – 'buried it in oblivion,' as Brailsford put it. Gollancz also invented 'technical reasons' for suppressing a book by Leonard Woolf, which contained some criticism of Stalin; but Woolf, who had his own presses, knew more about printing than Gollancz did, exposed the lie and threatened public trouble if his agreement was broken. Here again Gollancz gave way, though he made sure the book failed.

Left Book Club publications were, in fact, deliberately conceived to promote the CP line by deception. As Gollancz wrote to the editor of the Club's educational books, the Left Home University Library, 'The treatment should *not* of course be aggressively Marxist.' Volumes should be written 'in such a way that, while the reader will at any point draw the right conclusions, the uninitiated would not be put off by feeling, "Why, more of this Marxist stuff!"' At times the links with the CP hierarchy were extremely close: the records show Gollancz transferring sums of money to Pollitt in cash – 'I wonder whether you could let me have the money some time this morning in pound notes. Sorry to trouble you, Victor, but you know how things are.'[27] CP censorship went down to very small details; thus J. R. Campbell, later the *Worker* editor, was responsible for having removed from the bibliography of one volume works by Trotsky and other 'non-persons'.

Gollancz's behaviour, though indefensible and documented by what his biographer calls 'a mass of incriminating material', must be seen in context. Even more than the other decades of our century, the 1930s was the age of the lie, both big and small. The Nazi and Soviet governments lied on a colossal scale, using vast financial resources and employing thousands of intellectuals. Honourable institutions, once celebrated for their devotion to truth, now suppressed it deliberately. In London, Geoffrey Dawson, editor of *The Times*, 'kept out of the paper', as he put it, material from his own correspondents which might damage Anglo-German relations. In Paris, Félicien Challaye, a leading member of the famous Ligue des Droits de l'Homme, created to establish the innocence of Dreyfus, felt obliged to resign from it in protest at the shameless manner in which it helped to conceal the truth about Stalin's atrocities.[28] The Communists ran professional lie-organizations whose specific purpose was to deceive fellow-travelling intellectuals through various front organizations, such as the League Against Imperialism. One such was run first from Berlin, then after Hitler's advent to power from Paris, by the German Communist Willi Muenzenberg, described by the *New Statesman*'s editor, Kingsley Martin, as 'an inspired propagandist'. His right-hand man, the Czech Communist Otto Katz, 'a fanatical and ruthless commissar' as Martin calls him, recruited various British intellectuals to help.[29] They included the former *London Times* journalist Claud Cockburn, editor of the left-wing scandal sheet *The Week*, who helped Katz to concoct entirely imaginary news stories, such as an 'anti-Franco revolt' in Tetouan. When Cockburn subsequently published his account of these exploits, he was attacked by R.H.S.Crossman MP in the *News Chronicle* for his shameless delight in his lies. Crossman had been officially involved in British government 'disinformation activities' (i.e. lying) in the 1939–45 war. He wrote: 'Black propaganda may be necessary in war, but most of us who practised it detested what we were doing.' Crossman who, as it happened, was a typical intellectual who always put ideas before people and lacked a strong sense of the truth, was rebuked by Cockburn, who described the Crossman view as 'a comfortable ethical position if you can stop laughing. To me at least there seems something risible in the spectacle of a man firing off his own propaganda-lies . . . but keeping his conscience clear by "detesting" his own activities.' To Cockburn, a cause for which a man 'is fighting is worth lying for'.[30] (Some cause! Both Muenzenberg and Katz were murdered by Stalin for 'treason', Katz on the grounds that he had consorted with such 'Western imperialists' as Claud Cockburn.)

Gollancz's dishonesties should be judged against this background. The most notorious of them was his refusal to publish George Orwell's exposure of Communist atrocities against the Spanish anarchists, *Homage to Catalonia*. He was not alone in rejecting Orwell. Kingsley Martin refused to publish a series of articles by Orwell dealing with the same theme, and three decades later he was still defending his decision: 'I would no more have thought of publishing them than of publishing an article by Goebbels during the war against Germany.' He also persuaded his literary editor, Raymond Mortimer, to turn down a 'suspect' book review by Orwell, an episode which Mortimer later regretted bitterly.[31] Gollancz's relations with Orwell were protracted, complex, sour and mean. He published *The Road to Wigan Pier*, which was critical of the British left, before the Left Book Club was started, and when he decided to bring out an LBC edition he wanted to suppress the objectionable part. Orwell would not let him. So Gollancz published it with a mendacious introduction by himself, trying to explain away Orwell's 'errors' by saying he wrote as 'a member of the lower-upper-middle class'. As he was, if anything, a member of that class himself (though of course immensely richer than Orwell) and as, unlike Orwell, he had had virtually no contact with working people, this introduction was peculiarly dishonest. Gollancz was later deeply ashamed of it and furious when an American publisher reprinted it.[32]

By the time the Orwell row was at its height, Gollancz himself was already having second thoughts about his Communist connections. There were a number of reasons for this. One may have been the belief that he was damaging his commercial prospects. Secker & Warburg had eagerly snatched up *Homage to Catalonia*, as well as other books and authors who might naturally have been published by Gollancz but for CP objections. Gollancz's CP line, in fact, created a formidable rival for his firm. A second reason was Gollancz's limited attention-span. Books, authors, women (except Ruth), religions, causes could never retain his enthusiasm indefinitely. For a time Gollancz enjoyed the LBC and the immense rallies the CP helped to organize on its behalf in the Albert Hall, at which the Dean of Canterbury would intone: 'God Bless the Left Book Club!' He discovered he had considerable gifts as a public speaker. But it was always the CP stars, above all Pollitt himself, who got the most applause from the well-drilled audience, and Gollancz did not like that. By the autumn of 1938 he was showing signs of impatience and boredom with the whole thing.

In this mood he was more inclined to be open-minded. During a Christmas holiday in Paris he read a detailed account of the Moscow trials which convinced him they were fraudulent. Back in London he told Pollitt that the LBC could no longer peddle the Moscow line on that issue at least. In February he went so far as to admit in *LBC News* that there were 'certain barriers against full intellectual freedom in the Soviet Union'. Orwell was astonished, in the spring, by Gollancz's decision to bring out his novel, *Coming Up for Air*, a sure sign of a changed line. By the summer Gollancz was clearly anxious to have done with Moscow and he greeted the Hitler–Stalin Pact in August, if not exactly with relief – it meant war was inevitable – then as a heaven-sent opportunity to complete the break with the CP. He immediately began writing anti-Moscow propaganda, pointing out a large number of instances of evil behaviour which most sensible people had been aware of for years. As Orwell commented to Geoffrey Gorer: 'It's frightful that people who are so ignorant should have such influence.'[33]

The Left Book Club was never the same after Gollancz's break with Moscow. Its own staff were divided. Sheila Lynd, Betty Reid and John Lewis clung to the Communist Party. Gollancz decided not to sack Lewis and Lynd (who was now no longer his mistress). But he characteristically made commercial use of the occasion to demote them, reduce their salaries and shorten their period of notice.[34] Unlike Kingsley Martin, who uneasily defended his thirties' fellow-travelling to the end of his life, or Claud Cockburn who cynically boasted of his behaviour, Gollancz decided to go the whole hog and make a virtue of repentance. In 1941 he edited a volume, which included contributions from Laski and Stra-chey as well as Orwell, called *The Betrayal of the Left: An Examination and Refutation of Communist Policy*. In this he made a formal confession of the sins of the LBC:

> I accepted manuscripts about Russia, good or bad, because they were 'orthodox'; I rejected others, by *bona fide* socialists and honest men, because they were not . . . I published only books which justified the Trials, and sent the socialist criticism of them elsewhere . . . I am sure as a man can be – I was sure at the time in my heart – that all this was wrong.

How genuine and transforming was Gollancz's change of heart and admission of guilt it is hard to say. He certainly went through a dark night of the soul during the middle of the war, culminating in the physical crisis already described. But then, up in Scotland, and unusually for an intellectual, he heard the voice of God, which told him He would 'not despise' a 'humble and contrite heart'. Thus reassured, he acquired a new religion, in the shape of his own version of Christian socialism, a new mistress and a new zest for publishing, which took the form of the enthusiastic promotion of the Labour Party in a series of volumes called 'The Yellow Perils'. But he was soon up to his old tricks. In April 1944 he rejected Orwell's devastating satire *Animal Farm*: 'I could not possibly publish a general attack [on Russia] of this nature.' That too went to Secker & Warburg which also, in consequence, secured Orwell's famous best-seller, *Nineteen Eighty-Four*, obliging the furious and remor-seful Gollancz to dismiss it as 'enormously overrated'.[35] He was haunted by Orwell's honesty – as indeed was Kingsley Martin – for the rest of his life, and driven, in exasperation, to attacks on him which do not make much ethical, or indeed any other, sense. He could not accept, he wrote, 'that [Orwell's] intellectual honesty was impeccable . . . in my opinion he was too desperately anxious to be honest to be really honest . . . Didn't he have a certain *simplicité*, which in a man of as high intelli-gence as he is, is really always a trifle dishonest? I think so myself.'[36]

Gollancz lived on until 1967 but he never again exerted quite the power and influence that was his in the 1930s. Many held him responsible, along with the *New Statesman* and the *Daily Mirror*, for the Labour Party's historic election victory in 1945, creating the political framework in post-war Britain and much of Western Europe which lasted right up to the Thatcher era. But Prime Minister Attlee did not offer him the peerage he felt he deserved; indeed he got nothing at all until Harold Wilson, a more generous man, gave him a knighthood in 1965. The trouble with Gollancz's vanity was that it persuaded him he was more famous or notorious than he actually was. In 1946, when the ship on which he was taking a holiday docked in the Canary Isles, he had a sudden bout of the terrors and shouted that Franco's police intended to seize and torture him as soon as he landed. He insisted the British consul come on board to protect him. The consul sent his clerk to assure him that nobody on the islands had ever heard of him; indeed, a disappointed Gollancz reported, 'he had never heard of me himself'.

Gollancz's post-war career, in fact, was a dying fall. He wrote some highly successful books, but his own business was gradually edged out of its market-leader position. He did not keep up with the times or recognize the new intellectual stars. When Ludwig Wittgenstein wrote to him in September 1945 pointing out a weakness in one of his public arguments, he responded with a one-line note: 'Thank you for your letter, which I am sure was very well meant'; he misspelt the philosopher's name, believing him to be an obscure don.[37] He lost some of his best authors and missed getting some important books. He hailed Nabokov's *Lolita* as 'a rare masterpiece of spiritual understanding', failed to buy it, furiously decided it was 'a thoroughly nasty book, the literary value of which has been grossly overestimated' and finally denounced it to the *Bookman* as 'pornographic'. He played an important part in one hugely successful campaign, to abolish capital punishment – a cause which engrossed him for longer than any other and which was probably closest to his heart – but his role in this venture was overshadowed by Arthur Koestler, whom he hated, and by the elegant and eloquent Gerald Gardiner, who carried off the honours. Worse still, Gollancz failed to be given the top place in the Campaign for Nuclear Disarmament when it was formed in 1957. He was away at the time and mortified to return and find he had not even been asked to join its committee. He regarded it, he said, as a 'devastating insult' which had left him 'broken-hearted'. At first he blamed his old friend Canon John Collins, who had been made chairman in what Gollancz saw as his rightful place. In fact Collins had fought a losing battle to get him included. Then Gollancz held J.B. Priestley responsible, attributing his enmity to a dispute they had had over Priestley's *English Journey* back in the early 1930s. In fact Priestley was only one of many among the founders who said that they would not work with Gollancz at any price.

In the end, almost all men found Gollancz's self-centred vanity insupportable, especially as it often found expression in unpleasant outbursts of rage. In 1919 he had told his brother-in-law that he could not decide whether to become headmaster of Winchester or prime minister.[38] In fact he was fortunate that his business acumen enabled him to create a private autocracy where no one could challenge him and his inability to make other men like him did not matter so much. Ruth Dudley Edwards quotes a characteristic letter from the Gollancz files which conjures up the man better than any description. He had been asked, and had agreed, to give one of the Memorial Lectures in honour of Bishop Bell, the only man who had spoken out strongly against the area bombing of Germany. But some more attractive engagement had appeared and Gollancz had cancelled his appearance. The organizer, one Pitman, was understandably annoyed and had written Gollancz a reproachful letter. Gollancz replied at length and in furious terms. He rebuked Pitman for 'writing before the sun had gone down on your wrath', explained in immense detail the appalling burden of the commitments which had led him to cancel the lecture, objected in the strongest terms to Pitman's assertion he was 'under a moral obligation' to give it and then, warming to his task, went on: 'In fact, I am beginning to lose my temper as I dictate and I must say that such a remark is patently absurd.' Then followed two more paragraphs accusing Pitman of being 'grossly impertinent', and finally: 'I am conscious of the fact that I have started this letter in a moderate tone but have ended it in an immoderate one. I am also conscious of the fact that, in spite of my advice to you, I do not at the moment feel like letting the sun go down on my wrath and am therefore instructing my secretary to post the letter immediately.' This egotistical tirade might have been penned, *ceteris paribus*, by a Rousseau, a Marx or a Tolstoy. But is it possible to detect a tiny, self-mocking hint of irony? We must hope so.

12

Lies, Damned Lies and Lillian Hellman

*I*F Victor Gollancz was an intellectual who tampered with the truth in the interests of his millenarian aims, Lillian Hellman seems to have been one to whom falsehood came naturally. Like Gollancz she was part of that great intellectual conspiracy in the West to conceal the horrors of Stalinism. Unlike him, she never admitted her errors and lies, except in the most perfunctory and insincere fashion; indeed she went on to pursue a career of even more flagrant and audacious mendacity. It may be asked: why bother with Lillian Hellman? Was she not an imaginative artist to whom invention was a necessity and the worlds of reality and fantasy inevitably overlapped? As in the case of Ernest Hemingway, another notorious liar, is it fair to expect absolute truth from a contriver of fictions? Unfortunately for Hellman, disregard for the truth came to occupy a central place in her life and work; and there are two reasons why it is difficult to ignore her. She was the first woman to achieve international status as a playwright, and as such became a symbolic figure to educated women all over the world. Second, during the last decades of her life, and partly in consequence of her deceptions, she achieved a position of prestige and power in the American intellectual scene which has seldom been equalled. Indeed the Hellman case raises an important general question: to what extent do intellectuals as a class expect and require truth from those they admire?

Lillian Hellman was born on 20 June 1905 of middle-class Jewish parents. Like Gollancz, though for political as much as for personal reasons, she sought in her autobiographical writings to downgrade her mother and exalt her father. Her mother came from the rich and prolific families of Newhouses and Marxes who had flourished in American capitalism. Isaac Marx, following a common pattern in Jewish immigration, had come to America from Germany in the 1840s, started as a travelling peddler, settled down as a merchant and achieved wealth during the Civil War; his son had founded the Marx Bank, first in Demopolis, then in New York. Hellman described her mother, Julia Newhouse, as a fool. In fact she seems to have been cultured and well-educated, and it is likely that she was the source of Hellman's gifts. But Hellman found it politically desirable to dismiss the Newhouses and the Marxes, and she almost tried to pretend that her mother's family was Gentile.[1]

By contrast, her father Max was her hero. Hellman was an only child and Max spoilt her, frustrating such discipline as her mother tried to impose. She presented him as a radical, whose parents had fled as political refugees to the US in 1848. She exaggerated his education and intellectual gifts. In fact he seems to have been just as anxious to make capitalism work for him as the Marxes and Newhouses, but not so good at it. His business failed in 1911 (Hellman later blamed a non-existent business partner) and thereafter he lived mainly off his rich in-laws, ending as a mere salesman. There is no evidence of his radicalism, other than Hellman's assertions. She described, in an article about race relations, how he had saved a black girl from being raped by two white men. But then she also told a tale about how, when she was eleven or twelve, she insisted that she and her black nurse Sophronia sit down in the 'whites only' section of a streetcar, and about how they were evicted after a noisy protest. This anticipation, by forty years, of Rosa Parks's famous act of defiance in 1955 seems improbable, to put it mildly.[2]

Max's sisters kept a boarding-house, where Hellman was actually born and where she spent much time, a lonely but lively and sharp-eyed child, watching the inmates and telling stories to herself about them. She got a lot of material from the boarders; and later in Manhattan, she and Nathanael West, who managed the hotel where she lived, used secretly to open the guests' letters – the source of his book *Miss Lonelyhearts*, as well as incidents in her plays. She described herself as 'a prize nuisance child', which we can well believe, who smoked, went joy-riding in New Orleans, and ran away, undergoing amazing adventures, which seem less credible. When her father moved to New York for work, she attended New York University, cheated at exams, and emerged as a five foot four, 'rather homely' girl, with the possibility of becoming a successful *jolie laide*; she seems, even as a teenager, to have had an assertive sexual personality.

Hellman's early career, like her childhood, has been traced by her careful and fair-minded biographer, William Wright, though he did not find it easy to disentangle it from her highly unreliable autobiography.[3] At nineteen she got a job at the publishers Boni and Liveright which, under Horace Liveright, was then the most enterprising firm in New York. She later claimed she had discovered William Faulkner and was responsible for the publication of his satirical novel *Mosquitoes*, set in New Orleans; but facts prove otherwise. She had an abortion and then, pregnant again, married the theatrical agent Arthur Kober, left publishing and took up reviewing. She had an affair with David Cort, subsequently foreign editor of *Life*; in the 1970s he proposed to publish her letters, some with erotic drawings in the margins, and she took legal action to prevent him – when he died, destitute, the letters were accidentally destroyed. Married to Kober, Hellman made trips to Paris, Bonn (in 1929), where she considered joining the Nazi Youth, and Hollywood. She worked briefly as a play-reader for Anne Nichols and later claimed she had discovered Vicki Baum's *Grand Hotel*; but this was not true either.[4] In Hollywood, where Kober had a staff-writing job, she read scripts for Metro-Goldwyn-Mayer at $50 a week.

Hellman's radicalism began with her involvement in the trade-union side of the motion picture industry, where writers were bitter at their treatment by the big studios. But the crucial event in her political as well as her emotional life occurred in 1930 when she met Dashiell Hammett, the mystery-writer. As she subsequently romanticized both him and their relationship, it is necessary to be clear about what kind of man he was.[5] He came from an old, genteel-poor Maryland family. He left school at thirteen, did odd jobs, fought in the First World War and was wounded, then gained his inside knowledge of police work as a Pinkerton detective. At the agency he had worked for the lawyers employed by Fatty Arbuckle, who was broken by the court case in which the film comedian was accused of raping Virginia Rappe, who died afterwards. According to what detectives told him, the woman died not of the rape but of venereal disease, and the case seems to have given him a cynical dislike for authority generally (and also a fascination for fat villains, who figure largely in his fiction). When he met Hellman he had published four novels and was in the process of becoming famous through *The Maltese Falcon*, his best.

Hammett was a very serious case of alcoholism. The success the book enjoyed was perhaps the worst thing that could have happened to him; it brought him money and credit and meant he had little need to work. He was not a natural writer and seems to have found the creative act extraordinarily daunting. He did, after many efforts, finish *The Thin Man* (1934) which brought him even more money and fame, but after that he wrote nothing at all. He would hole up in a hotel with a crate of Johnnie Walker Red Label and drink himself into sickness. Alcohol brought about moral collapse in a man who seems to have had, at times, strong principles. He had a wife, Josephine Dolan, and two children, but his payments to them were haphazard and arbitrary; sometimes he was generous, usually he just forgot them. Pathetic letters to his publisher, Alfred A. Knopf, survive: 'For the past seven months Mr Hammett has sent me only one hundred dollars and has failed to write and explain his troubles – right now I am desperate – the children need clothing and are not getting the right food – and I am unable to find work – living with my parents who are growing old and can't offer us any more help . . .' Hammett, with a script contract, was to be found in Bel Air, drinking. The studio secretary assigned to him, Mildred Lewis, had nothing to do as he would not write but lay in bed; she described how she heard prostitutes, summoned by phone from Madame Lee Francis's – they were usually black or oriental women – creeping up and down the stairs; she would turn her back so she could not see them.[6] He probably made over two million dollars from his books but often contrived to be penniless and in debt, and would sneak out of hotels in which he had run up large bills (the Pierre in New York, for instance, where he owed $1000) wearing his clothes in layers.

Alcohol also made Hammett abusive and violent, not least to women. In 1932 he was sued for assault by the actress Elise de Viane. She claimed he got drunk at his hotel and when she resisted his attempts to make love to her, beat her up. Hammett made no effort to contest the suit and $2500 damages were awarded against him. Shortly after he met Hellman, he hit her on the jaw at a party and knocked her down. Their relationship can never have been easy. In 1931 and again in 1936 he contracted gonorrhoea from prostitutes, and the second time had great difficulty in getting cured.[7] There were constant rows over his women. Indeed it is not clear whether, and if so for how long, they ever actually lived together, though both eventually divorced their respective spouses. Years later, when her lying about many other things had been thoroughly exposed, Gore Vidal asked cynically: 'Did anybody ever see them together?'

Clearly Hellman exaggerated their relationship for her own purposes of self-publicity. Yet there was substance to it. In 1938, by which time she had moved to New York, where she had a town house and a farm at Pleasantville, Hammett was reported to be lying hopelessly drunk in the Beverly Wilshire Hotel, where he had run up a bill of $8000. Hellman had him brought by air to New York; he was met by an ambulance and taken to hospital. Later he lived for some time at her house.

But he made a habit of visiting Harlem brothels, which were much to his taste. So there were more rows. In 1941, while drunk, he demanded sex with her and she refused; after that he never made or attempted to make love with her again.[8] But their relationship continued, if in tenuous form, and for the last three years of his life (he died in 1958) he led a zombie-like existence in her New York home. This was an unselfish act on her part for it meant sacrificing the work-room she adored. She would say to guests: 'Please keep it down. There's a dying man upstairs.'[9]

What is clear about their friendship is that Hellman, as a writer, owed a great deal to Hammett. In fact there is a curious, and some would say suspicious, asymmetry about their writing careers. Not long after he met Hellman, Hammett's writing dwindled to a trickle, then dried up altogether. She, by contrast, began to write with great fluency and success. It was as though the creative spirit moved from one into the other, remaining in her until his death; once he had gone, she never wrote another successful play. All this may be pure coincidence; or not. Like almost everything to do with Hellman, it is hard to get at the truth. What is certain is that Hammett had a good deal to do with her first hit *The Children's Hour*. Indeed, he may be said to have thought of it. The subject of lesbianism on the stage had been a Broadway issue ever since 1926, when the New York police had closed down *The Captive*, a translation of a play about lesbianism by Edouard Bourdet. When Hellman began to work for Herman Shumlin as a reader, and started to write plays herself, Hammett drew her attention to a book by William Roughhead, *Bad Companions*. It dealt with an appalling case in Scotland in 1810 in which a black mulatto girl, from unprovoked malice and by skilful lying, ruined the lives of two sisters who ran a school and whom she accused of lesbianism. It is a curious fact that the devastation caused by lies, particularly female lies, fascinated both Hellman and Hammett; the lies of the woman are the threads which link together the brilliant complexities of *The Maltese Falcon*. When drunk, Hammett lied like any other alcoholic; when sober, he tended to be a stickler for exactitude, even if it was highly inconvenient. While he was around, he tended to exercise some restraint on Hellman's fantasies. She, by contrast, was both obsessed by lies and perpetrated them. She often lied about the origin of *The Children's Hour*, and the circumstances surrounding the first night. Moreover she did not indicate her indebtedness to Roughhead's book and when the play appeared one critic, John Mason Brown, accused her of plagiarism – the first of many such charges she had to face.[10]

All the same, it was a brilliant play, the changes made to the original story being the key to its excitement and dynamism. How large a part, if any, Hammett played in the actual writing is now impossible to establish. One dramatic gift Hellman had in abundance (like George Bernard Shaw) was the ability to provide her most reprehensible or unsympathetic characters with plausible speeches. It is the chief source of the powerful tension her plays generate. *The Children's Hour* was bound to arouse controversy because of its subject matter. Its eloquence and verbal edge exacerbated the hostility of its opponents and enthused its defenders. In London it was refused a licence by the Lord Chamberlain, and it was banned in Chicago and many other cities (in Boston the ban remained in force for a quarter of a century). But the police made no move against it in New York, where it was an instant critical and box-office success, running for 691 performances. Moreover, the daring of the theme and the brilliance of the treatment – and, most of all, the fury it provoked among the *bien-pensants* – immediately gave Hellman a special place in the affections of progressive intellectuals, something she kept until her death. When it failed to win the Pulitzer Prize for the best play of the 1934–35 season because one of the judges, the Reverend William Lyon Phelps, objected to its topic, the New York Drama Critics Circle was formed to create a new award precisely so it could be given to her.

The play's success also brought her a contract to write screenplays in Hollywood at $2500 a week, and for the next decade she alternated between screenplays and Broadway. The achievement was mixed but impressive on the whole. Her play *Days to Come*, dealing with strikes, was a disaster. It opened on 15 December 1936 and closed six days later. On the other hand *The Little Foxes* (1939), which deals with the lust for money in the South, *circa* 1900, and was based on people she knew as a child, was another big success, running for 410 performances. Thanks to some brutal but constructive criticism by Hammett, it is the best-written and constructed of her plays and the most frequently revived. Moreover, it is worth pointing out that it succeeded against strong competition: the 1939 season included Maxwell Anderson's *Key Largo*, Moss Hart and George S. Kaufman's *The Man Who Came to Dinner*, William Saroyan's *The Time of Your Life*, Philip Barry's *The Philadelphia Story*, Cole Porter's *Leave It to Me*, *Life with Father* and some hot British imports. She followed it with another hit, *Watch on the Rhine*, two years later. Meanwhile, of her six Hollywood screenplays, half became classics. The film of *The Children's Hour* she wrote for Sam Goldwyn, who persuaded her to retitle it *These Three* and remove the lesbian element, was a big success; so was her brilliant *Dead End* (1937). She also won a notable victory over the Hays Office in writing the screenplay of *Watch on the Rhine*. The left-wing hero of the play, the German anti-Nazi, Kurt Muller, eventually contrives to kill the villain, Count Teck. The Hays Office protested that, under their rules, murderers must be punished. Hellman countered that it was right to murder Nazis or fascists and, it being wartime, she gained her point. Indeed, the movie was chosen for a benefit performance before President Roosevelt himself. That was a sign of the times. Another was her writing, for Sam Goldwyn, a straight, pro-Soviet propaganda movie, *North Star* (1942), about a delightful Soviet collective farm, one of only three CP-line films made in Hollywood (the others were *Mission to Moscow* and *Song of Russia*).[11]

The themes of Hellman's theatre and screen plays suggest a close involvement with the radical left from the mid-1930s. The notion of her being recruited to the Communist Party by Hammett, however, is probably mistaken. In the first place, she tended to be more aggressively political than he was. If anything, it was she who drew him into serious and regular political activities. Moreover, while she continued to have intermittent sexual relations with Hammett until 1941 (1945 according to her account), there were many affairs with men: with the magazine manager Ralph Ingersol, with two Broadway producers, and with the Third Secretary of the US Embassy in Moscow, John Melby, among others. Hellman was notorious for taking the sexual initiative with men. Moreover she enjoyed considerable success. As a friend put it, 'It was simple. She was sexually aggressive at a time when no women were. Others were promiscuous, God knows, but they wouldn't make the first move. Lillian never hesitated, and she cleaned up.'[12] Not always, of course. Martha Gellhorn claimed she made an unsuccessful pass at Hemingway in Paris in 1937. Arthur Miller attributed her bitter enmity towards him to the fact that he turned her down: 'Lillian came on with every man she met. I wasn't interested and she never forgave me for it.'[13] In late middle age she had to use her money to buy the companionship of good-looking young men. But her successes were common enough to give her an unusual reputation on which rumour fed. It was said, for instance, that she attended all-male poker parties at the home of Frederick Vanderbilt Field, the winner taking Hellman into a bedroom. Her memoirs, otherwise boastful, make no mention of her conquests.

A woman with such a reputation and tastes is unlikely to have been greatly trusted by the American Communist Party of the 1930s, a highly doctrinaire body. But her name was certainly useful to them. Was she ever an actual member of the Party? Her strike play, *Days to Come*, was not a Marxist-inspired work. *Watch on the Rhine* ran counter to the Party line of August 1939–early June 1941, which was to back the Hitler–Stalin Pact. On the other hand, Hellman was very active in the Hollywood Screen Writers Guild, which was dominated by the CP, especially during the bitter battles of 1936–37. It would have been logical for her to have joined the Party in 1937, as she said Hammett did. It was the peak year for CP membership, when the party was backing Roosevelt's New Deal and Popular Front policies everywhere. Whereas the early thirties' converts had tended to be serious idealists, who had read Marx and Lenin (like Edmund Wilson), and were slipping away again by 1937, the Popular Front line made the CP briefly fashionable and attracted a good many recruits from the show-biz milieu, who knew little of politics but were anxious to be in the intellectual swim.[14] Hellman fitted into this category; but the fact that she continued to support Soviet policies over many years and did not renege when the fashion faded strongly suggests that she became an actual Party *apparatchik*, though not a senior one. She herself always denied being a Party member. Against this, Martin Berkeley testified that in June 1937, Hellman, along with Hammett, Dorothy Parker, Donald Ogden Stewart and Alan Campbell, attended a meeting at his house with the precise purpose of forming a Hollywood branch of the CP; later, Hellman took the Fifth Amendment rather than answer questions about this meeting. Her interrogation by the House Un-American Activities Committee strongly suggests she was a member, 1937–49. Her FBI file, nearly 1000 pages long, though full of the usual rubbish and very repetitious, contains much solid fact. Apart from Berkeley, Louis Budenz, former managing editor of the *Daily Worker*, likewise stated she was a Party member, as did two other informants; others testified to her taking an active role at Party meetings.[15]

What seems most probable is that, for a variety of reasons, including her sexual promiscuity, the CP found it more convenient to have her as a secret rather than an open member, and to keep her under control as a fellow-traveller, though allowing her some latitude. This is the only explanation which fits all her acts and attitudes during the period. Certainly she did everything in her power, quite apart from her plays and scripts, to assist the CP's penetration of American intellectual life and to forward the aims of Soviet policy. She took part in key CP front groups. She attended the tenth National Convention of the CP in New York, June 1938. She visited Russia in October 1937, under the tuition of the pro-Stalin *New York Times* correspondent, Walter Duranty. The trials were then at their height. On her return she said she knew nothing about them. As for the attacks on the trials by Western libertarians, she said she was unable to distinguish 'true charges from the wild hatred' and 'fact from invention when it is mixed with blind bitterness about a place and people'. But the next year her name was among the signatories (together with Malcolm Cowley, Nelson Algren, Irwin Shaw and Richard Wright) to an advertisement in *New Masses* which approved the trials. Under the auspices of the notorious Otto Katz, she paid two visits to Spain in 1937 and contributed, along with other writers, $500 to the pro-CP propaganda film with which Hemingway was also connected. But her account of what she did in Spain is plainly full of lies – it was refuted in detail by Martha Gellhorn – and it is difficult now to establish exactly what she did there.

Like most intellectuals, Hellman engaged in rancorous quarrels with other writers. These envenomed and complicated her political positions. Her anxiety to support the Soviet line in Spain drew her into a row with William Carney, the *New York Times* correspondent there who persisted in publishing facts which did not fit in with Moscow's version. She accused him of covering the war from the safety and comfort of the *Côte d'Azur*. Again, she backed the 1939 Soviet invasion of Finland, stating: 'I don't believe in that fine, lovable little Republic of Finland that everyone gets so weepy about. I've been there and it looks like a pro-Nazi little republic to me.' That brought her into conflict with Tallulah Bankhead, who had starred in the stage version of *The Little Foxes* and was already an enemy for a number of reasons (mainly sexual jealousy). Bankhead had done a benefit show for Finnish relief agencies. Hellman accused her of refusing an invitation to do a similar benefit for Spain. Bankhead retorted that the accusation was 'a brazen invention'. There is no evidence, as it happens, that Hellman ever went to Finland, and her biographer thinks it improbable.[16] But she continued to attack Bankhead in various publications, even after the actress was dead. She wrote of Bankhead's drunken family, use of drugs and described her making passes at black waiters; she told a repellent story (in her autobiographical *Pentimento*) of Bankhead insisting on showing a visitor her husband's gigantic erect penis.

The quarrel between Hellman and Bankhead was really about who was on the side of 'the workers'. The truth is that neither knew anything about the working class beyond occasionally taking a lover from its ranks. Hellman once did an investigation in Philadelphia for the New York liberal evening paper, *PM*, which involved talking to one cab-driver, two men in a shop and two black children; from which she concluded that America was a police state. But she had no friends among the workers with the exception of a longshoreman, Randall Smith, whom she made friends with at Martha's Vineyard after the war. He had served in the Lincoln Brigade in Spain and was certainly not typical of the American proletariat. Moreover, he grew to dislike Hellman, Hammett and their wealthy radical friends. 'As a former Communist,' he said, 'I used to resent their attitude – so lofty and intellectual. I doubt if either of them ever went to a meeting or did any work. They were like officers, I had been an enlisted man.' He particularly disliked Hammett's habit, in company, of showing off his power over women by 'taking his cane and lifting the skirt of his current girlfriend'.[17] The life Hellman led was certainly remote from what she liked to call 'the struggle'. At her house on East 82nd Street and on her 130-acre Westchester farm she lived like the New York rich, with a housekeeper, butler, secretary and personal maid. She went to the most fashionable psychiatrist, Gregory Zilboorg, who charged $100 an hour. Her stage and screen plays brought her deference as well as wealth. In September 1944 she went to Moscow at the invitation of the Soviet government and stayed in Ambassador Harriman's house, where she had her affair with the diplomat Melby; but she kept rooms at the Metropole and National Hotels as well as at the Embassy. This trip produced the usual crop of lies. She said she was in Russia five months; Melby, a more reliable witness, said three. Hellman published two quite different accounts of her Russian experiences, in *Collier's* magazine in 1945, and in her first autobiographical volume, *An Unfinished Woman*, in 1969. The magazine article made no mention of seeing Stalin. The autobiography stated that, although she had not asked to see Stalin, she was informed he had granted her an interview. She politely declined, on the grounds that she had nothing of importance to say and did not wish to take up his valuable time. This preposterous tale is contradicted by what Hellman said at the time of her return, when she told a New York press conference that she had asked to see Stalin but had been told he was 'too busy with the Poles'.[18]

In the 1930s and early 1940s, Hellman was a left-wing success heroine, a feted celebrity. In the late 1940s her life went into a new phase, subsequently glorified in radical legend as a time of martyrdom. For a time her political activities continued. Along with other members of the far left, she backed Wallace for president in 1948. In 1949 she was among the organizers of the Soviet-backed Cultural and Scientific Conference for World Peace, held at the Waldorf. But her troubles were beginning. The post-war plays did less well than their predecessors. A sequel to *The Little Foxes*, written about the same family and called *Another Part of the Forest*, opened in November 1947 and ran for 191 performances, but got some poor notices. It was notable for the appearance of her wayward father Max, who sat in the stalls noisily counting new dollar bills throughout the first act, then announced in the interval: 'My daughter wrote this play. It gets better.' Six months later, on the advice of her psychiatrist, she had him committed for senile dementia. There were difficulties over her next play, *The Autumn Garden*. Hellman later said that, after Hammett criticized the first draft, she tore it up; but an entire manuscript marked 'First Draft' survives in the University of Texas library. When it opened in March 1951 it ran for only 101 performances.

Meanwhile the House Un-American Activities Committee had been combing through the movie industry. The so-called Hollywood Ten, who refused to answer the Committee's questions about political activities, were cited for contempt. In November 1947 the studio producers agreed to sack any writers who fell into this category. The magazine of the Screen Writers' Guild attacked the decision in an editorial entitled 'The Judas Goats' written by Hellman, which contained the astonishing statement: 'There has never been a single line or word of communism in any American picture at any time.' The mills of the law continued to grind slowly. Hammett contributed to the bail fund for the screenwriters indicted for contempt. Three of them jumped bail and disappeared. The FBI believed Hammett knew where they were and a team arrived at Hellman's farm to search the place. Hammett himself was brought to court on 9 July 1951 and asked to help find the missing men by giving the names of other contributors to the fund. Instead of saying he did not know them (which was true) he stubbornly refused to answer at all, and was jailed. Hellman claimed she had to sell her farm to pay his legal expenses of $67,000.

She herself had been blacklisted in Hollywood in 1948, and four years later, on 21 February 1952, she got a summons to appear before the dreaded Committee. It was at this point she snatched victory from the jaws of defeat. She had always been good at public relations; it was a skill she shared with many of her intellectual contemporaries such as Brecht and Sartre. Brecht, as we have seen, managed to turn his appearance before the Committee into a propaganda score for himself. Hellman's achievement was even more remarkable and laid the foundation for her subsequent fame as the martyr-queen of the radicals. As with Brecht, she was helped by the stupidity of Committee members. Before appearing, she took very careful legal advice from her counsel, Joseph Rauh. There is no doubt she understood the legal position, which was complex. Her instructions to Rauh were that she would not name names; on the other hand, she did not want to go to jail under any circumstances. Thirdly, she did not wish to plead the Fifth Amendment if by doing so she appeared to be protecting herself, as this would be seen as an admission of guilt (the phrase at the time was 'a Fifth Amendment communist'). She would, however, be prepared to plead the Fifth if, by doing so, she appeared to be protecting only others. Therein, however, lay Rauh's difficulty, for the Fifth Amendment protects the witness only against self-incrimination. How could Hellman be saved from jail by the Fifth while at the same time presented as an innocent saving others? He said later there was never any question of her going to jail. 'It was like an algebra problem,' he said. 'But then I began to see it as primarily a public relations problem. I knew that if the headline in the *New York Times* the next day read "Hellman Refuses to Name Names", I had won. If it said "Hellman Pleads the Fifth", I had lost.'

Hellman solved the problem for him by writing a cunning and mendacious letter to John S. Wood, the chairman of the Committee, on 19 May 1952. She argued that she had been advised she could not plead the Fifth about herself, then refuse to answer about others. Then came the big lie: 'I do not like subversion and disloyalty in any form and if I had ever seen any I would have considered it my duty to have reported it to the proper authorities.' There followed a brilliant debating trick, which upended the true legal position, making it appear that Hellman would be happy to go to jail if only her own freedom was at stake but was taking the Fifth to protect other, quite blameless people: 'But to hurt innocent people whom I knew many years ago in order to save myself is, to me, inhuman, indecent and dishonourable. I cannot and will not cut my conscience to fit this year's fashions, even though I long ago came to the conclusion that I was not a political person and could have no comfortable place in any political group.' To the fury of the chairman, who seems to have understood the trick Hellman was playing, a member of the Committee who had failed to grasp the legal point moved that the letter be read into the record, and this enabled a delighted Rauh to issue immediate copies to the press. The next day he got exactly the headline he wanted. In her autobiographical volume about these events, *Scoundrel Time*, Hellman subsequently embroidered the story, inventing various details, including a man shouting from the gallery, 'Thank God somebody finally had the guts to do it.' But she need not have bothered. Her letter was the only 'fact' that emerged from the hearing into the history books. It went into the anthologies too, as a moving plea for freedom of conscience by a selfless and heroic woman.[19]

This was the core of the later Hellman legend. But a collateral myth was that she had been ruined financially by a combination of the blacklist and by the huge legal bills with which she and Hammett were faced as a result of the witch-hunt. But there is no evidence she was ruined at all. *The Children's Hour* was revived in 1952 and brought in a handsome income. She kept her New York townhouse until, in old age, she moved into a more convenient apartment. It is true she sold her farm, but in 1956 she bought a fine property on Martha's Vineyard, which had by then become a smarter place for wealthy intellectuals to relax in than the fringes of New York. Hammett's financial troubles sprang from many causes. When he finally stopped drinking he did not begin to work but merely sat glued to the television set. He had also been recklessly generous. There was no danger of that in Hellman's case but she shared with Hammett another habit – a failure to pay income tax. As the cases of Sartre and Edmund Wilson suggest, there is a common propensity among radical intellectuals to demand ambitious government programmes while feeling no responsibility to contribute to them.

Hammett's failure to pay income tax went right back to the 1930s and did not come to light merely as a result of his going to jail. Indeed his habit was noted by the FBI before the war. But of course his sentence spurred the Internal Revenue Service, along with other creditors – of whom there were many – to press claims. On 28 February 1957 a federal court entered a default judgment against him for $104,795, and this was just for the years 1950–54. The authorities were not particularly harsh, a court deputy reporting that no money was to be had: 'In my opinion after my investigation, I was speaking to a broken man.' By the time of his death the sum owing, including interest, had risen to $163,286.[20] Hellman's debts to the tax-man were even greater: in 1952 they were estimated at between $175,000 and $190,000 – enormous sums in those days. She later claimed she was so broke she had to take a job in Macy's department store; but there was no truth in this either.

Hellman lay low in the 1950s, a difficult decade for radicals. But by 1960 she was on the upsurge again. Her play *Toys in the Attic*, based on an idea by Hammett and using her childhood memories of the boarding-house, opened in New York on 25 February 1960 with a superb cast. It ran for 556 performances, won the Circle award again and made Hellman a lot of money. But it was her last serious play, and the death of Hammett the next year suggested to many that, without him, she could not write another. Be that as it may, she had a second career to pursue. Radicalism was reviving throughout the 1960s, and by the end of the decade it was almost as strong as in her thirties' heyday. A trip to Russia produced another batch of fibs and the assertion that Khrushchev's Secret Session speech, confirming Stalin's crimes, had been to stab his old patron in the back.[21] Hellman, sniffing the wind of opinion in America, decided it was time to write her memoirs.

They became one of the great publishing successes of the century and brought Hellman more fame, prestige and intellectual authority even than her plays. They were, indeed, a canonization of herself while she was still alive, an apotheosis by the printed word and the public-relations machine. *An Unfinished Woman*, published in June 1969, was a best-seller and won the National Book Award for Arts and Letters. *Pentimento*, out in 1973, was on the best-seller list for four months. The third, *Scoun-drel Time* (1976), was on the best-seller list for no less than twenty-three weeks. She was offered half a million dollars for the film rights of her life. She found herself with a completely new reputation as a master of prose style and was asked to take writing seminars at Berkeley and MIT. The awards and honours came rolling in. New York University made her Woman of the Year, Brandeis gave her its Theatre Arts Medal, Yeshiva its Achievement Award. She got the MacDowell Medal for Con-tributions to Literature and honorary PhDs from Yale, Columbia and many other universities. By 1977 she was back at the top of Hollywood society, presenting at the Academy Awards. The same year a section of her memoirs appeared as the much-praised movie *Julia*, which won many awards in its turn. On the East Coast, she was the queen of radical chic and the most important single power-broker among the progressive intelligentsia and the society people who seethed round them. Indeed in the New York of the 1970s she dispensed the same kind of power which Sartre had wielded in Paris, 1945–55. She promoted and selected key committees. She compiled her own blacklists and had them enforced by scores of servile intellectual flunkies. The big names of New York radicalism scurried to do her bidding. Part of her power sprang from the fear she inspired. She knew how to make herself unpleasant, in public or in private. She would spit in a man's face, scream abuse, smash him on the head with her handbag. At Martha's Vineyard the fury with which she assailed those who crossed her garden to the beach was awe-some. She was now very rich and employed posses of lawyers to attack the slightest opposition or infringement of her rights. Sycophants who thought they were merely worshipping at her shrine might get a nasty shock. When Eric Bentley, Brecht's friend, put on an off-Broadway anti-witch-hunt play, *Are You Now or Have You Ever Been?*, which involved actresses reading from her letters, Hellman demanded royalties and said she would close the show down unless the owners complied. She was a fast woman with a writ. Most people preferred to settle. She is said to have got a million-dollar buy-out to avoid a lawsuit over a 1981 revival of *The Little Foxes*. Supposedly powerful institutions jumped to do her bidding, often before she had issued her commands. Thus Little, Brown of Boston cancelled a book by Diana Trilling when she refused to remove a passage critical of Hellman. Mrs Trilling, who was merely trying to defend her late husband Lionel from one of Hellman's vicious attacks in *Scoundrel Time*, said of her: 'Lillian was the most powerful woman I've ever known, maybe the most powerful *person* I've ever known.'

The basis of Hellman's authority was the extraordinary myth she had created about herself in her autobiographies. In a way it was comparable to the self-canonization of Rousseau in his *Confessions*. As has been shown repeatedly, the memoirs of leading intellectuals – Sartre, de Beauvoir, Russell, Hemingway, Gollancz are obvious examples – are quite unreliable. But the most dangerous of these intellectual self-glorifications are those which disarm the reader by what appears to be shocking frankness and admission of guilt. Thus Tolstoy's diaries, honest though they appear to be, in fact hide far more than they reveal. Rousseau's *Confessions*, as Diderot and others who really knew him perceived at the time, are an elaborate exercise in deception, a veneer of candour concealing a bottomless morass of mendacity. Hellman's memoirs conform to this cunning pattern. She often admits to vagueness, confusion and lapses of memory, giving readers the impression that she is engaged in making a constant effort to sift the exact truth from the shadowy sands of the past. Hence when the books first appeared many reviewers, including some of the most perceptive, praised her truthfulness.

But amid the chorus of praise and the din of flattery by Hellman courtiers during the 1970s, dissenting voices were raised by those who had personally experienced her lies. In particular, when *Scoundrel Time* appeared, her account was challenged by such weighty figures as Nathan Glazer in *Commentary*, Sidney Hook in *Encounter*, Alfred Kazin in *Esquire* and Irving Howe in *Dissent*.[22] But these writers concentrated on exposing her shocking distortions and omissions. They were largely unaware of her inventions. Their attacks were part of the continuing battle between the democratic liberals and the hard-line Stalinists; as such they aroused comparatively little attention and did Hellman no serious damage.

But then Hellman made a catastrophic error of judgment. It was an uncharacteristic one, in an area where she was usually very much at home, public relations. She had long had a feud with Mary McCarthy. This dated from the Stalinist–Trotskyist split among the American left of the 1930s. It had been kept alive by a row at a Sarah Lawrence College seminar in 1948, when McCarthy had detected Hellman lying about John Dos Passos and Spain, and by further exchanges over the 1949 Waldorf Conference. McCarthy had since repeatedly accused Hellman of lying on a large scale, but had apparently done her no harm. Then in January 1980, on *The Dick Cavett Show*, McCarthy repeated her most comprehensive charge about Hellman's lies: 'I once said in an interview that every word she writes is a lie, including "and" and "the".' Hellman was watching. Her fury and her taste for litigation overcame her prudence. She began a suit for $2,225,000 in damages, and pursued it with great persistence and vigour.

What followed was a classic proof of the contention that to sue for defamation merely draws attention to the charge. Earlier accusations of lying had left Hellman unscathed. Now the public pricked up its ears. It scented a hunt – possibly a kill. Litigation was bad public relations anyway, for writers who sue other writers are never liked. It was known that Hellman was rich, whereas McCarthy would have to sell her house to meet the cost of the suit. Friends on both sides piled in with money and advice, and the case, with its preliminary hearings, became a major story, thus drawing further attention to the question of Hellman's veracity. More seriously, the case promoted a new intellectual game: detecting Hellman's inventions. McCarthy, who rapidly had to fork out $25,000 in fees, had no alternative but to lead the pack, as she faced financial ruin. As Hellman's biographer, William Wright, put it: 'By suing McCarthy, Hellman forced one of the country's sharpest and most energetic minds to pore through the entire Hellman *oeuvre* in search of lies.'[23] Others were happy to join in. In the Spring 1981 issue of the *Paris Review*, Martha Gellhorn listed and documented eight major Hellman lies about Spain. And Stephen Spender drew McCarthy's attention to the curious case of Muriel Gardiner.

Spender had had a brief affair with Muriel, a wealthy American girl who had once been married to an Englishman, Julian Gardiner. She had gone to Vienna to study psychiatry and had there become involved in the anti-Nazi underground, under the alias of 'Mary', smuggling out messages and people. She had fallen in love with another anti-Nazi, an Austrian socialist called Joe Buttinger, and married him. After war broke out in 1939 they left Europe and settled in New Jersey. Hellman never met Muriel, but she heard all about her, her husband and their activities underground from her New York lawyer. The idea of a rich American heiress married to a Central European socialist resistance leader is the starting point for *Watch on the Rhine*, which Hellman started to write five months after the Buttingers came to New Jersey, though the actual plot has little to do with them. When Hellman came to write *Pentimento*, she again made use of Muriel's experiences, calling her 'Julia'; but this time she brought herself into the story, in a heroic and flattering light, as Julia's friend. Moreover, she presented it all as autobiographical fact.

When the book appeared, no one challenged Hellman's account. But Muriel read it and wrote a perfectly friendly letter to Hellman, pointing out the similarities. She got no reply and Hellman later denied ever having received such a letter. Since she had never actually met Muriel, her contention had to be that there were two American underground agents, 'Julia' as well as 'Mary'. Who, then, was Julia? She was dead, said Hellman. What was her real name, then? Hellman could not reveal it: her mother was still alive and would be persecuted as a 'premature anti-Nazi' by German reactionaries. When the controversy over Hellman's lies got going, Muriel gradually abandoned belief in Hellman's good faith. In 1983 she got Yale University to publish her own memoirs, *Code Name Mary*. When it was published reporters from the *New York Times* and *Time* magazine began to ask awkward questions about *Pentimento* and the movie *Julia*. The Director of the Austrian Resistance Archives, Dr Herbert Steiner, confirmed there was only one 'Mary'. Either Julia was Mary or she was an invention and in either event Hellman was exposed as a liar on a massive scale. McCarthy, in touch with Muriel, filed much of this material for the preliminary proceedings of the libel action. Then, in June 1984, *Commentary* published a remarkable article by Samuel McCracken of Boston University, 'Julia and Other Fictions by Lillian Hellman'. He had done a great deal of police-type research into train times, boat-sailing schedules, theatre programmes and other checkable facts which made up the detail of Hellman's account of Julia in *Pentimento*. Nobody with an open mind, reading this article, could be left with any doubt that the Julia episode was a piece of fiction, based upon the true experiences of a woman Hellman had never met.

McCracken's investigation also lifted the cover on another murky corner of Hellman's life: her pursuit of money. She had always been avaricious, and the propensity increased with age. Most of her lawsuits had had a financial object. After Hammett died, she formed a liaison with a rich Philadelphian, Arthur Cowan. He advised her on investments. He also put her up to a dodge to acquire Hammett's copyrights, held by the US government in lieu of his tax debts. As very little was coming in royalties, Cowan persuaded the government to put the rights up to auction, setting a minimum bid of $5000. Hellman persuaded Hammett's daughters to agree to the sale, telling them, falsely, that otherwise they would themselves be liable for Hammett's debts. Cowan and Hellman were the only bidders, at $2500 each, and got the rights. Hellman then began to work this literary property vigorously and it was soon bringing in hundreds of thousands of dollars – $250,000 for one television adaptation of a Hammett story alone. When Cowan died in turn, he left no will, according to Hellman's account in *Pentimento*. McCracken established that he did leave a will, and Hellman got nothing, suggesting they had a quarrel before he died. But Hellman evidently persuaded Cowan's sister that it had been his intention she should get his share of the Hammett rights, as the sister wrote a letter relinquishing them to her. Thus Hellman enjoyed the increasingly valuable Hammett copyrights *in toto* until her death, and it was only then that she left something, in her will, to the impoverished Hammett daughters.[24]

Hellman died on 3 July 1984, a month after the publication of McCracken's article. By that time her fantasy world, on which she had built her reputation, was crashing down about her ears. From being the aggressive queen of the radical left, she was everywhere on the defensive. However, intellectual heroes, or heroines, are not disposed of so easily. Just as south Italian peasants continue to make offerings and present petitions to their favourite saints long after their very existence has been exposed as an invention, so the lovers of progress too cling to their idols, feet of clay notwithstanding. Though Rousseau's monstrous behaviour was well known even in his own lifetime, the reason-worshippers flocked to his shrine and institutionalized the myth of his goodness. No revelations about Marx's private conduct or his public dishonesty, however well-documented, seem to have disturbed the faith of his followers in his righteousness. Sartre's long decline and the unrelieved fatuity of his later views did not prevent 50,000 Parisian cognoscenti turning out for his obsequies. Hellman's funeral, in Martha's Vineyard, was also well attended. Among the notables who paid homage were Norman Mailer, James Reston, Katherine Graham, Warren Beatty, Jules Feiffer, William Styron, John Hersey and Carl Bernstein. She left nearly four million dollars, most of which went to two trusts. One was the Dashiell Hammett Fund, enjoined to make grants, 'guided by the political, social and economic beliefs, which of course were radical, of the late Dashiell Hammett, who was a believer in the doctrines of Karl Marx'. Despite all the revelations and exposures, the nailing of so many falsehoods, the Lillian Hellman myth industry continued serenely on its course. In January 1986, eighteen months after her death, the hagiographical play *Lillian* opened in New York, and was well attended. As the 1980s wane, votive candles to the goddess of reason are still being lit, secular Masses said. Will Lillian Hellman, like her hero Stalin, ever be finally buried in decent obscurity, or will she – fables and all – remain a fighting symbol of progressive thought? We shall see. But the experience of the last two hundred years suggests that there is plenty of life, and lies, in the old girl yet.

13

The Flight of Reason

*A*T the end of the Second World War, there was a significant change in the predominant aim of secular intellectuals, a shift of emphasis from utopianism to hedonism. The shift began slowly at first, then gathered speed. Its origins can best be studied by looking at the views and relationships of three English writers, all of them born in the year 1903: George Orwell (1903–50), Evelyn Waugh (1903–66) and Cyril Connolly (1903–74). They might be described as the Old Intellectual, the Anti-Intellectual and the New Intellectual. Waugh began a cautious friendship with Orwell only when the latter was already stricken by fatal illness. Waugh and Connolly sparred together all their adult lives. Orwell and Connolly had known each other since their school days. Each writer kept a wary, sceptical and sometimes envious eye on the other two. Indeed Connolly, who felt himself the failure of the three, wrote a self-pitying couplet into a copy of Virgil he gave to the drama critic T. C. Worsley:

> At Eton with Orwell, at Oxford with Waugh
> He was nobody after and nobody before.[1]

But this was far from true. In some ways he was to prove the most influential of the three.

Orwell, whom we shall look at first, was an almost classic case of the Old Intellectual in the sense that for him a political commitment to a utopian, socialist future was plainly a substitute for a religious idealism in which he could not believe. God could not exist for him. He put his faith in man but, looking at the object of his devotion too closely,

306

lost it. Orwell, born Eric Blair, came from a family of minor empire-builders, and looked like one. He was tall, spare, with a short-back-and-sides haircut and a severely trimmed moustache. His paternal grandfather was in the Indian Army; his maternal grandfather was a teak merchant in Burma. His father worked in the Opium Department of the Indian Civil Service. He and Connolly attended the same fashionable private school and later both went to Eton. He received this expensive education because, like Connolly, he was a clever boy who was expected to get scholarships and bring credit to the school. In fact both boys later wrote entertaining but devastating accounts of the school which did it damage.[2] Orwell's essay, 'Such, Such Were the Joys', is uncharacteristically exaggerated and even mendacious. It was the belief of his Eton tutor, A.S.F. Gow, who knew the private school well, that Orwell had been corrupted by Connolly into producing this unfair indictment.[3] If so, it was the only occasion on which Connolly persuaded Orwell to embark on an immoral course, particularly one involving lying. Orwell, as Victor Gollancz observed through clenched teeth, was painfully truthful.

After Orwell left Eton he joined the Indian police, serving five years, 1922–27. As such, he saw the seamier side of imperialism, the hangings and floggings, and found he could not stomach it. In fact his two brilliant essays, 'A Hanging' and 'Shooting an Elephant', perhaps did more to undermine the empire spirit in Britain than any other writings.[4] He returned to England on home leave, resigned from the service, and determined to be a writer. He chose the name 'George Orwell' after considering various alternatives, including P.S. Burton, Kenneth Miles and H. Lewis Allways.[5] Orwell was an intellectual in the sense that he believed, at any rate when young, that the world could be reshaped by the power of intellect. He thus thought in terms of ideas and concepts. But his nature, and perhaps his police training, gave him a passionate interest in people. His policeman's instinct certainly told him that things were not what they seemed, and that only investigation and close scrutiny would yield the truth.

Hence, unlike most intellectuals, Orwell embarked on his career as a socialist idealist by examining working-class life at close quarters. In this respect he had something in common with Edmund Wilson, who shared his passion for exact truth. But he was far more persistent than Wilson in seeking knowledge of 'the workers' and for many years this quest for experience remained the central theme of his life. He first took rooms in Notting Hill, at that time a London slum. Then in 1929 he worked in Paris as a dishwasher and kitchen porter. But he became ill with pneumonia – he suffered from chronic weakness of the lungs, which killed him at the age of forty-seven – and the venture ended in a spell in a Paris charity hospital, an episode harrowingly described in *Down and Out in Paris and London* (1933). He then lived with tramps and hop-pickers, boarded with a working-class family in the Lancashire industrial town of Wigan, and kept a village store. All these activities had one aim: 'I felt I had got to escape not merely from imperialism but from every form of man's dominion over man. I wanted to submerge myself, to get right down among the oppressed, to be one of them and on their side against the tyrants.'[6]

Hence, when the Spanish Civil War broke out in 1936, Orwell not only gave moral support to the Republic, as did over 90 per cent of Western intellectuals, but – unlike virtually all of them – actually fought for it. Moreover, as luck would have it, he fought for it in what was itself to become the most oppressed and martyred section of the Republican Army, the anarchist (POUM) militia. This experience was critical to the rest of his life. Characteristically, Orwell wanted to go to Spain first and see the situation for himself, before deciding what he would do about it. But getting to Spain was difficult and entry was in effect controlled by the Communist Party. Orwell went first to Victor Gollancz, who put him on to John Strachey; and Strachey in turn referred him to Harry Pollitt, the Communist Party boss. But Pollitt would not provide Orwell with a letter of recommendation unless he first agreed to join the Communist Party-controlled International Brigade. This Orwell declined to do, not because he had anything against the Brigade – in fact he tried to join it in Spain the following year – but because it would have closed his options before he had examined the facts. So he turned to the left-wing faction known as the Independent Labour Party. They got him to Barcelona and put him in touch with the anarchists, and it was thus he joined the POUM militia. He was impressed by Barcelona, 'a town where the working class was in the saddle', and still more by the militia existence, in which 'many of the normal motives of civilized life – snobbery, money-grubbing, fear of the boss, etc. – had simply ceased to exist. The ordinary class division of society had disappeared to an extent that is almost unthinkable in the money-tainted air of England.'[7] He found the fighting, in which he was wounded, in some ways an uplifting experience, and wrote a letter of gentle reproach to Connolly, who had inspected the war, like most intellectuals, simply as a 'concerned' tourist: 'A pity you didn't come up to our position and see me when you were in Aragon. I would have enjoyed giving you tea in a dugout.'[8] Orwell described the militia on active service as 'a community where hope was more normal than apathy or cynicism, where the word "comrade" stood for comradeship and not, as in most countries, for humbug'. There, 'no one was on the make'; there was 'a shortage of everything but no privilege and no boot-licking'. He thought it 'a crude foretaste of what the opening stages of socialism might be like'. In conclusion, he wrote home, 'I have seen wonderful things and at last really believe in socialism, which I never did before.'[9]

There followed, however, the devastating experience of the Communist Party's purge of the anarchists on Stalin's orders. Thousands of Orwell's comrades were simply murdered or thrown into prison, tortured and executed. He himself was lucky to escape with his life. Almost as illuminating, to him, was the difficulty he found, on his return to England, in getting his account of these terrible events published. Neither Victor Gollancz, in the Left Book Club, nor Kingsley Martin, in the *New Statesman* – the two principal institutions whereby progressive opinion in Britain was kept informed – would allow him to tell the truth. He was forced to turn elsewhere. Orwell had always put experience before theory, and these events proved how right he had been. Theory taught that the left, when exercising power, would behave justly and respect truth. Experience showed him that the left was capable of a degree of injustice and cruelty of a kind hitherto almost unknown, rivalled only by the monstrous crimes of the German Nazis, and that it would eagerly suppress truth in the cause of the higher truth it upheld. Experience, confirmed by what happened in the Second World War, where all values and loyalties became confused, also taught him that, in the event, human beings mattered more than abstract ideas; it was something he had always felt in his bones. Orwell never wholly abandoned his belief that a better society could be created by the force of ideas, and in this sense he remained an intellectual. But the axis of his attack shifted from existing, traditional and capitalist society to the fraudulent utopias with which intellectuals like Lenin had sought to replace it. His two greatest books, *Animal Farm* (1945) and *Nineteen Eighty-Four* (1949), were essentially critiques of realized abstractions, of the totalitarian control over mind and body which an embodied utopia demanded, and (as he put it) 'of the perversions to which a centralized economy is liable'.[10]

Such a shift of emphasis necessarily led Orwell to take a highly critical view of intellectuals as such. This accorded well with his temperament, which might be described as regimental rather than bohemian. His work is scattered with such asides as (of Ezra Pound) 'one has the right to expect ordinary decency even of a poet'. Indeed it was an axiom of his that the poor, the 'ordinary people', had a stronger sense of what he called 'common decency', a greater attachment to simple virtues like honesty, loyalty and truthfulness, than the highly educated. When he died in 1950, his ultimate political destination was unclear and he still vaguely ranked as a left-wing intellectual. As his reputation rose, left and right fought, and in fact still fight, for the title to his allegiance. But in the forty years since his death, he has been increasingly used as a stick to beat the intellectual concepts of the left. Intellectuals who feel most solidarity with their class have long recognized him as an enemy. Thus, in her essay on Orwell, Mary McCarthy, sometimes confused in her political ideas but nothing if not caste-conscious, was severe: Orwell was 'conservative by temperament, as opposed as a retired colonel or a working man to extremes of conduct, dress or thought'. He was 'an incipient philistine. Indeed he was a philistine.' His socialism was 'an unexamined idea off the top of his head, sheer rant'. His pursuit of Stalinists was occasionally 'a mere product of personal dislike'. His 'political failure... was one of thought'. Had he lived he must surely have moved to the right, so 'it was a blessing for him probably that he died.'[11] (This last thought – better dead than anti-red – is a striking example of the priorities of archetype intellectuals.)

One reason why professional intellectuals moved away from Orwell was his growing conviction that, while it was right to continue to look for political solutions, 'just as a doctor must try to save the life of a patient who is probably going to die', we had to start 'by recognizing that political behaviour is largely non-rational', and therefore not as a rule susceptible to the kind of solutions intellectuals habitually sought to impose.[12] But while the intellectuals were becoming suspicious of Orwell, those of the contrary persuasion – the men of letters, if you like – tended to warm to him. Evelyn Waugh, for instance, was never a man to underrate the importance of the irrational in life. He began to correspond with Orwell and visited him in hospital; had Orwell lived, their friendship might well have blossomed. They first came together over a common desire that P. G. Wodehouse, a writer they admired, should not be persecuted for his foolish (but, compared with Ezra Pound's, quite innocuous) wartime broadcasts. This was a case where both men insisted that an individual person must take precedence over the abstract concept of ideological justice. Waugh quickly saw in Orwell a potential defector from the ranks of the intelligentsia. He noted in his diary on 31 August 1945: 'I dined with my communist cousin, Claud [Cockburn], who warned me against Trotskyist literature, so that I read and greatly enjoyed Orwell's *Animal Farm*.'[13] He likewise recognized the power of *Nineteen Eighty-Four*, though he found it implausible that the religious spirit would not have survived to take part in the resistance to the tyranny Orwell portrayed. In his last letter to Orwell, 17 July 1949, Waugh made this point, adding: 'You see how much your book excited me so that I risk preaching a sermon.'[14]

What Orwell came reluctantly and belatedly to accept – the failure of utopianism on account of the fundamental irrationality of human behaviour – Waugh had vociferously upheld for most of his adult life. Indeed no great writer, not even Kipling, ever gave a clearer statement of the anti-intellectual position. Waugh, like Orwell, believed in personal experience, in seeing for himself, as opposed to theoretical imagining. It is worth noting that while he did not seek deliberately, like Orwell, to live with the oppressed, he was an inveterate traveller, often in remote and difficult regions; he had seen a great deal of men and events and had a practical as well as a bookish knowledge of the world. When writing on serious matters he also had an unusual regard for truth. His one avowedly political work, a description of the Mexican revolutionary regime called *Robbery Under Law* (1939), is preceded by a warning to readers. He made it plain exactly what his credentials for writing on the subject were, and how inadequate they seemed to him. He drew their attention to books written by those with views different to his own, and warned them not to make up their minds about what was going on in Mexico simply on the basis of his account. He stressed that he deplored 'committed' literature. Many readers, he said, 'bored with the privilege of a free press', had decided 'to impose on themselves a voluntary censorship' by forming book clubs – he had in mind Gollancz's Left Book Club – so that 'they may be perfectly confident that whatever they read will be written with the intention of confirming their existing opinions'. Hence in fairness to *his* readers, Waugh thought it proper to summarize his own beliefs.

He was, he said, a conservative, and everything he had seen in Mexico strengthened his convictions. Man was, by his nature, 'an exile and will never be self-sufficient or complete on this earth'. He thought man's chances of happiness were 'not much affected by the political and economic conditions in which he lives', and that sudden changes in man's state usually made matters worse, 'and are advocated by the wrong people for the wrong reasons'. He believed in government: 'men cannot live together without rules' but these 'should be kept to the bare minimum of safety'. 'No form of government ordained from God' was 'better than another' and 'the anarchic elements in society' were so strong it was a 'whole-time job to keep the peace'. Inequalities in wealth and position were 'inevitable' and it was therefore 'meaningless to discuss the advantages of their elimination'. In fact men 'naturally arrange themselves in a system of classes', such being 'necessary for any cooperative work'. War and conquest were likewise inevitable. Art was also a natural function of man, and 'it so happens' that most great art had been produced 'under systems of political tyranny' though 'I do not think it has a connection with any particular system'. Finally, Waugh said he was a patriot in the sense that, while he did not think British prosperity was necessarily inimical to anyone else, if on occasions it was, then 'I want Britain to prosper and not her rivals.'

Thus Waugh described society as it was and must be, and his response to it. He did indeed have a personal, idealized vision; but, being an anti-intellectual, he freely conceded it was unrealizable. His ideal society, as described in an introduction to a book published in 1962, had four orders. At the top was 'the fount of honour and justice'. Immediately below were 'men and women who hold offices from above and are the custodians of tradition, morality and grace'. They had to be 'ready for sacrifice' but were protected from 'the infections of corruption and ambition by hereditary possession'; they were 'the nourishers of the arts, the censors of manners'. Below them were 'the classes of industry and scholarship', trained from childhood 'in habits of probity'. At the bottom were the manual labourers, 'proud of their skills and bound to those above them by common allegiance to the monarch'. Waugh concluded by asserting that the ideal society was self-perpetuating: 'In general a man is best fitted to the tasks he has seen his father perform.' But such an ideal 'has never existed in history nor ever will' and we were 'every year drifting further' away from it. All the same, he was not a defeatist. He did not believe, he said, in simply deploring, then bowing to, the spirit of the age: 'for the spirit of the age is the spirits of those who compose it, and the stronger the expressions of dissent from prevailing fashion, the higher the possibility of diverting it from its ruinous course'.[15]

Waugh constantly and to the best of his considerable ability did 'dissent from prevailing fashion'. But, holding the views he did, he naturally did not take part in politics as such. As he put it, 'I do not aspire to advise my sovereign in her choice of servants.'[16] Not only did he eschew politics himself. He deplored the fact that so many of his friends and contemporaries, not least Cyril Connolly, succumbed to the 1930s spirit of the age and betrayed literature by politicizing themselves. Connolly fascinated Waugh. He brought him into several of his books, in one way or another, and would annotate Connolly's own with fierce and perceptive marginal observations. Why this interest? There were two reasons. First, Waugh thought Connolly worth his notice because of his brilliant wit and because, in his writing, he was capable 'of phrase after phrase of lapidary form, of delicious exercises in parody, of good narrative, of luminous metaphors' and sometimes 'of haunting originality'.[17] Yet at the same time Connolly lacked a sense of literary structure, or architecture as Waugh preferred to call it, as well as persistent energy, and was therefore incapable of producing a major book. Waugh found this incongruity of great interest. Secondly and more importantly, however, Waugh saw Connolly as the representative spirit of the times, and therefore to be watched as one might a rare bird. In his copy of Connolly's *The Unquiet Grave*, now at the Humanities Research Center at the University of Texas, Austin, he made many notes about Connolly's character. He was 'the most typical man of my generation', with his 'authentic lack of scholarship', his 'love of leisure and liberty and good living', his 'romantic snobbery', 'waste and despair' and 'high gift of expression'. But he was 'strait-jacketed by sloth' and handicapped by his Irishry; however much he tried to conceal it he was 'the Irish boy, the immigrant, homesick, down-at-heel and ashamed, full of fun in the public house, a ready quotation on his lips, afraid of witches, afraid of the bog-priest, proud of his capers'; he had 'the Irishman's deep-rooted belief that there are only two realities – hell and the USA'.[18] In the 1930s he deplored the fact that Connolly wrote about 'recent literary history' not in terms of writers 'employing and exploring their talents in their own ways' but as 'a series of "movements", sappings, bombings and encirclements, or party racketeering and gerrymandering. It is the Irish in him perhaps.' He blamed him severely for 'surrendering' into 'the claws' of commitment, 'the cold dank pit of politics into which all his young friends have gone tobogganing.' He thought this 'a sorry end to so much talent; the most insidious of all the enemies of promise'.[19] He thought Connolly's obsession with politics could not last. He was capable of better, or at any rate other, things. In any case, how could someone like Connolly give advice to humanity on how to conduct its affairs?

How indeed! Without being in any way an evil man, Connolly exhibited the characteristic moral weaknesses of the intellectual to an unusual degree. In the first place, while professing egalitarianism, at least when it was fashionable, from 1930–50, he was a lifelong snob. 'Nothing infuriates me more than to be treated as an Irishman,' he complained, pointing out that Connolly was the only Irish surname among those of his eight great-grandparents. He came from a family of professional soldiers and sailors. His own father was an undistinguished army officer, but *his* father had been an admiral and his aunt had been Countess of Kingston. In 1953, in an anonymous profile in the *New Statesman*, the critic John Raymond pointed out that Connolly had doctored details of his biography in *Enemies of Promise*. While the original 1938 (and therefore 'proletarian') edition had suppressed his grand and landed connections, they had been elaborately resurrected in the revised edition of 1948, by which time intellectual fashions had changed. Connolly, Raymond noted, was always 'plumb on target' in getting such 'cultural trends' right: 'No one has a better knowledge of the poses, the rackets, the gimmicks of English literature over the last quarter of a century.'[20]

The snobbery started early. Like many leading intellectuals such as Sartre, Connolly was an only child. His mother, who adored him, called him Sprat. Spoilt, self-centred, ugly and no use at games, he found boarding-school tough. He survived first by enthusiastic flunkying of well-born boys. 'This term,' he exulted to his mother, 'we have an awful lot of nobility . . . one Siamese prince, the grandson of the Earl of Chelmsford, the son of Viscount Malden, who is the son of the Earl of Essex, another grandson of a Lord and the nephew of the Bishop of London.'[21] His second survival device was wit. Like Sartre, he quickly discovered that intellectual ingenuity, particularly the ability to make the other boys laugh, earned him a certain grudging acceptance. He later recorded that 'the word would go round' – "Connolly's being funny" and 'soon I would have collected a crowd'. As court jester to more powerful boys, Connolly got on even at Eton, though there he expanded into the wisdom field: 'I am becoming quite a Socrates in the Lower Half of the College.' Known as 'the tug who's been kicked in the face by a mule', Connolly used his intellectual gifts to get himself into the coveted 'Pop', and an Oxford 'Open' followed almost as a matter of course. His contemporary Lord Jessel said to him: 'Well, you've got a Balliol scholarship and you've got into Pop – you know, I shouldn't be surprised if you never did anything else the rest of your life.'

There was an appalling danger, as Connolly was aware, of this prediction coming true. He was always very perceptive about himself, as about others. He early recognized that he was by nature a hedonist; his aim he designated as not so much perfection as 'perfection in happiness'. But how could he be happy if, having no inherited money of his own, he was forced to be energetic? Waugh was right to point to his laziness. Connolly himself admitted 'that sloth by which I have been disabled'. He did little work at Oxford and took a Third. He then landed an easy job as amanuensis to the wealthy writer Logan Pearsall Smith, who assigned him few duties and in effect gave him an allowance of £8 a week, quite a lot in those days. Smith was hoping for a Boswell but doomed to disappointment since Boswellizing requires energetic assiduity. Besides, Connolly soon married a well-off woman, Jean Bakewell, with £1000 a year. He seems to have been fond of her but the couple were too selfish to have a child. There was a botched abortion in Paris, forcing Jean to have a further operation which meant she could never have children; it affected her glands, she became overweight and her husband lost interest in her. Connolly never seems to have developed a mature attitude to women. He admitted that for him 'love' took the form of 'the exhibitionism of the only child'. It meant 'a desire to lay my personality at someone's feet as a puppy deposits a slobbery ball'.[22] Meanwhile, Jean's money was sufficient to remove the need for regular work. Connolly's diaries, which he kept from 1928–37, record the consequences: 'Idle morning.' 'Extremely idle morning, lunch about two.' 'I am lying on the sofa trying fo imagine a yellow slab of sunshine spread thickly over a white wall.' 'Too much leisure. With so much leisure one leans too hard on everyone and everything, and most of them give way.'[23]

In fact Connolly was not quite as idle as he liked to make out. He completed his sharp critique of literary fashions, *Enemies of Promise*, which, when eventually published (in 1938) proved one of the most influential books of the decade. It suggested that he had a natural gift for leading at any rate the more gregarious intellectuals of his generation. When the Spanish war came, he duly politicized himself and paid three visits there; rather like the Grand Tour, it was compulsory among intellectuals of a certain class, the cerebral equivalent of big-game hunting. Connolly had the mandatory letter of authorization from Harry Pollitt, which came in useful when his companion, W. H. Auden, was arrested in Barcelona for urinating in the Monjuich public gardens, a serious offence in Spain.[24]

Connolly's accounts of these visits, mainly in the *New Statesman*, are acute and a refreshing contrast to the field-grey committed prose most other intellectuals were producing at that time. But they indicate the strain he found in carrying the Left Man's Burden. 'I belong,' he introduced himself, 'to one of the most non-political generations the world has ever seen . . . we would no sooner have attended a political meeting than we would have gone to church.' The 'more realistic' of them – he instanced Evelyn Waugh and Kenneth Clark – had grasped that 'the kind of life they lived depended on close cooperation with the governing class.' The rest had 'wavered' until the Spanish war erupted: 'they have [now] become politically-minded entirely, I think, through foreign affairs.' But he was quick to note that many on the left had been motivated by careerism or because 'they hated their father or were unhappy at public school or insulted at the Customs or worried about sex'.[25] He drew pointed attention to the importance of literary as well as political merit and commended Edmund Wilson's *Axel's Castle* as 'the only left-wing critical book to accept aesthetic as well as economic standards'.[26]

What Connolly was hinting was that politicized literature did not work. In due course, as soon as it was intellectually safe, he proclaimed the demise of 'commitment' openly. In October 1939 a wealthy admirer, Peter Watson, devised the perfect role for Connolly: editing a monthly magazine of new writing, *Horizon*, whose specific purpose was to uphold literary excellence in the teeth of the all-enveloping wartime spirit. From the start it was a striking success and confirmed Connolly's position as a leading power-broker among the intelligentsia. By 1943 he felt he could afford to write off the 1930s as a mistake: 'The literature most typical of those ten years was political and it failed both ways, for it accomplished none of its political objects, nor did it evoke any literary work of lasting merit.'[27] Instead, Connolly began the process of replacing the intellectual search for utopia by the pursuit of enlightened hedonism. He did this both in the columns of *Horizon* and in another highly influential book, a collection of escapist thoughts about pleasure, *The Unquiet Grave* (1944). In youth Connolly had described his ideology as seeking 'perfection in happiness'; in the proletarian 1930s he had called it 'aesthetic materialism'; now it was 'the defence of civilized standards'.

However, it was not until June 1946, when the war was over, that Connolly actually got down to defining his programme in detail in a *Horizon* editorial.[28] Characteristically it was the sharp-eyed Evelyn Waugh who drew attention to this statement. He had been following Connolly's doings with close attention, despite all the wartime distractions; later, in his *Sword of Honour* trilogy he was to lampoon the wartime Connolly as Everard Spruce, his magazine as *Survival* and his pretty girl intellectual assistants – in real life Lys Lubbock, who shared Connolly's bed, and Sonia Brownell, who became the second Mrs Orwell – as Frankie and Coney. Now he drew the attention of the Catholic readers of the *Tablet* to the enormity of Connolly's programme.[29] This list of ten aims, described by Connolly as 'the major indications of a civilized society', was as follows: (1) abolition of the death penalty; (2) penal reform, model prisons and rehabilitation of prisoners; (3) slum clearance and 'new towns'; (4) light and heating subsidized and 'supplied free like air'; (5) free medicine, food and clothes subsidies; (6) abolition of censorship, so that everyone can write, say and perform what they wish, abolition of travel restrictions and exchange control, the end of phone-tapping or the compiling of dossiers on people known for their heterodox opinions; (7) reform of the laws against homosexuals and abortion, and the divorce laws; (8) limitations on property ownership, rights for children; (9) the preservation of architectural and natural beauty and subsidies for the arts; (10) laws against racial and religious discrimination.

This programme was, in fact, the formula for what was to become the permissive society. Indeed, if we leave out some of Connolly's more impractical economic ideas, virtually everything he called for was to be enacted into law in the 1960s, not only in Britain but in the United States and most other Western democracies. These changes, affecting almost every aspect of social, cultural and sexual life, were to make the 1960s one of the most crucial decades in modern history, akin to the 1790s. Waugh was understandably alarmed. He suspected that doing what Connolly proposed involved the virtual elimination of the Christian basis of society and its replacement by the universal pursuit of pleasure. Connolly saw it as the final attainment of civilization; to others it would end in a pandaemonium. What it unquestionably showed, however, was how much more influential intellectuals are when they turn from political utopias to the business of eroding social disciplines and rules. This was demonstrated by Rousseau in the eighteenth century and again by Ibsen in the nineteenth. Now it was proved again: whereas the political 1930s, as Connolly remarked, had been a failure, the permissive 1960s, from the intellectuals' point of view anyway, were a spectacular triumph.

Connolly himself, having set the agenda, played little part in carrying through the revolution, though he survived till 1974. He was not made for long campaigns or heroic endeavours. The spirit might have been willing, at any rate at times, but the flesh was always weak. He coined the phrase, apropos of himself, 'imprisoned in every fat man a thin one is wildly signalling to be let out.'[30] But the thin Cyril never did emerge. He was an anti-hero long before the word was coined. Greed, selfishness and petty depredations ever marked his steps. As early as 1928, an unpaid laundry bill led Desmond MacCarthy to denounce him as an opportunist and a sponger. Indeed most people who had given Connolly hospitality had cause to regret it. One found what was described as 'bathroom detritus' in the bottom of his grandfather clock. Lord Berners discovered a mouldering tub of potted shrimps amid his Chippendale. Somerset Maugham detected Connolly stealing two of his prize avocados and forced him to unpack his suitcases and disgorge. Half-eaten meals were retrieved, weeks later, from bedroom drawers, or bits of spaghetti and bacon rashers marking his place in the host's books. Then there was 'the cigar ash dropped with absent malice into the culinary triumph proffered by the wife of a celebrated American intellectual'.[31] Or the unchivalrous behaviour during a V-bomb raid over London in 1944 when Connolly – like Bertrand Russell thirty years before – was in bed with a lady of quality. She was possibly the Lady Perdita (later, in real life, Mrs Annie Fleming) described by Evelyn Waugh as figuring in his interests at that time. But whereas Russell leapt out of Lady Constance Malleson's bed in a gesture of generous indignation at man's inhumanity, in Connolly's case the leap was dictated by panic, redeemed by a *bon mot*: 'Perfect fear casteth out love.'

Clearly such a man could not lead a crusade for civilization, even had the energy been there. But of course it was not. Sloth, boredom and self-disgust caused Connolly to kill *Horizon* in 1949: 'We closed the long windows over Bedford Square, the telephone was taken, the furniture stored, the back numbers went to their cellar, the files rotted in the dust. Only contributions continued inexorably to be delivered, like a suicide's milk.' He finally divorced poor Jean and married a beautiful intellectual's moll called Barbara Skelton. But the union (1950–54) did not prosper. Each watched the other warily. Both, in the tradition of Tolstoy and Sonya, and many denizens of Bloomsbury, kept competitive diaries for future publication. After it broke up, Connolly complained to Edmund Wilson bitterly about Skelton's diary, which described her relations with him and might at any moment appear as a novel. Meanwhile, Wilson records him saying, 'she had confiscated and hidden a diary that he has kept about his relations with her. He knew where it was, however, and was going to break in and get it sometime when she was away.'[32] Evidently this did not happen, as no such diary has yet surfaced. But Skelton's was eventually published in 1987 and Connolly was right to be apprehensive about its contents. It provides an unforgettable portrait of the comatose intellectual in supine posture.

Thus, on 8 October 1950 she records: '[Cyril] sinks back into the bed like a dying goose, still in his dressing-gown ... Sinks further into the pillow and closes his eyes, with an expression of resigned suffering ... An hour later I go into the bedroom. Cyril is lying with his eyes closed.' 10 October: '[Cyril] has a long session in the bath while I do the laundry. Go into the bedroom later to find him standing naked in an attitude of despair staring into space ... return to bedroom, find C. still gazing into space ... write a letter, return to bedroom, C. still with his back to the room propped against the window ledge.' A year later, 17 November 1951: '[Cyril] wouldn't come down to breakfast but lay in his bed sucking the sheet-ends ... He sometimes lies for an hour with folds of sheet pouring from his mouth like ectoplasm.'[33]

Nonetheless, this upholder of civilized values had laid the egg of permissiveness rather in the same way that Erasmus laid the egg of the Reformation. But the hatching was the work of others, and in the process a new and disturbing element, which Connolly certainly did not foresee and would have deplored if he had, was added to the brew: the cult of violence. It is a curious fact that violence has always exercised a strong appeal to some intellectuals. It goes hand in hand with the desire for radical, absolutist solutions. How else can we explain the taste for violence in Tolstoy, Bertrand Russell and so many other nominal pacifists? Sartre, too, was a man fascinated by violence, dabbling in it beneath an obfuscating cloud of euphemism. Thus he argued: 'when youth confronts the police our job is not only to show that it is the police who are the violent ones but to join youth in counter-violence.' Or again, for an intellectual not to engage in 'direct action' (i.e. violence) on behalf of blacks 'is to be guilty of murder of the blacks – just as if he actually pulled the triggers that killed [the Black Panthers] murdered by the police, by the system'.[34]

The association of intellectuals with violence occurs too often to be dismissed as an aberration. Often it takes the form of admiring those 'men of action' who practise violence. Mussolini had an astonishing number of intellectual followers, by no means all of them Italian. In his ascent to power, Hitler consistently was most successful on the campus, his electoral appeal to students regularly outstripping his performance among the population as a whole. He always performed well among teachers and university professors. Many intellectuals were drawn into the higher echelons of the Nazi Party and participated in the more gruesome excesses of the SS.[35] Thus the four *Einsatzgruppen* or mobile killing battalions which were the spearhead of Hitler's 'final solution' in Eastern Europe contained an unusually high proportion of university graduates among the officers. Otto Ohlendorf, who commanded 'D' Battalion, for instance, had degrees from three universities and a doctorate in jurisprudence. Stalin, too, had legions of intellectual admirers in his time, as did such post-war men of violence as Castro, Nasser and Mao Tse-tung.

Encouragement or tolerance of violence by intellectuals has sometimes been the product of characteristic loose thinking. Auden's poem which dealt with the Civil War, 'Spain', published in March 1937, contained the notorious line: 'The conscious acceptance of guilt in the necessary murder'. This was criticized by Orwell, who liked the poem as a whole, on the grounds that it could only have been written by someone 'to whom murder is at most a word'. Auden defended the line by arguing that, 'if there is such a thing as a just war, then murder can be necessary for the sake of justice' – but he nonetheless cut the word 'necessary'.[36] Kingsley Martin, who served in the Quaker Ambulance Unit in the First World War and shrank from actual violence in any shape, sometime muddled himself into defending it theoretically. In 1952, applauding the final triumph of Mao in China, but nervous about reports that one and a half million 'enemies of the people' had been disposed of, he asked foolishly in his *New Statesman* column: 'Were these executions really necessary?' Leonard Woolf, a director of the journal, forced him to publish a letter the following week in which he asked the pointed question: would Martin 'give some indication... under what circumstances the execution of 1.5 million persons by a government *is* "really necessary"?' Martin, of course, could give no answer and his wriggling efforts to get off the hook on which he had impaled himself were painful to behold.[37]

On the other hand, some intellectuals do not find even the fact of violence abhorrent. The case of Norman Mailer (1923–) is particularly illuminating because he is in many ways so typical of the intellectual type we have been examining.[38] The first-born only son of a matriarchal family, he was from the start the centre of an admiring female circle. It consisted of his mother Fanny, whose own family, the Schneiders, were well-to-do and who ran a successful business herself, and her various sisters. Later Mailer's own sister joined the circle. The boy was a model Brooklyn child, quiet, well-behaved, always first in class, at Harvard by sixteen; his progress was enthusiastically applauded by the females. 'All the women in the family thought that Norman was the cat's miaow.' That was the comment of his first wife, Beatrice Silverman, who also noted: 'Fanny just didn't want her little genius to be married.' 'Genius' was a word often on Fanny's lips in relation to Mailer; she would inform reporters, at one of his many court appearances, 'My boy's a genius.' Sooner or later Mailer's wives became disagreeably conscious of the Fanny Factor. The third, Lady Jean Campbell, complained: 'All we ever did was go to dinner with his mother.' The fourth, a blonde actress who called herself Beverley Bentley, was censured (and indeed physically attacked) for making anti-Fanny remarks. However, the wives were themselves an adult substitute for the childhood circle of femininity, since Mailer continued to be on terms with all except one of them after their divorces, arguing: 'When you're divorced from a woman, the friendship can then start because one's sexual vanity is not in it any longer.' There were six wives altogether, who between them produced eight children, the sixth wife, Norris Church, being the same age as Mailer's eldest daughter. There were also many other women, the fourth wife complaining: 'When I was pregnant, he had an airline stewardess. Three days after bringing home our baby, he began an affair.' The progression from one woman to another was strongly reminiscent of Bertrand Russell, while the harem atmosphere recalls Sartre. But Mailer, while from a matriarchal background, had strongly patriarchal notions. His first marriage broke up because his wife wanted a career, Mailer dismissing her as 'a premature Woman's Liberationist'. He complained of the third: 'Lady Jean gave up $10 million to marry me but she would never make me breakfast.' He finished with the fourth when she, in turn, had an affair. One of his women complained: Norman just won't have anything to do with a woman who has a career.' Reviewing one of his books in 1971, V.S. Pritchett argued that the fact Mailer had so many wives (only four at that date) showed he was 'clearly not interested in women but in something they had got'.[39]

The second characteristic Mailer has in common with many intellectuals is a genius for self-publicity. The brilliant promotion of his outstanding war novel, *The Naked and the Dead* (1948), was a highly professional job by his publishers Rinehart, among the most elaborate and certainly the most successful campaigns of the post-war period. But, once launched, Mailer took over his own public relations, which for the next thirty years were a wonder and a warning to all – work, wives, divorces, views, quarrels and politics being woven skilfully into one seamless garment of self-advertisement. He was the first intellectual to make effective use of televison to promote himself, staging on it memorable and sometimes alarming scenes. He early grasped television's insatiable appetite for action, as opposed to mere words, and accordingly turned himself into the most hyperactive of intellectuals, following a course already piloted by Hemingway. What was all this self-promotion designed to serve? Vanity and egoism, of course: it cannot be stressed too strongly that many of the activities of men like Tolstoy, Russell and Sartre, though superficially rationalized, can only be adequately explained by the desire to draw attention to themselves. Then too there was the more mundane purpose of earning money. Mailer's patriarchal tastes proved expensive. Taken to court in 1979 by his fourth wife, Mailer argued that he could not afford to pay her $1000 a week; he was, he said, paying $400 a week to his second, $400 to his fifth and $600 to his sixth wife; he was $500,000 in debt, owing $185,000 to his agent and unpaid taxes of $80,500, leading the Internal Revenue Service to put a $100,000 lien on his house. His self-advertising was clearly designed to attract readers, and did so handsomely. To give only one example, his long essay 'The Prisoner of Sex', attacking feminism and much-canvassed as a result of his marital escapades, appeared in *Harper's* in March 1971 and sold more newsstand copies than any other issue in the magazine's 120-year history.

However, Mailer's self-promotion also had a serious purpose, to promote the concept which became the dominant theme of his work and life – the need for man to throw off some of the constraints which inhibit the use of personal force. Hitherto, most educated people had identified such inhibitions with civilization – the poet Yeats, for instance, had defined civilization precisely as 'the exercise of self-restraint'. Mailer questioned this assumption. Might not personal violence sometimes, for some people, be necessary and even virtuous? He reached this position by a devious route. As a young man he was a standard fellow-traveller, making eighteen speeches on behalf of Wallace in the 1948 presidential campaign.[40] But he broke with the Communist Party at the notorious 1949 Waldorf Conference and thereafter his political views, while sometimes merely reflecting the liberal-left consensus, became more idiosyncratic and original. In particular, his novel-writing and journalism led him to explore the position of blacks and black cultural assumptions in the life of the West. In the 1957 summer issue of Irving Howe's magazine *Dissent* he published a thesis *The White Negro*, which proved to be his most influential piece of writing, indeed a key document of the post-war epoch. In this he analysed 'hip consciousness', the behaviour of young, self-assertive and confident blacks, as a form of counterculture; he explained and justified it, indeed urged its adoption by radical whites. There were, Mailer argued, many aspects of black culture progressive intellectuals should be prepared to examine carefully: anti-rationalism, mysticism, the sense of the life force and, not least, the role of violence and even revolution. Consider, Mailer wrote, the actual case of two young men beating to death a sweetshop owner. Did it not have its beneficial aspect? 'One murders not only a weak, fifty-year-old man but an institution as well, one violates private property, one enters into a new relation with the police and introduces a dangerous element into one's life.' Since rage, when turned inwards, was a danger to creativity, was not violence, when used, externalized and vented, itself creative?

This was the first carefully considered and well-written attempt to legitimize personal violence – as opposed to the 'institutionalized violence' of society – and it aroused understandable anger in some quarters. Indeed Howe himself later admitted he should have cut the passage about the sweetshop murder. At the time Norman Podhoretz attacked it as 'one of the most morally gruesome ideas I have ever come across', which showed 'where the ideology of hipsterism can lead'.[41] But very large numbers of young people, white as well as black, were waiting for just such a lead and a rationalization. *The White Negro* was the authenticating document for much of what took place in the 1960s and 1970s. It gave intellectual respectability to many acts and attitudes hitherto regarded as outside the range of civilized behaviour, and added some significant and baleful items to the permissive agenda Cyril Connolly had proposed a decade before.

The message had all the more impact in that Mailer reinforced and publicized it by his own actions, in public and private. On 23 July 1960 he was tried for his part in a police-station brawl in Provincetown, being found guilty of drunkenness but not of 'disorderly conduct'. On 14 November he was again charged with disorderly conduct at a Broadway club. Five days later he gave a big party at his New York home to announce his candidacy for mayor of New York. By midnight he was fighting drunk down on the street outside his apartment house, engaging in fisticuffs with various acquaintances, such as Jason Epstein and George

Plimpton, as they left his party. At 4.30 am he came in from the street with a black eye, swollen lip and blood-stained shirt. His second wife, Adele Morales, a Spanish-Peruvian painter, remonstrated with him; whereupon he took out a penknife with a 2½-inch blade and stabbed her in the abdomen and back; one wound was three inches deep. She was fortunate not to die. The legal processes which followed this incident were complex; but Adele refused to sign a complaint and it ended a year later in Mailer being given a suspended sentence and probation. His comments in the meantime did not strike any particular note of remorse. In an interview with Mike Wallace he said: 'The knife to a juvenile delinquent is very meaningful. You see, it's his sword, his manhood.' There ought, he added, to be annual gang-jousts in Central Park. On 6 February 1961 he gave a reading of his verses at the Young Men's Hebrew Association Poetry Center, including the lines 'So long as you use a knife/there is some love left'; the Director rang down the curtain on grounds of obscenity. After the whole episode was over, Mailer summed up: 'A decade's anger made me do it. After that I felt better.'[42]

There were also Mailer's more calculated public efforts to push the counter-culture. One of those who had been inspired by *The White Negro* was the Yippie Jerry Rubin, and Mailer was the star speaker at the huge anti-Vietnam rally Rubin staged at Berkeley on 2 May 1965. He said that President Lyndon Johnson's 'Great Society' was moving 'from camp to shit' and he exhorted 20,000 students not just to criticize the President but to insult him by sticking his picture upside down on walls. One of those who heard him was Abbie Hoffman, soon to be the counter-culture's high priest. Mailer showed, he argued, 'how you can focus protest sentiment effectively by aiming not at the decisions but at the guts of those who make them'.[43] Two years later Mailer took a flamboyant part in the big march on the Pentagon, on 21 October 1967, entertaining and provoking the vast audience with obscenities, telling them, 'We're going to try to stick it up the government's ass, right up the sphincter of the pentagon,' and getting himself arrested and sentenced to thirty days in jail (twenty-five suspended). Released, he told reporters: 'You see, dear fellow Americans, it is Sunday and we are burning the body and blood of Christ in Vietnam' – defending his allusion by saying that, though not a Christian himself, he was now married to one. This was wife number four.

In effect Mailer brought into politics the language of hip, the voice of the street. He eroded the statesman's hieratic and a lot of the assumptions that went with it. In May 1968, at the height of the student unrest, a writer in the *Village Voice*, analysing Mailer's appeal, wrote: 'How could they not dig Mailer? Mailer, who preached revolution before there was a movement. Mailer, who was calling LBJ a monster while slide-rule liberals were still writing speeches for him. Mailer, who was into Negroes, pot, Cuba, violence, existentialism... while the New Left was still a twinkle in C.Wright Mills's eye.'[44] But, while certainly lowering the tone of political discourse, it is not clear that Mailer raised the content. His impact on literary life was similar. His rows with fellow authors rivalled, even outstripped, those of Ibsen, Tolstoy, Sartre and Hemingway. He quarrelled, privately and publicly, with William Styron, James Jones, Calder Willingham, James Baldwin and Gore Vidal, among others. As with Hemingway, these rows sometimes took violent forms. In 1956 he was reported fighting on the flowerbeds outside Styron's house. His opponent was Bennett Cerf, whom he told: 'You're not a publisher, you're a dentist.' In 1971 there was face-slapping and head-butting with Gore Vidal before a Dick Cavett television show; at a 1977 party the script read: Mailer to Vidal: 'You look like a dirty old Jew.' Vidal: 'Well, *you* look like a dirty old Jew.' Mailer throws drink in Vidal's face; Vidal bites Mailer's finger.[45] The television debate which followed the face slapping, which also involved the harmless and ladylike *New Yorker* Paris correspondent, Janet Flanner, developed into an angry Vidal–Mailer discussion of buggery. Then:

Flanner: 'Oh, for goodness sake!' [laughter]
Mailer: 'I know you've lived in France for many years but believe me, Janet, it's possible to enter a woman another way.'
Flanner: 'So I've heard.' [more laughter]
Cavett: 'On that classy note we will end the show.'

Mailer epitomized the interthreading of permissiveness with violence which characterized the 1960s and 1970s, and miraculously survived his own antics. Others were not so lucky or resilient. Indeed there were some sad casualties of the switch of intellectual thrust away from old-style utopianism and towards the new, vertiginous and increasingly brutal hedonism. When Cyril Connolly published his manifesto in June 1946, Kenneth Peacock Tynan had just completed his first year at Magdalen College, Oxford, and has already established himself as the leader of intellectual society there. Four months later, when the new term opened, I was an awestruck freshman-witness to his arrival at the Magdalen lodge. I stared in astonishment at this tall, beautiful, epicene youth, with pale yellow locks, Beardsley cheekbones, fashionable stammer, plum-coloured suit, lavender tie and ruby signet-ring. I was trundling my solitary regulation school trunk. He seemed to fill the lodge with his possessions and servitors whom he ordered about with calm and imperious authority. One sentence particularly struck me: 'Have a care for that box, my man – it is freighted with golden shirts!' Nor was I the only one to be struck by this elegant cameo performance. In 1946 Tynan and I were among the few undergraduates who had come straight to university from school. The great majority had been in the war; some had held senior ranks and had witnessed or perhaps perpetrated scenes of appalling slaughter. But they had seen nothing like this. Beefy majors from the Foot Guards were struck speechless. Bomber pilots who had killed thousands in the Berlin firestorms simply goggled. Lieutenant-Commanders who had sunk the *Bismarck* gazed in awe. With super timing, having dominated his self-created scene, Tynan swept out followed by his toiling bearers.

Behind this strange young man there was (though he did not then know it) an even stranger story. It might have come straight from the pages, not indeed of those Magdalen alumni and heroes, Oscar Wilde or Compton Mackenzie, but from Arnold Bennett. The facts about Tynan's life have all been carefully gathered by his second wife, Kathleen, and published in a tender and sorrowing biography, a model of its kind.[46] Tynan was born (in 1927) and brought up in Birmingham, attended its famous grammar school and flourished mightily there, playing the lead in *Hamlet* and winning a demyship to Oxford. He thought himself the only, much-adored and indulged child of Rose and Peter Tynan, a draper. His father gave him £20 spending money every fortnight, a great deal in those days. In fact Tynan was illegitimate and his father was what Bennett called a 'card', leading a double life. Half the week he was Peter Tynan in Birmingham. The other half, in his tailcoat, top hat, grey spats and handmade silk shirts, he was Sir Peter Peacock, Justice of the Peace, successful entrepreneur, six times mayor of Warrington, and with a Lady Peacock and many small Peacocks in train. The deception only came to light in 1948 at the end of Tynan's Oxford career, when Sir Peter died, the indignant legitimate family swept down from Warrington to claim the body, and Tynan's tearful mother was banned from the funeral. It is not unknown for Oxford undergraduates suddenly to discover they are illegitimate – it happened to another Magdalen man, the putative baronet Edward Hulton, who was obliged to have the 'Sir' removed from his nameplate – and Tynan responded by instantly inventing the story that his father had been financial adviser to Lloyd George. But the discovery hurt. He dropped 'Peacock' as his middle name. Moreover, his mother's guilt feelings at what she had done to her son help to explain why, from the beginning, she over-protected and spoiled him; indeed, he always treated her like some kind of superior servant.

It was always Tynan's habit to order people around; he had the touch of the master. At Oxford he dressed in princely style at a time when clothes-rationing was still strictly enforced. In addition to the plum-coloured suit and golden shirts there was a cloak lined in red silk, a sharp-waisted doeskin outfit, another suit in bottle green, said to be made of billiard-cloth, and green suede shoes; he used make-up – 'just a little crimson lake around the mouth'.[47] He thus restored Oxford's reputation for aesthetic extravagance. During the whole of his time there he was easily the most talked-of person in the city. He produced and acted in plays. He spoke brilliantly at the Union. He wrote for or edited the magazines. He gave sensational parties, attended by London show-biz celebrities.* He kept a court of young women and admiring dons. He was burnt in effigy by envious bloods. He seemed to bring to life the pages of *Brideshead Revisited*, then a recent best-seller, and to add fresh ones.

Moreover, unlike virtually all those who cause a splash at Oxford, he made good in the real world. He produced plays and reviews. He acted alongside Alec Guinness. More important, he quickly established himself as the most audacious literary journalist in London. His motto was: 'Write heresy, pure heresy.' He pinned to his desk the exhilarating slogan: 'Rouse tempers, goad and lacerate, raise whirlwinds.' He followed these injunctions at all times. They quickly brought him the coveted shop-window as drama critic of the *Evening Standard*, and in due course the still more influential drama post on the *Observer*, then the best paper in Britain. Readers goggled, as the students had once done in Magdalen lodge, at this amazing phenomenon, who seemed to know all world literature and used words such as esurient, cateran, cisisbeism and eretheism.[48] He became a power in the London theatre, which regarded him with awe, fear and hatred. He turned Osborne's *Look Back in Anger* into a hit and launched the legend of the Angry Young Men. He introduced Britain to Brecht. Not least, he campaigned powerfully for the subsidized theatre which made the Brechtian drama effective. When Britain's own National Theatre was created he became its first literary manager, 1963–73, and established it with a strong cosmopolitan repertoire: of the seventy-nine plays done there during his regime, most of them his ideas, half were hits, an astonishing record. He was already well-known in the United States, thanks to some superb reviews in the *New Yorker*, 1958–60; at the National Theatre he established a global reputation. Indeed at times in the 1960s he probably had more influence than anyone else in world theatre; and, as I have argued in this book,

---

* Charging an entrance fee of ten shillings.

the theatre ultimately has more effect on behaviour than any other art.

Nor was Tynan without a serious purpose. Like Connolly, and equally vaguely, he linked hedonism and permissiveness with socialism. In *Declaration* (1957), the manifesto of the 'Angries', he contributed his one considered statement of aim. Art, he insisted, had to go 'on record; it must commit itself'. But equally socialism ought to mean 'progress towards pleasure', and be 'a gay international affirmation' (this was in the age before the hijacking of 'gay').[49] Writing in the year Mailer published *The White Negro*, he had in part the same aim, of breaking down linguistic inhibitions on the stage and in print. No one in Britain played a bigger role in destroying the old system of censorship, formal and informal. His efforts to do so were punctuated by more traditional political gestures, though even these had a permissive aspect. In 1960, after much manoeuvring, he succeeded in getting a four-letter word into the *Observer*. The next year he organized a pro-Castro demonstration in Hyde Park, with the help of scores of pretty girls. On 13 November 1965 he achieved his masterpiece of calculated self-publicity when he uttered the word 'fuck' on a BBC television late-night satire programme. For a time it made him the most notorious man in the country. On 17 June 1969 he put organized nudity on the general stage with his review *Oh! Calcutta!* It was eventually performed all over the world and grossed over $360 million.

Yet in destroying censorship, Tynan was also destroying himself. His actual death in 1980 was caused by emphysema, the product of habitual smoking on a weak chest he had inherited from his father. But some time before that he had irreparably damaged himself as a moral being by what can only be called a self-immolation on the altar of sex. His sexual obsessions began early. He later claimed he masturbated from the age of eleven and often vaunted the joys of this activity; towards the end of his life, in a haunting self-characterization, he described himself as a dying species, *Tynanosaurus homo masturbans*. He also, as a boy, did his best to collect pornography, no easy matter in wartime Birmingham. When he performed his schoolboy *Hamlet*, he induced James Agate, then the leading critic and a notorious homosexual, to write a notice of the show. Agate did so, and also invited the youth to his London flat and put his hand on his knee: 'Are you homo, my boy?' 'I'm afraid not.' 'Ah well, I thought we'd get that out of the way.'[50] Tynan spoke the truth. He sometimes enjoyed dressing up in women's clothes and did not particularly discourage the common notion that he was homosexual, believing it gave him easier access to women. But at no point did he have a homosexual experience, 'never even a mild grope', as he put it.[51] He was, however, interested in sado-masochism. Agate, having discovered this, gave him the key to his own extensive pornographic collection, and that completed Tynan's corruption.

Thereafter he built up a store of his own. In due course, various landladies and both his wives stumbled across it and were profoundly shocked. This is curious because Tynan never troubled to hide his sexual interests and sometimes proclaimed them. He announced during an Oxford Union debate: 'My theme is – just a thong at twilight.' He formed relationships with large numbers of young women at Oxford and usually asked them to present him with a pair of their knickers to hang alongside the whip which graced his walls. He liked voluptuous Jewish girls, especially those with strict fathers and used to corporal punishment. He told one of them that the word 'chastise' had 'a good Victorian ring of retribution'. He added: 'the word spank is very potent and has the correct schoolgirlishness ... Sex means spank and beautiful means bottom and always will.'[52] He did not expect either of his wives to submit to such doings, which he associated with sin and wickedness to be guiltily enjoyed. But once he was a power in the theatre there was no difficulty in finding out-of-work actresses who cooperated in return for some help.

Women seem to have objected less to his sadism, which took only a mild form, than to his vanity and authoritarianism. One left him when she noticed that, entering a restaurant, he always blocked her efforts to use the looking-glass. Another reported: 'The moment you left him you were out of his mind.' He treated women as possessions. He had in many ways a sweet nature and could be perceptive and understanding. But he expected women to revolve around men like moons around a planet. His first wife, Elaine Dundy, had ambitions of her own and eventually wrote a first-rate novel.* This led to quarrels of a spectacular, rather stagey kind, with screams, smashed crockery and cries of 'I'll kill you, you bitch!' Mailer, no mean judge of marital rows, rated the Tynans highly: 'They'd hit each other shots that you'd just sit there and applaud like you would at a prizefight.' Tynan, while reserving the unqualified right to be unfaithful himself, expected loyalty from his spouse. On one occasion, returning from his current mistress to his London flat, he found his first wife in the kitchen with a naked man. Tynan saw that the man was a poet, a BBC producer and therefore wet, and took the bold step of seizing his clothes from the bedroom and hurling them down the lift-shaft. Usually he was not so courageous. After divorcing his first wife, he induced Kathleen Gates, who was to become his second, to leave her husband and live with him. The husband

---

* Asked if it would be any good, Cyril Connolly replied: 'Oh I shouldn't think so. Just another wife trying to prove she exists.'

broke in through Tynan's front door, while Tynan himself cowered behind the sofa. Later the husband caught up with him near her mother's house in Hampstead; there was a scrimmage and tufts of Tynan's golden hair, now greying, were pulled out before he got to safety in the house. The second wife's account continued: 'For a time Ken and I managed to hold out in my mother's house. Then we crept into the night. Some distance down the road, Ken swore we were being followed and climbed into a nearby dustbin.'[53] Tynan did not relish this forced evocation of Samuel Beckett, a playwright he originally discounted.

The second marriage crumbled like the first on Tynan's insistence on total sexual latitude for himself, fidelity for his wife. He formed a permanent liaison with an unemployed actress with whom he enacted elaborate sado-masochistic fantasies, involving himself dressing up as a woman, the actress as a man, and sometimes prostitutes as extras. He told Kathleen he intended to continue with these sessions twice a week, 'although all common sense and reason and kindness and even camaraderie are against it... It is my choice, my thing, my need... It is fairly comic and slightly nasty. But it is shaking me like an infection and I cannot do anything but be shaken until the fit has passed.'[54] This was bad enough. Still worse was Tynan's decision to push aside his career in order to become a pornographer, and an unsuccessful one at that. As far back as 1958 his engagement book contained the notes: 'Write play. Write pornographic book. Write autobiography.' In 1964 he formed a relationship with *Playboy* magazine, though oddly enough they resisted his attempts to provide them with erotic material. It seems to have been Tynan's optimistic belief that he could turn pornography into a serious art form, encouraged by the meretricious success of *Oh! Calcutta!* In the early 1970s he tried to enlist a number of distinguished writers in compiling an anthology of masturbation fantasies, but got humiliating rebuffs from, among others, Nabokov, Graham Greene, Beckett and Mailer. Thereafter he became increasingly involved in long-drawn-out but ultimately abortive attempts to produce a pornographic film. For one thing, he could never raise the finance. Unlike most intellectuals, he was not avaricious. Quite the contrary: he was generous, even reckless, a quality he shared with Sartre. When his mother died she left him a sizeable sum of old Sir Peter's money, which Tynan spent as quickly as he could. When eased out of the National Theatre, the payoff he received was niggardly. He signed such foolish contracts for *Oh! Calcutta!* that he received barely $250,000 from this hugely successful show. Much of his time in his last years was spent trying to raise funds for a project his wiser friends viewed with contempt and despair. He had doubts himself. He wrote to Kathleen from Provence: 'What am I doing here churning out pornography? It is very shaming.' At St Tropez he dreamed of a naked girl, covered with dust and excrement, her hair shaved off, with dozens of drawing pins nailed into her head. At that point, he recorded, 'I woke up filled with horror. And at once dogs in the hotel grounds began to bark pointlessly, as they are said to do when the King of Evil, invisible to man, passes by.'[55] Tynan's last years, a sinister counterpoint of sexual obsession and physical debility, are movingly told by his widow and make appalling reading to those who knew and admired the man. They recall Shakespeare's arresting phrase, 'the expense of spirit in a waste of shame'.[56]

A still more striking casualty of permissiveness, with a stronger note of violence, was Rainer Werner Fassbinder, perhaps the most gifted film director even Germany has produced. He was a child of defeat, born in Bavaria on 31 May 1945 in the immediate aftermath of Hitler's suicide; he was the adolescent beneficiary, and victim, of the new freedoms which intellectuals like Connolly, Mailer and Tynan were seeking to bestow on civilized humanity. In the 1920s the German cinema had led the world. The coming of the Nazis had created a diaspora of its talents, Hollywood getting the lion's share; and when the Nazi regime fell, the American occupying authorities planted Hollywood cinema on German soil. This epoch ended in 1962 when twenty-six young German film writers and directors issued a declaration of German cinematic independence known as the Oberhausen Manifesto. Fassbinder left school two years later and by the age of twenty-one he had shot two short films. In a German arts world dominated by the shadow of Brecht, he formed a small production collective known as Antitheatre. In its first successful production he himself played Mac the Knife in Brecht's *Threepenny Opera*. Although the Antitheatre was egalitarian in theory, in practice it was a hierarchical, structured tyranny, with himself as despot, and run (it has been said) in the way Louis XIV ran Versailles.[57] He used this outfit to make his first successful film, *Love is Colder than Death*, shot over a mere twenty-four days in April 1969.

Fassbinder turned himself into not only the leading but also the symbolic film maker of the permissive age with astonishing speed. He had great will and authority, and an enviable power of taking fast, firm decisions; this enabled him to make high-quality movies quickly and economically. Critical approval came soon. He did not achieve world box-office success until *Fear Eats the Soul* (1974), but this was already his twenty-first movie. Indeed in the twelve months beginning in November 1969 he made nine full-length features. One of the most highly rated, both critically and commercially, *The Merchant of Four Seasons* (1971), had 470 scenes and was shot in twelve days. By the age of thirty-seven he had made forty-three films, one every hundred days for thirteen years.[58] There were no days off. He always worked, and made others work, on Sundays. In a professional sense he had a fanatical and sustained self-discipline. He would say: 'I can sleep when I am dead.'

This prodigious output was achieved against a background of personal self-indulgence and self-abuse which makes the flesh creep. His father was a doctor who left his family when Fassbinder was six, giving up medicine to write poetry and supporting himself by running cheap properties. His mother was an actress, later appearing in some of his films. After her divorce she married a short-story writer. The background to Fassbinder's childhood and adolescence was bohemian, literary, uncertain, amoral and irresponsible. He read a great deal and quickly began to produce creative work, stories and songs. He absorbed the new permissive culture with the same speed and certainty he did everything else. He was, in the new hip terminology, streetwise. By fifteen he was helping his father to collect the rents of his slum flats. He announced he was in love with a butcher's boy. The father replied – it was characteristically German – 'If you want to go to bed with men, can't it be with someone from the university?'[59] Thereafter Fassbinder pursued with relentless ferocity one of the three great themes of the new sixties' culture: the uninhibited exploitation of sex for pleasure. As his power in the theatre and cinema world grew, so did his demands and ruthlessness. Most of his lovers were male. Some were married and with children and there were appalling scenes of family distress. Almost from the start there was a hint of sado-masochism and extremism. He drew men from the working class and turned them into actors as well as lovers. One, whom he called 'my Bavarian negro', seems to have specialized in wrecking expensive cars. Another, a former North African male prostitute, was homicidal and created moments of terror for Fassbinder's associates, and indeed for himself. A third, a butcher-turned-actor, committed suicide. But Fassbinder was also interested in women and talked patriarchally of 'producing a traditional family'. His attitude to women was proprietorial. He liked to control them. In his early days, to raise money for his movies, he used women he controlled to service immigrant 'guest-workers', as the Germans call them. In 1970 he married an actress called Ingrid Caven, who believed she could transform him into a heterosexual. The wedding feast turned into a predictable orgy. The bride found her bedroom door locked, and the groom and the best man in her bed. Divorced, Fassbinder eventually went through another marriage with one of his film editors, Juliane Lorenz, but continued his ostentatious homosexual life in bars, hotels and brothels. Yet curiously enough he demanded fidelity from her. During the filming of the novel *Berlin Alexanderplatz* (1980), he discovered she had spent a night with one of the electricians. He created a jealous scene and called her a whore; she tore up her wedding certificate in his face.

Fassbinder also reflected, in his films and lifestyle, the second great theme of the new culture: violence. As a very young man, he seems to have been close to Andreas Baader, who helped to create one of West Germany's most notorious terrorist gangs, and to Horst Sohnlein, who made incendiaries for the Baader-Meinhof group. His actor friend Harry Baer said Fassbinder often said he was tempted to go into terrorism but told himself that 'making films would be more important for "the cause" than going out onto the street'.[60] When Baader and other members of his gang committed suicide in the Stammheim Prison, in October 1977, Fassbinder shouted angrily: 'They have murdered our friends.' His subsequent film *The Third Generation* (1979) produced the argument that the threat of terrorism was being exploited by the authorities to make Germany totalitarian again, and it aroused fury. In Hamburg a mob beat the cinema projectionist unconscious and destroyed the film. In Frankfurt youths hurled acid-bombs at the cinema that was showing it. Fassbinder usually got state subsidies for his films – that too was a sign of the times – but he made this one out of his own funds: it was a labour of love, or hate.

By this time however he had embraced, and was in the process of being crushed by, a third theme of the new culture: drugs. Tolerance, acceptance of drugs had always been an implicit assumption of the permissive society, especially in its hip vernacular. In the 1960s it became standard practice for intellectuals to sign petitions demanding the liberalization of drug laws. As a young man Fassbinder had earned money by driving stolen cars across frontiers but he does not seem to have become involved in drugs then. He was, naturally, part of the German hip scene. Like Brecht he designed a suitable uniform for himself: carefully torn jeans, check shirt, scuffed patent-leather shoes, a thin, wispy beard. He chain-smoked a hundred cigarettes a day. He ate a great deal of rich food and by his thirties had began to look bloated and frog-like, claiming: 'Growing ugly is your way of sealing yourself off... Your stout, fat body, a monstrous bulwark against all forms of affection.'[61] He also drank heavily: in the United States he would finish a fifth of Jim Beam bourbon and sometimes a second fifth every day. When he decided to sleep he would consume large quantities of pills, such as Mandrax. He does not seem to have taken up hard drugs until he made his film *Chinese Roulette* in 1976, when he was thirty-one. But then, having tried cocaine, he became convinced of its creative power and used it regularly and in rising doses. Indeed, when filming *Bolswiser* (1977), he forced one of the actors to play his part drugged.

Thus events moved to an inevitable climax. In February 1982 he won the Golden Bear at the Berlin Film Festival; he hoped to do the hat trick by getting the Golden Palm at Cannes and the Golden Lion at Venice. But he did not get the Cannes prize; instead he spent 20,000 marks there on cocaine and made over the distribution rights of his next film for a guaranteed future supply. He had recently developed an erratic habit of violence towards women. At one point, drunk or drugged, he had become enraged and, for no reason, had hacked a script girl on the shins. At his birthday party on 31 May, a quasi-public event, he had handed his former wife Ingrid an enormous plastic dildo, saying it should keep her happy for a while. His schedule of work and interviews continued but his consumption of drugs, drink and banned sleeping pills mounted. On the morning of 10 June Juliane Lorenz found him dead in bed, the TV-video machine still switched on. A funeral of sorts took place but the coffin was empty as the police were still examining his body for drugs. The moral was so plain and emphatic that it was worth nobody's while to draw it, though many did. To honour the departed artist a death-mask was taken in the manner of Goethe or Beethoven, and at the Venice Film Festival that September there circulated among the café tables on the Piazza San Marco pirated copies of this gruesome object.

Tynan and Fassbinder might be described as victims of the cult of hedonism. There were those, too, who fell victim to the intellectual legitimization of violence as well. Among them was James Baldwin (1924–88), the most sensitive and in some ways the most powerful of twentieth-century black American writers. His was a case of a man who might have led a happy and fulfilled life by virtue of his achievements, which were considerable. But instead he was rendered intensely miserable by the new intellectual climate of his time, which persuaded him that the message of his work must be hate – a message he delivered with angry enthusiasm. It is further evidence of the curious paradox that intellectuals, who ought to teach men and women to trust their reason, usually encourage them to follow their emotions; and, instead of urging debate and reconciliation on humanity, all too often spur it towards the arbitration of force.

Baldwin's own account of his childhood and youth is unreliable, for reasons we will come to in a moment. But from the work of his biographer Fern Marja Eckman and other sources, it is possible to give a reasonably accurate summary.[62] Baldwin's background in 1920s Harlem was in some ways deprived. He was, in effect, the eldest of eight children. His mother did not get married until he was three. His grandfather was a Louisiana slave. His stepfather was a Sunday preacher, a Holy Roller, who worked in a bottling plant for very low wages. But despite the poverty, Baldwin was well, if strictly, brought up. His mother said that he always had one of his little brothers or sisters in one arm, but a book in the other. The first book he read right through was *Uncle Tom's Cabin*, and he read it again and again; its influence on his work, despite his attempts to erase it, was striking. His parents recognized his gifts and encouraged them; so did everyone else. In the 1920s and 1930s, there was no race-conscious defeatism in Harlem schools. The belief was that blacks, if they worked hard enough, could excel. Poverty was never accepted as an excuse for not learning. The academic standards were high. Children were expected to meet them, or be punished. Baldwin thrived in this atmosphere. At Public School 24 his head teacher, Gertrude Ayer, then the only black school head in New York City, was excellent; another teacher, Orilla Miller, took him to his first play and encouraged his writing. At the Frederick Douglas Junior High School, he published his first short story when he was thirteen in the school magazine, the *Douglas Pilot*, which he later edited; he was helped by two outstanding black teachers, the poet Countee Cullen, who taught French, and Herman Porter. As a teenager he wrote with extraordinary grace and exulted in his progress. A year after he left he contributed an article to the magazine applauding the school's 'spirit of good will and friendship' which make it 'one of the greatest junior high schools in the country'.[63] As well as being an accomplished writer he was also by now an outstanding teenage preacher, described as 'very hot'. He was admired, encouraged and greatly befriended by senior blacks in the pentacostal-tabernacle circuit. He went on to a celebrated New York academy, the De Witt Clinton High School in the Bronx, which produced, among others, Paul Gallico, Paddy Chayevsky, Jerome Weideman and Richard Avedon. Again he published his fiction in the school's superb magazine, *The Magpie*, and went on to edit it. Again he was befriended by some first-rate teachers, who gave his obvious talents every encouragement in their power.

Baldwin's later articles in *The Magpie* reflect his loss of faith. He left the Church. He worked as a porter and elevator boy, then on a construction site in New Jersey, writing at night. Again, there are many examples of his being helped and encouraged by his elders, both black and white. The then leading black writer, Richard Wright, got him the Eugene F. Saxton Memorial Trust Award, which enabled him to travel to Paris. He was published in the *Nation* and the *New Leader*. His ascent was not sensational but steady and methodical. People who knew him then testified to his hard-working earnestness, his dutiful support of his family, to whom he·sent every penny he could spare. He gave every sign of being happy. His breakthrough came in 1948 when he published a much-applauded article, 'The Harlem Ghetto', in the Jewish intellectual monthly *Commentary*.[64] A lot of people lent him money to get on with his imaginative work. A loan from Marlon Brando enabled him to finish his novel of Harlem church life, *Go Tell It On the Mountain*, which was published to great applause in 1953. He led the life of a cosmopolitan intellectual, leaping straight from Harlem to Greenwich Village and the Paris Left Bank, bypassing the black bourgeoisie completely. He ignored the South. The Negro Problem was not a primary issue to him. Indeed, from much of his earlier and best writing it is impossible to tell that he was black. He stood for integration, in his life no less than in his work. Some of his finest essays were contributed to the integrationist *Commentary*.[65] Its editor Norman Podhoretz later remarked: 'He was a Negro intellectual in almost exactly the same sense as they were Jewish intellectuals.'[66]

But in the second half of the 1950s, Baldwin sensed the emerging new intellectual climate, of permissiveness on the one hand and approved hatred on the other. He was, or believed himself to be, a homosexual, and his second novel, *Giovanni's Room* (1956), dealt with this theme. It was rejected by his own publisher and he was forced to turn to another, which (he convinced himself) paid him too little. The experience filled him with rage against the American publishing industry. Moreover, he discovered that rage, at any rate from a 'deprived' person like himself, was becoming topical, fashionable and just. He expanded it, to include people and institutions he had once held in respect. He turned on Richard Wright and many other older blacks who had helped him.[67] He began to pass collective judgments on the white race. He rewrote his entire personal history, to a great extent unconsciously, no doubt. He became yet another intellectual whose autobiographical writings, beneath a deceptive veneer of exhibitionist frankness, are dangerously misleading.[68] He discovered he had been a very unhappy child. His father had told him he was the ugliest boy he had ever seen, as ugly as the devil's son. Of his father he wrote: 'I do not remember, in all those years, that one of his children was ever glad to see him come home.' He claims, when his father died, he heard his mother sigh: 'I'm a widow forty-one years old with eight small children I never wanted.' He discovered he was savagely bullied at school. He described Frederick Douglas Junior High School with horror. When he revisited it in 1963 he told the pupils: 'White people have convinced themselves the Negro is happy in this place. It's your job not to believe it one moment longer.'[69] He declared of his high school that only the whites were happy in it, a claim his contemporary Richard Avedon strenuously denied. He said of the English teacher who helped him: 'We *hated* each other.' He violently and repeatedly denounced books he had once loved, such as *Uncle Tom's Cabin*. He attacked the whole concept of the would-be integrationist, middle-class Negro.[70] He investigated the South and in the late 1950s became associated with the civil-rights movement, two phenomena he had hitherto ignored. But he was not interested in the Gandhi-like tactics of a Martin Luther King. Nor did he care for the sinewy reasoning of such black intellectuals as Bayard Rustin, who put the strictly rational case for equality with masterly skill. In the atmosphere generated by Mailer's *The White Negro*, Baldwin played, with ever-growing vehemence, the emotional card – not least against Mailer himself, telling him he would rather spend his time with a white racist than a white liberal, since then at least he knew where he stood.

In fact Baldwin spent plenty of time with white liberals, in America and in Europe. There was, in fact, nothing he liked more, or for longer, than white liberal hospitality. In the grand old intellectual tradition of Rousseau, he turned its enjoyment into a princely favour. He condescended to accept it. As his biographer Fern Eckman wrote in 1968: 'While in the throes of creativity, Baldwin regularly progresses from one house to another, rather like a medieval king travelling through his realm, dispensing royal favour by granting honoured subjects the privilege of serving as his host.'[71] He invited his friends too, and transformed his host's establishment into an informal clubhouse, then would leave on the grounds (as he said to one of them) that 'Your house has become too public.' As one host put it, in respectful admiration rather than anger: 'Having Jimmy at your house isn't like having a guest, it's like entertaining a caravan.' The more hatred he generated, the more subservience he received. The echoes of Rousseau are uncanny.

The hatred was very widely dispersed, black liberals receiving even more of it than the white variety. As one of them complained, 'No matter how free you think you are, Jimmy makes you feel you've still got a little bit of Uncle Tom in you.' At the beginning of the 1960s, Podhoretz asked Baldwin to investigate the new black violence being preached by Malcolm X and his Black Muslims, and offered to publish his findings in *Commentary*. Baldwin did so but sold the piece to the *New Yorker* for considerably more money.[72] Accompanied by a description of his youthful experiences of racism, it appeared in book form in 1963 under the title *The Fire Next Time*. For forty-one consecutive weeks it was among the top five titles on the American best-seller list and was translated all over the world. In one respect it was the logical successor to Mailer's

The White Negro, and would perhaps have been impossible without it. But it was a far more influential work, both in the United States and elsewhere, because it was a statement, by a leading black intellectual – hitherto operating within the literary conventions and mode of discourse of Western culture – of black nationalism conceived on a racial basis. Baldwin now gave his rage formal literary expression, institutionalized it, defended and propagated it. In doing so, he set up a new kind of racial asymmetry. No white intellectual could conceivably assert that all whites hated blacks, let alone defend such hatred. Yet Baldwin now insisted that blacks hated whites, and the implication of his work was that they were justified in doing so. Hence he gave intellectual respectability to a new and rapidly spreading form of black racism, which was taking over the leadership of black communities all over the world.

Whether Baldwin really believed in the inevitability of black racism and of a seemingly unbridgeable chasm dividing the races is open to doubt. The young James Baldwin would have denied it strongly. It conflicted with his actual experience – which is why the older Baldwin had to rewrite his personal history. The last twenty years of Baldwin's life were thus founded on a falsehood, or at least a culpable confusion. In fact he lived most of it abroad, remote from any struggle. But his work was vandalized by the fire he had lit himself, and ceased to be effective. What lived on was the spirit of *The Fire Next Time*. It reinforced the message of Frantz Fanon's frenzied polemic, *Les Damnés de la terre*, and Sartre's rhetoric, that violence was the legitimate right of those who could be defined, by race, class or predicament, to be the victims of moral iniquity.

Now here we come to the great crux of the intellectual life: the attitude to violence. It is the fence at which most secular intellectuals, be they pacifist or not, stumble and fall into inconsistency – or, indeed, into sheer incoherence. They may renounce it in theory, as indeed in logic they must since it is the antithesis of rational methods of solving problems. But in practice they find themselves from time to time endorsing it – what might be called the Necessary Murder Syndrome – or approving its use by those with whom they sympathize. Other intellectuals, confronted with the fact of violence practised by those they wish to defend, simply transfer the moral responsibility, by ingenious argument, to others whom they wish to attack.

An outstanding practitioner of this technique is the linguistic philosopher Noam Chomsky. In other respects he is very much an old-style utopian, rather than a new-style hedonist intellectual. He was born in Philadelphia in December 1928 and rapidly achieved economic eminence at a number of leading universities: the Massachussetts Institute of Technology, Columbia, Princeton, Harvard and so forth. In 1957, the same year as Mailer published *The White Negro*, Chomsky produced a masterly volume called *Syntactic Structures*. This was a highly original, and seemed at the time a decisive, contribution to the ancient but continuing debate on how we acquire knowledge and, in particular, how we acquire so much. As Bertrand Russell put it, 'How comes it that human beings, whose contacts with the world are brief and personal and limited, are nevertheless able to know as much as they do know?'[73] There are two rival explanations. One is the theory that men are born with innate ideas. As Plato put it in his *Meno*: 'There are, in a man who does not know, true opinions concerning that which he does not know.' The most important contents of the mind are there from the beginning, though external stimulation or experience, acting on the senses, is required to bring this knowledge into consciousness. Descartes held that such intuitive knowledge is more dependable than any other, and that all men are born with a residuum of it, though only the most reflective realize its full potentiality.[74] Most Continental European philosophers take this view to some degree.

As against this there is the Anglo-Saxon tradition of empiricism, taught by Locke, Berkeley and Hume. It argues that, while physical characteristics can be inherited, the mind is at birth a *tabula rasa* and mental characteristics are all acquired through experience. These views, usually in a highly qualified form, are generally held in Britain, the United States and other countries which follow their culture.

Chomsky's study of syntax, which is the principles governing the arrangements of words or sounds to form sentences, led him to discover what he called 'linguistic universals'. The world's languages are much less diverse than they superficially appear because all share syntactic universals which determine the hierarchical structure of sentences. All the languages he, and later his followers, studied conformed to this pattern. Chomsky's explanation was that these unvarying rules of intuitive syntax are so deep in the human consciousness that they must be the result of genetic inheritance. Our ability to use language is an innate rather than an acquired ability. Chomsky's explanation of his linguistic data may not be correct. But so far it is the only plausible one produced, and it puts him firmly in the Cartesian or 'Continental' camp.[75]

It also aroused considerable intellectual excitement, not merely in academic circles, and made Chomsky something of a celebrity, as Russell became after his work on mathematical principles, or Sartre when he popularized existentialism. The temptation for such celebrities is to use the capital they have acquired from eminence in their own discipline to acquire a platform for their views on public issues. Both Russell and Sartre succumbed to this temptation, as we have seen; and so did Chomsky. Throughout the 1960s, intellectuals in the West, but especially in the United States, became increasingly agitated by American policy in Vietnam, and by the growing level of violence with which it was executed. Now therein lay a paradox. How came it that, at a time when intellectuals were increasingly willing to accept the use of violence in the pursuit of racial equality, or colonial liberation, or even by millenarian terrorist groups, they found it so repugnant when practised by a Western democratic government to protect three small territories from occupation by a totalitarian regime? There is really no logical manner in which this paradox can be resolved. The explanations intellectuals offered, that they were objecting to 'institutionalized violence' on the one hand, and justifying individual, personal, counter-violence on the other (and many variations on the same) had to suffice. They certainly sufficed for Chomsky, who became and remained the leading intellectual critic of US Vietnam policy. From explaining how mankind acquired its capacity to use language, he turned to advising it on how to conduct its geopolitics.

Now it is a characteristic of such intellectuals that they see no incongruity in moving from their own discipline, where they are acknowledged masters, to public affairs, where they might be supposed to have no more right to a hearing than anyone else. Indeed they always claim that their special knowledge gives them valuable insights. Russell undoubtedly believed that his philosophical skills made his advice to humanity on many issues worth heeding – a claim Chomsky endorsed in his 1971 Russell Lectures.[76] Sartre argued that existentialism was directly relevant to the moral problems raised by the Cold War and our response to capitalism and socialism. Chomsky in turn concluded that his work on linguistic universals was itself primary evidence of the immorality of American policy in Vietnam. How so? Well, Chomsky argued, it depends on which theory of knowledge you accept. If the mind at birth is indeed a *tabula rasa*, and human beings are, as it were, pieces of plasticine which can be modelled into any shape we please, then they are fit subjects for what he calls 'the "shaping of behaviour" by the state authority, the corporate manager, the technocrat or the Central Committee'.[77] If, on the other hand, men and women possess innate structures of mind and have intrinsic needs for cultural and social patterns which for them are 'natural', such state efforts must fail in the end, but in the process of failing they will hinder our development and involve terrible cruelty. The attempt of the United States to impose its will, and particular patterns of social, cultural and political development, on the peoples of Indo-China was an atrocious instance of such cruelty.

It required a peculiar perversity, with which anyone who studies the careers of intellectuals becomes depressingly familiar, to reach this conclusion. Chomsky's argument from innate structures, if valid, might fairly be said to constitute a general case against social engineering of any kind. And indeed, for a variety of reasons, social engineering has been the salient delusion and the greatest curse of the modern age. In the twentieth century it has killed scores of millions of innocent people, in Soviet Russia, Nazi Germany, Communist China and elsewhere. But it is the last thing which Western democracies, with all their faults, have ever espoused. On the contrary. Social engineering is the creation of millenarian intellectuals who believe they can refashion the universe by the light of their unaided reason. It is the birthright of the totalitarian tradition. It was pioneered by Rousseau, systematized by Marx and institutionalized by Lenin. Lenin's successors have conducted, over more than seventy years, the longest experiment in social engineering in history, whose lack of success does indeed confirm Chomsky's general case. Social engineering, or the Cultural Revolution as it was called, produced millions of corpses in Mao's China, and with equal failure. Though applied by illiberal or totalitarian governments, all schemes of social engineering have been originally the work of intellectuals. Apartheid, for instance, was worked out in its detailed, modern form in the social psychology department of Stellenbosch University. Similar systems elsewhere in Africa – Ujaama in Tanzania, 'Consciencism' in Ghana, Négritude in Senegal, 'Zambian Humanism', etc. – were cooked up in the political science or sociology departments of local universities. American intervention in Indo-China, imprudent though it may have been, and foolishly conducted as it undoubtedly was, was originally intended precisely to save its peoples from social engineering.

Chomsky ignored such arguments. He showed no interest in totalitarian attempts to suppress or change innate characteristics. He argued that the liberal democracy, the laissez-faire state, was just as objectionable as the totalitarian tyranny, since the capitalist system, of which it is necessarily an organic part, supplies the elements of coercion which produce the same denial of self-fulfilment. The Vietnam War was the outstanding case of capitalist-liberal oppression of a small people who were trying to respond to their own intuitive urges; of course it was bound to fail but unspeakable cruelty was meanwhile being inflicted.[78]

The arguments of intellectuals like Chomsky undoubtedly played a major part in reversing what was originally a strong determination on the part of the United States to ensure that a democratic society had the chance to develop in Indo-China. When the American forces withdrew, the social engineers promptly moved in, as those who supported American intervention had all along predicted they would. It was then that the unspeakable cruelties began in earnest. Indeed in Cambodia, as a direct result of American withdrawal, one of the greatest crimes, in a century of spectacular crimes, took place in 1975. A group of Marxist intellectuals, educated in Sartre's Paris but now in charge of a formidable army, conducted an experiment in social engineering ruthless even by the standards of Stalin or Mao.

Chomsky's reaction to this atrocity is instructive. It was complex and contorted. It involved the extrusion of much obfuscating ink. Indeed it bore a striking resemblance to the reactions of Marx, Engels and their followers to the exposure of Marx's deliberate misquotation of Gladstone's Budget speech. It would take too long to examine in detail but the essence was quite simple. America was, by Chomsky's definition, which by now had achieved the status of a metaphysical fact, the villain in Indo-China. Hence the Cambodian massacres could not be acknowledged to have taken place at all until ways had been found to show that the United States was, directly or indirectly, responsible for them.

The response of Chomsky and his associates thus moved through four phases.[79] (1) There were no massacres; they were a Western propaganda invention. (2) There may have been killings on a small scale; but the 'torment of Cambodia has been exploited by cynical Western humanitarians, desperately eager to overcome the "Vietnam Syndrome"'. (3) The killings were more extensive than at first thought, and were the result of the brutalization of the peasants by American war crimes. (4) Chomsky was finally driven to quoting 'one of the handful of authentic Cambodian scholars' who, by skilful shifting of the chronology, was able to 'prove' that the worst massacres occurred not in 1975 but 'in mid-1978', and took place not for Marxist but for 'traditionalist, racist, anti-Vietnamese reasons'. The regime had by then 'lost any Marxist coloring it had once had' and had become 'a vehicle for hyperchauvinist poor peasant populism'. As such it 'at last' won the approval of the CIA, who moved from exaggerating the massacres for propaganda purposes to actively perpetrating them. In short Pol Pot's crime was in fact America's, *quod erat demonstrandum*.

By the mid-1980s, Chomsky's focus of attention had shifted from Vietnam to Nicaragua, but he had moved himself well beyond the point at which reasonable people were still prepared to argue with him seriously, thus repeating the sad pattern of Russell and Sartre. So yet another intellect, which once seemed to tower over its fellows, plodded away into the wasteland of extremism, rather as old Tolstoy set off, angry and incoherent, from Yasnaya Polyana. There seems to be, in the life of many millenarian intellectuals, a sinister climacteric, a cerebral menopause, which might be termed the Flight of Reason.

*   *   *

We are now at the end of our enquiry. It is just about two hundred years since the secular intellectuals began to replace the old clerisy as the guides and mentors of mankind. We have looked at a number of individual cases of those who sought to counsel humanity. We have examined their moral and judgmental qualifications for this task. In particular, we have examined their attitude to truth, the way in which they seek for and evaluate evidence, their response not just to humanity in general but to human beings in particular; the way they treat their friends, colleagues, servants and above all their own families. We have touched on the social and political consequences of following their advice.

What conclusions should be drawn? Readers will judge for themselves. But I think I detect today a certain public scepticism when intellectuals stand up to preach to us, a growing tendency among ordinary people to dispute the right of academics, writers and philosophers, eminent though they may be, to tell us how to behave and conduct our affairs. The belief seems to be spreading that intellectuals are no wiser as mentors, or worthier as exemplars, than the witch doctors or priests of old. I share that scepticism. A dozen people picked at random on the street are at least as likely to offer sensible views on moral and political matters as a cross-section of the intelligentsia. But I would go further. One of the principal lessons of our tragic century, which has seen so many millions of innocent lives sacrificed in schemes to improve the lot of humanity, is – beware intellectuals. Not merely should they be kept well away from the levers of power, they should also be objects of particular suspicion when they seek to offer collective advice. Beware committees, conferences and leagues of intellectuals. Distrust public statements issued from their serried ranks. Discount their verdicts on political leaders and important events. For intellectuals, far from being highly individualistic and non-conformist people, follow certain regular patterns of behaviour. Taken as a group, they are often ultra-conformist within the circles formed by those whose approval they seek and value. That is what makes them, *en masse*, so dangerous, for it enables them to create climates of opinion and prevailing orthodoxies, which themselves often generate irrational and destructive courses of action. Above all, we must at all times remember what intellectuals habitually forget: that people matter more than concepts and must come first. The worst of all despotisms is the heartless tyranny of ideas.

Notes

Chapter One: Jean-Jacques Rousseau: 'An Interesting Madman'

1. See Joan Macdonald, *Rousseau and the French Revolution* (London, 1965).
2. J.H.Huizinga, *The Making of a Saint: The Tragi-Comedy of Jean-Jacques Rousseau* (London, 1976), pp. 185 ff.
3. Ernst Cassirer, *The Philosophy of the Enlightenment* (Princeton, 1951), p. 268.
4. Jean Chateau, *Jean-Jacques Rousseau: Sa Philosophie de l'éducation* (Paris, 1962), pp. 32 ff.
5. Lester G. Crocker, *Jean-Jacques Rousseau: The Quest, 1712–1758* (New York, 1974), p. 263.
6. Ibid., pp. 238–39, 255–70.
7. For Rousseau's early life see ibid., pp. 7–15; the account he gives in his *Confessions* is quite unreliable.
8. Rousseau's letters are published in R.A.Leigh, *Correspondence Complète de Jean-Jacques Rousseau* (Geneva, 1965 ff) and in T.Dufour and P.P.Plan, *Correspondance Générale de Jean-Jacques Rousseau* (20 vols., Paris, 1924–34).
9. Crocker, vol. i, pp. 160 ff.
10. Quoted in Huizinga, p. 29.
11. The *Discours* is published in G.R.Havens (ed.), *Discours sur les sciences et les arts* (New York, 1946).
12. For Rousseau's works see Bernard Gagnebin and Marcel Raymond (eds), *Oeuvres complètes* (3 vols., Paris, 1959–64).
13. Macdonald.
14. Quoted in Huizinga, pp. 16–17.
15. Crocker, vol. i, p. 16; see also pp. 194 ff.
16. Quoted by Huizinga, p. 50. The passage occurs in an unposted letter to Monsieur de Mirabeau, 1767.
17. J.Y.T.Greig (ed.), *Letters of David Hume* (Oxford, 1953), vol. ii, p. 2.
18. Huizinga, pp. 15–16.
19. Such *obiter dicta*, and many similar, are collected in Huizinga.
20. Crocker, vol. ii: *The Prophetic Voice, 1758–1783* (New York, 1973), pp. 28–29.
21. P.M. Masson, *La Réligion de Jean-Jacques Rousseau* (3 vols., Paris, 1916).
22. Crocker, vol. i, pp. 146–47.
23. C.P.Duclos: *Considérations sur les moeurs de ce siècle* (London, 1784), quoted in Huizinga.

24. Crocker, vol. ii, pp. 208, 265–302.
25. Huizinga, pp. 56–57, 112.
26. W.H.Blanchard, *Rousseau and the Spirit of Revolt* (Ann Arbor, 1967), p. 120.
27. Quoted in Huizinga, p. 119.
28. E.C.Mossner, *Life of David Hume* (Austin, 1954), p. 528–29.
29. Crocker, vol. ii, pp. 300–2.
30. Ibid., pp. 318–19, 339–41.
31. *Confessions*, Everyman edition (London, 1904), vol. i, p. 13.
32. Ronald Grimsley, *Jean-Jacques Rousseau: A Study in Self-Awareness* (Bangor, 1961), pp. 55 ff.
33. *Confessions*, vol. i, pp. 58 ff.
34. See Crocker's excellent analysis of this technique, vol. i, pp. 57–58.
35. Huizinga, p. 75.
36. Crocker, vol. i, pp. 340 ff.
37. *Confessions*, vol. i, p. 31.
38. Ibid., vol. i, p. 311.
39. Ibid.
40. While Thérèse was still alive, Madame de Charrière wrote *Plainte et défense de Thérèse Levasseur* (Paris, 1789). A powerful modern defence of her is I.W. Allen's Ph.D. thesis, *Thérèsê Levasseur* (Western Reserve University, Cleveland), cited in Crocker, vol. i, p. 172. Other works dealing with Rousseau's relations with Thérèse include Claude Ferval, *Jean-Jacques Rousseau et les femmes* (Paris, 1934).
41. See F.A.Pottle (ed.), *Boswell on the Grand Tour, Germany and Switzerland 1764* (London, 1953), pp. 213–58.
42. Printed in ibid., pp. 335–37.
43. Greig, vol. ii, pp. 14–15.
44. Quoted in Crocker, vol. i, p. 186.
45. Ibid., pp. 178 ff.
46. The main defences are in the *Confessions*, vol. i, pp. 314 ff, vol. ii, pp. 8 ff.
47. For the General Will, etc., see L.G.Crocker, *Rousseau's Social Contract: An Interpretive Essay* (Cleveland, 1968).
48. Printed in C.R.Vaughan (ed.), *The Political Writings of Rousseau* (2 vols., Cambridge, 1915), vol. ii, p. 250.
49. Sergio Cotta, 'La Position du problème de la politique chez Rousseau', *Études sur le Contrat social de J.J.Rousseau* (Paris, 1964), pp. 177–90.
50. I.W.Allen, quoted in Crocker, vol. i, p. 356, note 6.
51. See Huizinga, Introduction.
52. Judgments for and against Rousseau are listed in Huizinga, pp. 266 ff.
53. Quoted by Crocker, vol. i, p. 353; the remark is recorded in Henri Guillemin, *Un Homme, deux ombres* (Geneva, 1943), p. 323.

Chapter Two: Shelley, or the Heartlessness of Ideas

1. P.B.Shelley to Elizabeth Hitchener, in F.L.Jones (ed.), *Letters of Percy Bysshe Shelley* (2 vols., Oxford, 1964), vol. i, pp. 116–17.
2. See text in D.L.Clark (ed.), *Shelley's Prose* (New Mexico, rev. ed. 1966).
3. For a clear analysis of the essay see M.H.Scrivener, *Radical Shelley* (Princeton, 1982), pp. 249 ff.

4. An interesting analysis of these poems is in Art Young, *Shelley and Non-Violence* (The Hague, 1975).

5. *Essays in Criticism, Second Series: Byron*, reprinted in Matthew Arnold, *Selected Prose* (Harmondsworth, 1982), pp. 385–404.

6. Byron to John Murray, 3 August 1822; to Thomas Moore, 4 March 1822; both in Leslie A. Marchand (ed.), *Byron's Letters and Journals* (11 vols., London, 1973–82), vol. ix, pp. 119, 189–90.

7. The best biography of Shelley, a pioneering work, is Richard Holmes, *Shelley: The Pursuit* (London, 1974). This should be supplemented by Holmes's essay on Shelley in his *Footsteps: Adventures of a Romantic Biographer* (London, 1985).

8. For Sir Timothy Shelley, see R. C. Thorne (ed.), *History of Parliament: House of Commons 1790–1820* (London, 1986), vol. v, *Members Q–Y*, pp. 140–41.

9. For the radicalization of the young Shelley see Holmes, pp. 25 ff; and K. M. Cameron, *The Young Shelley: Genesis of a Radical* (New York, 1950).

10. N. Mackenzie (ed.), *Secret Societies* (London, 1967), p. 170; Nesta Webster, *Secret Societies and Subversive Movements* (London, 1964), pp. 196–268.

11. Marie Roberts, *British Poets and Secret Societies* (London, 1986), deals with Shelley in Chapter 4, pp. 88–101.

12. Shelley, *Letters*, vol. i, p. 54; Paul Dawson, *The Unacknowledged Legislator: Shelley and Politics* (Oxford, 1980), pp. 157 ff.

13. Sylvia Norman, *The Flight of the Skylark: The Development of Shelley's Reputation* (London, 1954), p. 162.

14. Thomas Jefferson Hogg, *Life of Shelley*, quoting Helen.

15. Holmes, pp. 36, 48.

16. Ibid., pp. 50–51.

17. Ibid., p. 57.

18. Letter to John Williams, in *Letters*, vol. i, p. 330.

19. Ibid., pp. 139–40, 146–47, 148–49.

20. Ibid., p. 155.

21. Ibid., pp. 156, 163.

22. Ibid., p. 165.

23. Ibid., pp. 205–6.

24. F. L. Jones (ed.), *Mary Shelley's Journal* (London, 1947), p. 17.

25. N. I. White, *Shelley* (2 vols., New York, 1940), vol. i, pp. 547–52.

26. See Louis Schutz Boas, *Harriet Shelley: Five Long Years* (Oxford, 1962).

27. Letters of 14 July, 27 August, 15 September and 16 September 1814, in *Letters*, vol. i, pp. 389–90, 391–92, 394, 396.

28. Letter of 26 September 1814, in *Letters*, vol. i, pp. 396–97.

29. Letter of 3 October 1814, in *Letters*, vol. i, p. 403.

30. Letters of 3 and 25 October 1814, in *Letters*, vol. i, pp. 400, 410.

31. Letter of 14 November 1814, in *Letters*, vol. i, p. 421.

32. *Letters*, vol. i, p. 520, footnote.

33. See the account of Harriet's last phase in Boas, Chapter vii, pp. 183 ff.

34. Letter of 16 December 1814, *Letters*, vol. i, pp. 519–21. The authenticity of this letter was later challenged by Shelley's Victorian apologists, but there seems no reason to doubt it. See Holmes, p. 353 and footnote.

35. *Letters*, vol. i, pp. 511–12.

36. Letter of 10 December 1812, *Letters*, vol. i, p. 338.

37. For Fanny Imlay, see Holmes, pp. 347 ff.

38. Letter to Godwin, *Letters*, vol. i, p. 311.
39. *Letters*, vol. i, p. 196.
40. *Letters*, vol. i, p. 314.
41. Holmes, p. 216.
42. *Letters*, vol. i, p. 530.
43. *Letters*, vol. ii, pp. 264–65.
44. Holmes, pp. 442–47; see also Ursula Orange: 'Shuttlecocks of Genius', *Keats–Shelley Memorial Bulletin*, clixv.
45. See letters of Byron to Hoppner, 10 September and 1 October 1820, in *Byron's Letters and Journals*, vol. 7, pp. 174, 191.
46. Byron to Douglas Kinnaird, 20 January 1817, in *Byron's Letters and Journals*, vol. 5, pp. 160–62.
47. Claire Clairmont to Byron, 6 May 1816, Murray Mss, quoted in Doris Langley Moore, *Lord Byron: Accounts Rendered* (London, 1974), p. 302.
48. The case that the mother was the nurse, Elise, is argued in Ursula Orange, 'Elise, Nursemaid to the Shelleys', *Keats–Shelley Memorial Bulletin*, 1955. Richard Holmes, though Shelley's best biographer, is implausible on this issue, and in fact takes two different views, one in *Shelley: The Pursuit* and another in *Footsteps*.
49. August 1821; quoted in Moore.
50. See Byron's letters to J.B. Webster, 8 September 1818, and to John Cam Hobhouse and Douglas Kinnaird, 19 January 1819, printed in *Byron's Letters and Journals*, vol. vi, pp. 65, 91–92.
51. *Letters*, vol. i, p. 323.
52. Letter to Byron, 14 September 1821, quoted in Moore.
53. Haydon wrote these comments in the margin of his copy of Medwin's *Conversations with Lord Byron* (now at Newstead Abbey, Roe–Byron Collection); quoted in Moore, pp. 301–2.
54. *Letters*, vol. i, p. 423, note 1; Shelley's letters to Hogg, 1 January and 26 April 1815, vol. i, pp. 423, 426; eleven letters of Mary to Hogg survive.
55. Robert Ingpen and W.E. Peck (eds.), *Complete Works of P.B. Shelley* (New York, 1926–30), vol. vii, p. 43.
56. Letter of 10 January 1812, in *Letters*, vol. i, pp. 227 ff.
57. For details of Shelley's financial transactions with Godwin, see Holmes, pp. 223–38, 250, 269–70, 284, 307, 311–21, 346, 379, 407–13, 526.
58. Harriet to Mrs Nugent, 11 December 1814, in *Letters*, vol. i, p. 422, note.
59. Letter of 7 March 1841, in Thomas Pinney (ed.), *Letters of Thomas Babington Macaulay* (6 vols., Cambridge, 1974–81), vol. iii, p. 366.
60. Quoted in Ann Blainey, *Immortal Boy: A Life of Leigh Hunt* (London, 1985), p. 189.
61. *Letters*, vol. i, pp. 366, 379, note.
62. Holmes, p. 161.
63. For Roberts, see *Letters*, vol. i, p. 339, note 1 to Letter 215; for Bedwell, *Letters*, vol. i, p. 362; for the Williamses, *Letters*, vol. i, pp. 360 and note, 386–87; for Evans, *Letters*, vol. i, pp. 332–33, 339.
64. For the booksellers, see Shelley to John Slatter, 16 April 1811; Henry Slatter to Sir Timothy Shelley, 13 August 1831; letter from Shelley, 23 December 1814; *Letters*, vol. i, pp. 438, note 1, 411.
65. *Letters*, vol. i, pp. 362–63.
66. A.M.D. Hughes, *The Nascent Mind of Shelley* (Oxford, 1947), pp. 131 ff.

67. Such as Art Young, see note 4 above.
68. See Scrivenor, *Radical Shelley* (Princeton, 1982), pp. 198–210.
69. See Edward Duffy, *Rousseau in England: The Context for Shelley's Critique of the Enlightenment* (Berkeley, 1979).
70. Claire Clairmont to Edward Trelawney, 30 September 1878, printed in the Carl H. Pforzheimer Library Bulletin iv, pp. 787–88.
71. Shelley to John Gisborne, 18 June 1822, in *Letters*, vol. ii, pp. 434–37.
72. Holmes, p. 728; *Letters*, vol. ii, p. 433.
73. F.L.Jones (ed.), *Maria Gisborne and Edward E. Williams: Their Journals and Letters* (London, 1946), p. 149.
74. Holmes, p. 729; Edward Dowden, *Life of P. B. Shelley* (2 vols., London, 1886), vol. ii, pp. 534 ff.

Chapter Three: Karl Marx: 'Howling Gigantic Curses'

1. Edgar von Westphalen, quoted in Robert Payne, *Marx* (London, 1968), p. 20.
2. See the excellent essay on Marx in Robert S. Wistrich: *Revolutionary Jews From Marx to Trotsky* (London, 1976).
3. Letter to Engels, 11 April 1868, *Karl Marx–Friedrich Engels Werke* (East Berlin, 1956–68), vol. xxxii, p. 58.
4. For Marx's poetry see Payne, pp. 61–71.
5. *Marx–Engels Werke*, vol. iii, pp. 69–71.
6. Payne, pp. 166 ff.
7. Text in Marx–Engels, *Selected Correspondence 1846–95* (New York, 1936), pp. 90–91.
8. *Capital*, Everyman edition (London, 1930), p. 873.
9. T.B.Bottomore (trans. and ed.), *Karl Marx: Early Writings* (London, 1963), pp. 34–37; the essays on the Jews are also in *Karl Marx–Engels Collected Works* (London, 1975 ff), vol. iii, pp. 146–74.
10. The decisive stage in Marx's writings was reached in *A Contribution to the Critique of Hegel's Philosophy of Law* (1844), *The Economic and Philosophical Manuscripts of 1844* (first published in 1932), and *The German Ideology* (1845–46).
11. For a valuable discussion of these writings, see Payne, pp. 98 ff.
12. Payne, p. 86.
13. Payne, pp. 134–36.
14. *Marx–Engels Werke*, vol. xxx, p. 259.
15. Karl Jaspers, 'Marx und Freud', *Der Monat*, xxvi (1950).
16. Geoffrey Pilling, *Marx's Capital* (London, 1980), p. 126.
17. Louis Althusser, *For Marx* (trans. London, 1969), pp. 79–80.
18. Printed in *Engels on Capital* (London, 1938), pp. 68–71.
19. *Capital*, pp. 845–46.
20. *Capital*, pp. 230–311.
21. *Capital*, p. 240, note 3.
22. W.O.Henderson & W.H.Challoner (trans. and eds.), *Engels's Condition of the Working Class in England* (Oxford, 1958).
23. Engels to Marx, 19 November 1844, *Marx–Engels Gesamt-Ausgabe* (Moscow, 1927–35), 1 part iii (1929).
24. Henderson & Challoner, Appendix v, from Dr Loudon's *Report on the Operation of the Poor Laws*, 1833, gives characteristic examples of Engels's

methods of misquotation which have the effect of seriously distorting Loudon's meaning.

25. *Nationalökonomie der Gegenwart und Zukunft*, i (Frankfurt, 1848), pp. 155–61, 170–241.
26. For a general analysis of Marx's methods see Leslie R. Page, *Karl Marx and the Critical Examination of his Works* (London, 1987).
27. As reported in seven London newspapers, 17 April 1863.
28. See David F. Felix, *Marx as Politician* (London, 1983), pp. 161–62, 269–70.
29. Ibid., p. 147.
30. For this see Page, pp. 46–49.
31. See also Felix, and Chushichi Tsuzuki: *The Life of Eleanor Marx, 1855–98: A Socialist Tragedy* (London, 1967).
32. Payne, p. 81.
33. Ibid., p. 134.
34. Geinzen's account was published in Boston in 1864; quoted in Payne, p. 155.
35. *Marx–Engels Gesamt–Ausgabe*, vol. vi, pp. 503–5.
36. *Marx–Engels Gesamt–Ausgabe*, vol. vii, p. 239.
37. Payne, p. 475 note.
38. Stephan Lukes, *Marxism and Morality* (Oxford, 1985), pp. 3 ff.
39. Quoted in David McLellan, *Karl Marx: His Life and Thought* (London, 1973), p. 455.
40. Payne, pp. 50 ff.
41. *Marx–Engels, Collected Works*, vol. ii, pp. 330–31.
42. Marx, *On Britain* (Moscow, 1962), p. 373.
43. Payne, pp. 251 ff; Michael Bakunin, *Oeuvres* (Paris, 1908).
44. E.g., *Marx–Engels Gesamt–Ausgabe*, vol. xxxiii, p. 117.
45. *Marx–Engels Gesamt–Ausgabe*, vol. xxxi, p. 305.
46. It appears as a footnote in *Capital*, vol. i, ii, vii Chapter 22.
47. Quoted in Payne, p. 54.
48. *Marx–Engels Gesamt–Ausgabe*, vol. xxvii, p. 227.
49. *Marx–Engels Gesamt–Ausgabe*, vol. xxx, p. 310; Engels's reply is in vol. xxx, p. 312.
50. *Marx–Engels Gesamt–Ausgabe*, vol. xxxi, p. 131.
51. For further information on Marx's finances, see David McLellan, *Karl Marx: Interviews and Recollections* (London, 1981) and his *Karl Marx: The Legacy* (London, 1983); Fritz J. Raddatz, *Karl Marx: A Political Biography* (trans., London, 1979).
52. *Marx–Engels Gesamt–Ausgabe*, vol. xxvii, p. 500.
53. *Marx–Engels Gesamt–Ausgabe*, vol. xxvii, p. 609.
54. Printed in *Archiv für Geschichte des Socialismus* (Berlin, 1922), pp. 56–58; in Payne, pp. 251 ff.
55. *Marx–Engels Gesamt–Ausgabe*, pp. 102–3.
56. *Marx–Engels Gesamt–Ausgabe*, vol. iii, pp. 4, 569.
57. For Marx's family, see H. F. Peters, *Red Jenny: A Life with Karl Marx* (London, 1986); Yvonne Kapp, 'Karl Marx's Children: Family Life 1844–55' in *Karl Marx: 100 Years On* (London, 1983), pp. 273–305, and her *Eleanor Marx* (2 vols., London, 1972).
58. Payne, p. 257.
59. The Soviet authorities, having published a bowdlerized version, have the

surviving manuscript locked up in the Marx–Engels–Lenin Institute in Moscow. Another version, possibly also censored, was published in Leipzig in 1965.

60. For this and other dates in Marx's life, see the chronological survey by Maximilien Rubel in *Marx: Life and Works* (trans., London, 1980); the existence of the illegitimate son was first revealed in W. Blumenberg, *Karl Marx: An Illustrated Biography* (1962, English trans. London, 1972).
61. See Payne, pp. 538–39.

Chapter Four: Henrik Ibsen: 'On the Contrary!'

1. 17 May 1814.
2. See Brian W. Downs, *Ibsen: The Cultural Background* (Cambridge, 1948) and the introduction to John Northam (trans. and ed.), *Ibsen's Poems* (Oslo, 1986).
3. 'Memories of Childhood', written in January 1881, printed in Evert Sprinchorn (ed.), *Ibsen: Letters and Speeches* (London, 1965), pp. 1–6.
4. For the facts of Ibsen's life I have relied mainly on Michael Meyer's biography: *Henrik Ibsen: i. The Making of a Dramatist, 1828–64* (London, 1967); *ii. The Farewell to Poetry, 1864–82* (London, 1971); *iii. The Top of a Cold Mountain, 1886–1906* (London, 1971). However, for the convenience of readers my notes usually refer to the abridged edition, *Henrik Ibsen* (London, 1974).
5. Meyer, p. 197 note.
6. *Rhymed Letter to Fru Heiberg.*
7. Some of George Brandes's views are in 'Henrik Ibsen: Personal Reminiscences and Remarks about his Plays', *Century Magazine*, New York, February 1917.
8. Quoted in Meyer, pp. 775–76.
9. Quoted in Bergliot Ibsen, *The Three Ibsens: Memories of Henrik I, Suzannah I and Sigurd I* (trans., London, 1951), pp. 17–18.
10. Meyer, p. 432; Paulsen's memoirs were published in Copenhagen in 1903.
11. Halvdan Koht, *Life of Ibsen* (2 vols., trans., London, 1931), vol. ii, p. 111.
12. Jaegar's notes about Ibsen were published in 1960; see Meyer, p. 603.
13. Quoted in Meyer, p. 592.
14. Bergliot Ibsen, p. 92.
15. Meyer, pp. 339, 343–44.
16. Hans Heiberg, *Ibsen: A Portrait of the Artist* (trans., London, 1969), p. 177.
17. Meyer, pp. 689–90.
18. Meyer, pp. 575–76.
19. Meyer, p. 805.
20. Meyer, pp. 277–78.
21. Meyer, p. 500.
22. Meyer, p. 258.
23. Letter of 9 December 1867, in Meyer, pp. 287–88.
24. Heiberg, pp. 20–22.
25. For Else see Meyer (3 vols.), vol. i, pp. 47–48.
26. Heiberg, p. 34.
27. The episode is related in Meyer (3 vols.), vol. iii, p. 206.
28. Heiberg, p. 241.
29. Meyer, p. 55.
30. Quoted in Meyer, pp. 304–5.

31. Meyer, pp. 293–94.
32. Printed in *Letters and Speeches*, pp. 315–16.
33. Bergliot Ibsen, pp. 84–85.
34. Quoted in Meyer, pp. 287–88.
35. Quoted in Meyer, p. 332.
36. Preface to *Cataline* (1875 edition).
37. 'Resignation' is included in John Northam's collection.
38. Meyer, p. 659.
39. Janson's diaries were published in 1913.
40. Quoted in Meyer, p. 531.
41. Heiberg, pp. 245–46; see Ibsen's speech to the working men of Trondhjem, 14 June 1885, in *Letters and Speeches*, pp. 248–49.
42. *Letters and Speeches*, pp. 251–56.
43. Meyer, p. 703.
44. *Letters and Speeches*, pp. 337–38.
45. Meyer, pp. 815–16.
46. Meyer, pp. 636 ff.
47. E. A. Zucker, *Ibsen: The Master Builder* (London, 1929).
48. Quoted in Meyer, p. 646.
49. Meyer, pp. 653–54.
50. The letters to Emilie Bardach are in *Letters and Speeches*, pp. 279–98.
51. Meyer, p. 97.
52. Letter to Magdalene Thoresen, 3 December 1865.
53. Meyer, pp. 250–51.
54. Meyer, p. 131.
55. Bergliot Ibsen, pp. 61–62.
56. Bergliot Ibsen, pp. 52, 79, 82 etc.
57. Meyer, pp. 280–81, 295–97.
58. Meyer, p. 581.

Chapter Five: Tolstoy: God's Elder Brother

1. Quoted in George Steiner, *Tolstoy or Dostoievsky* (London, 1960).
2. Diary entries for 12 October, 2–3 November 1853; 7 July 1857; 18 July 1853 in Aylmer Maude (ed.), *The Private Diary of Leo Tolstoy 1853–57* (London, 1927), pp. 79–80, 37, 227, 17.
3. Maxim Gorky, *Reminiscences of Tolstoy, Chekhov and Andreev* (London, 1934), quoted in Steiner, p. 125.
4. 19 January 1898, in *Diary*.
5. Quoted in Henri Troyat, *Tolstoy* (trans., London, 1968), pp. 133–40.
6. Ilya Tolstoy, *Tolstoy, My Father* (trans., London, 1972).
7. Leo Tolstoy, 'Boyhood'.
8. Quoted in Aylmer Maude, *Life of Tolstoy* (London, 1929), p. 69.
9. 3 November 1853, in *Diary*, p. 79.
10. Maude, *Life*, p. 37.
11. Maude, p. 126.
12. Maude, p. 200; Troyat, p. 194.
13. R. F. Christian, *Tolstoy: A Critical Introduction* (Cambridge, 1956).
14. Edward Crankshaw, *Tolstoy: The Making of a Novelist* (London, 1974) is particularly good on Tolstoy's strengths and weaknesses as a writer.

15. Elizabeth Gunn, *A Daring Coiffeur: Reflections on War and Peace and Anna Karenina* (London, 1971).
16. Both passages quoted by Gunn.
17. Quoted in Steiner, p. 229.
18. Quoted in Crankshaw, p. 66.
19. Entries for 25 and 27 July, 1 August 1857, in *Diary*; see also Introduction, p. xxiii.
20. *Diary*, pp. 10, 158.
21. *Diary*, pp. 10–16; Crankshaw, p. 128.
22. Troyat, p. 63.
23. Quoted in Anne Edwards, *Sonya: The Life of Countess Tolstoy* (London, 1981), p. 43.
24. Troyat, p. 212.
25. Quoted in Valentin F. Bulgakov, *The Last Year of Leo Tolstoy* (trans., London, 1971), pp. 145 46.
26. *Diary*, Introduction, p. xxi.
27. Quoted in Ernest J. Simmons, *Leo Tolstoy* (London, 1949), pp. 621–22.
28. Letter to N. N. Strakhov, author of an article, 'The Feminine Question', refuting J. S. Mill. Quoted in Simmons.
29. Crankshaw, pp. 145–52.
30. Edwards, pp. 77–87; Crankshaw, pp. 196–204; Simmons, p. 270.
31. Edwards, p. 267.
32. For a specimen of Tolstoy's holograph mss see photo in Crankshaw, p. 247.
33. Crankshaw, p. 198.
34. Quoted in Troyat, pp. 525–26.
35. Leo Tolstoy, *Recollections*.
36. Troyat, p. 141.
37. Quoted in Maude, pp. 250–51.
38. Crankshaw, p. 172.
39. Simmons, p. 400.
40. The death of Levin's brother in *Anna Karenina*; the refusal to attend the funeral in *War and Peace*.
41. Note of 16 December 1890.
42. Quoted in Troyat, p. 133.
43. Troyat, p. 212.
44. Crankshaw, pp. 237–38.
45. Letter to her sister, quoted in Simmons, p. 429.
46. Simmons, p. 738.
47. Quoted in Isaiah Berlin, *The Hedgehog and the Fox: An Essay on Tolstoy's View of History* (London, 1953), p. 6.
48. See the example cited by Berlin: the character of Kutuzov (a real person) in *War and Peace* is gradually transformed in successive drafts from 'the sly, elderly, feeble voluptuary' which he was in historical fact to 'the unforgettable symbol of the Russian people in all its simplicity and intuitive wisdom', which is what Tolstoy needed him to be.
49. Simmons, pp. 317–18.
50. For a shrewd analysis of Tolstoy's Christianity, see Steiner, pp. 260–65.
51. Diary entry of August 1898, quoted in Steiner, p. 259.

52. These *obiter dicta* are taken mainly from George Steiner's Introduction to Bulgakov, and from Bulgakov's text.

53. Simmons, pp. 493 ff.

54. Diary entry of 17 December 1890. Countess Tolstoy's diaries are published as *The Diary of Tolstoy's Wife, 1860–1891* (London, 1928); *The Countess Tolstoy's Later Diaries, 1891–97* (London, 1929); *The Final Struggle: Being Countess Tolstoy's Diary for 1910* (London, 1936).

55. See Bulgakov's own introduction to his *The Last Year of Leo Tolstoy*, esp. pp. xxiii–iv.

56. Bulgakov, p. 162.

57. Bulgakov, pp. 166 ff, 170–71.

58. Bulgakov, p. 197.

Chapter Six: The Deep Waters of Ernest Hemingway

1. See Edward Wagenknecht, *Ralph Waldo Emerson: Portrait of a Balanced Soul* (New York, 1974), Chapter 6, 'Politics', pp. 158–201.

2. *Journals and Miscellaneous Notebooks of Ralph Waldo Emerson* (14 vols., Harvard, 1960–) vol. vii, p. 435.

3. Thomas Wentworth Higginson, *Every Saturday*, 18 April 1868.

4. For this see Joel Porte, *Representative Man: Ralph Waldo Emerson in His Time* (New York, 1979).

5. *Correspondence of Emerson and Carlyle* (New York, 1964), p. 14.

6. Entry for 25 April 1848 in Joel Porte (ed.), *Emerson in his Journals* (Harvard, 1982), p. 385.

7. Henry James, *The Art of Fiction*, pp. 223–24.

8. *Journals and Misc. Notebooks*, vol. viii, pp. 88–89, 242.

9. Ibid., vol. ix, p. 115.

10. Ibid., vol. vii, p. 544.

11. See the illuminating article by Mary Kupiec Cayton, 'The Making of an American Prophet: Emerson, his audience and the rise of the culture industry in nineteenth-century America', *American Historical Review*, June 1987.

12. See Paul Boyer, *Urban Masses and Moral Order in America, 1820–1920* (Harvard, 1978), p. 109.

13. Quoted in Wagenknecht, p. 170; cf. Lewis S. Feuer, 'Ralph Waldo Emerson's Reference to Karl Marx', *New England Quarterly*, xxxiii (1960).

14. For Grace Hemingway, see Max Westbrook, 'Grace under Pressure: Hemingway and the Summer of 1920' in James Nagel (ed.), *Ernest Hemingway: The Writer in Context* (Madison, Wisconsin, 1984), pp. 77 ff; the family is described in Marcelline Hemingway Sandford, *At the Hemingways: A Family Portrait* (Boston, 1961).

15. Madeleine Hemingway Miller, *Ernie* (New York, 1975), p. 92. Kenneth S. Lynn, *Hemingway* (New York, 1987), pp. 19–20, says that these daily religious services were held only when the Hemingways were living with their Grandfather Hall, Grace's father.

16. Lynn, p. 115.

17. Carlos Baker (ed.), *Ernest Hemingway: Selected Letters, 1917–61* (New York, 1981), p. 3.

18. For Hemingway's religion, see Jeffrey Meyers, *Hemingway: A Biography* (London, 1985), pp. 31–32, 178, etc.; Lynn, pp. 70, 249, 312–14, etc.

19. Quoted in Lynn, pp. 117–18.
20. Quoted in Bernice Kert, *The Hemingway Women* (New York, 1983), p. 27.
21. *Selected Letters*, pp. 670, 663.
22. Lynn, p. 233.
23. Lynn, pp. 234 ff; see also B.J.Poli, *Ford Madox Ford and the Transatlantic Review* (Syracuse, 1967), p. 106.
24. Lynn, p. 230.
25. Quoted in Meyers, p. 24.
26. Meyers, p. 94.
27. See *Paris Review*, Spring 1981.
28. Given in Meyers, p. 137.
29. William White (ed.), *By-Line: Ernest Hemingway: Selected Articles and Dispatches of Four Decades* (New York, 1967), p. 219.
30. Quoted in Meyers, pp. 74–75.
31. *New Yorker*, 29 October 1927.
32. Introduction to an anthology, *Men at War* (New York, 1942).
33. Herbert Matthews, *A World in Revolution* (New York, 1971), pp. 24–25.
34. Quoted in Meyers, p. 426.
35. Quoted in Michael S. Reynolds, *Hemingway's Reading 1910–40* (Princeton, 1981), p. 4.
36. For Hemingway's lies, see Meyers, pp. 9, 15–16, 27, etc; Lynn, pp. 74, etc.
37. For this subject, see Michael S. Reynolds, *Hemingway's First War* (Princeton, 1976).
38. Letter to Hadley Hemingway, 31 January 1938, quoted in Lynn, p. 447.
39. John Dos Passos, *Best Times* (New York, 1966), p. 141.
40. *The Green Hills of Africa* (New York, 1935), p. 71.
41. Letter of 9 February 1937, in *Selected Letters*, p. 458; letter to Harry Sylvester, 1 July 1938, quoted in Meyers, p. 303.
42. See Hugh Thomas, *The Spanish Civil War* (London, 1982 edition), p. 706 and note; Lynn, pp. 448–49; *Selected Letters*, p. 463; Meyers, p. 307.
43. 'Fascism is a Lie', *New Masses*, 22 June 1937.
44. The best description of this is in Meyers, Chapter 18, 'Our Man in Havana', pp. 367–88; see also Lynn, pp. 502 ff.
45. Spruille Braden, *Diplomats and Demagogues* (New York, 1971).
46. Meyers, p. 370.
47. Jacqueline Tavernier-Courbin, 'Ernest Hemingway and Ezra Pound', in James Nagel (ed.), *Ernest Hemingway: The Writer in Context*, pp. 179 ff.
48. Letter to Archibald MacLeish, August 1943, quoted in Meyers, p. 514; E. Fuller Tolley, *The Roots of Treason: Ezra Pound and the Secrets of St Elizabeth's* (London, 1984).
49. *A Moveable Feast* (New York, 1964), pp. 208–9.
50. Meyers, pp. 205–6; Ludington Townsend, *John Dos Passos: A Twentieth-Century Odyssey* (New York, 1980).
51. See Lynn, pp. 38–48.
52. Letter to Arthur Mizener, 2 June 1950, in *Selected Letters*, p. 697.
53. Kert, *The Hemingway Women*, p. 170; this work is the primary source of information about all Hemingway's wives and girlfriends.
54. Quoted in Lynn, p. 356.
55. Kert, pp. 296–97.

56. Carlos Baker, *Ernest Hemingway: A Life Story* (New York, 1969), p. 380.
57. Meyers, p. 353.
58. Kert, pp. 391–92.
59. *Selected Letters*, p. 576.
60. Gregory H. Hemingway, *Papa* (Boston, 1976), pp. 91–92.
61. Meyers, p. 416.
62. Quoted in Meyers, p. 394.
63. *Selected Letters*, p. 572; Meyers, p. 530.
64. Letter of Martha Gellhorn to Clara Spieghel, 17 May 1940, quoted in Meyers, p. 353.
65. Lynn, pp. 517, 577; Meyers, p. 426.
66. Gregory Hemingway, p. 109; Meyers, pp. 447 ff; Adriana's side is put in her book of reminiscences, *La Torre Bianca* (Milan, 1980), which she wrote before committing suicide.
67. Kert, p. 476.
68. Mary Welsh Hemingway, *How It Was* (New York, 1976), p. 602.
69. *By-Line*, p. 473.
70. Mary Welsh Hemingway, p. 607.
71. Ibid., pp. 280–81.
72. Kert, pp. 268 ff.
73. Meyers, p. 480; *Selected Letters*, p. 367; Gregory Hemingway, p. 100.
74. Meyers, p. 351.
75. Kathleen Tynan, *The Life of Kenneth Tynan* (London, 1987), pp. 164–66.
76. Letter of 11 November 1920, quoted in Lynn, pp. 127–28.
77. It is printed in Meyers, Appendix I, pp. 573-75.
78. For a full medical analysis see Lynn, pp. 528–31.
79. C.L. Sulzberger, *A Long Row of Candles* (New York, 1969), p. 612.

Chapter Seven: Bertolt Brecht: Heart of Ice

1. Under the *glasnost* policy of Mikhail Gorbachov, more details about Brecht's life are beginning to appear in Communist publications: see Werner Mittenzwei, *The Life of Bertolt Brecht* (2 vols., East Berlin, 1987).
2. The most useful account of Brecht is Ronald Hayman, *Bertolt Brecht: A Biography* (London, 1983), which gives his background, pp. 5 ff. I have also made extensive use of Martin Esslin's brilliant work, *Bertolt Brecht: A Choice of Evils* (London, 1959).
3. Bertolt Brecht: *Gesammelte Gedichte*, p. 76.
4. Quoted by Sergei Tretyakov in 'Bert Brecht', *International Literature*, Moscow, 1937; cf. his poem, 'The Legend of the Dead Soldier'.
5. Esslin, pp. 8–9.
6. Walter Benjamin, *Understanding Brecht* (trans., London, 1973).
7. Esslin, pp. 27–28.
8. Quoted in Esslin, p. 22.
9. Ruth Fischer, *Stalin and German Communism* (Harvard, 1948), p. 615; Esslin, Chapter Seven, 'Brecht and the Communists', pp. 133–76.
10. Quoted by Daniel Johnson, 'Mac the Typewriter', *Daily Telegraph*, 10 February 1988.
11. Lotte H. Eisner, 'Sur le procès de l'Opéra de Quat' Sous', *Europe* (Paris), January–February 1957.

12. Esslin, pp. 42–43.
13. See James K. Lyon, *Bertolt Brecht in America* (Princeton, 1980), passim.
14. For Brecht's part in the Congressional hearings, see Lyon, pp. 326 ff.
15. *Hearings Regarding the Communist Infiltration of the Motion Picture Industry* (Washington DC, 1947) gives the text of the Brecht exchanges, pp. 491–504.
16. Quoted in Esslin, p. 71.
17. Hayman, pp. 337–40.
18. For Nellhaus and Bentley, see Lyon, pp. 152 ff, 205.
19. Esslin, pp. 81–82.
20. Hayman, p. 245.
21. Hayman, p. 225.
22. Quoted in Lyon, p. 209.
23. Hayman, pp. 140–41.
24. Lyon, pp. 238–39.
25. *New York Times*, 2 November 1958; Lyon, p. 300; Humphrey Carpenter, *W.H. Auden* (London, 1981), p. 412.
26. Lyon, pp. 264–65.
27. Esslin, p. 79.
28. Sidney Hook, *Out of Step: An Unquiet Life in the Twentieth Century* (New York, 1987), pp. 492–93.
29. See the *New Leader*, 30 December 1968, 28 April 1969.
30. Hayman, p. 209.
31. Brecht: *Schriften zur Politik und Gesellschaft*, pp. 111 ff.
32. Brecht: *Versuche* xii 147.
33. Quoted in Esslin, p. 162.
34. Quoted by Daniel Johnson, *Daily Telegraph*, 10 February 1988.
35. *Neues Deutschland*, 22 March, 19 October 1951; Esslin, pp. 154 ff.
36. *Tagesanzeiger* (Zurich), 1 September 1956.
37. *Neues Deutschland*, 23 June 1953.
38. See his *Arbeitsjournal* for 20 August 1953.
39. For an excellent treatment of the uprising, see Hayman, Chapter 33, 'Whitewashing', pp. 365–78.
40. *Europe*, January–February 1957.
41. Quoted in Esslin, p. 136.

Chapter Eight: Bertrand Russell: A Case of Logical Fiddlesticks

1. For bibliography see Barry Feinberg and Ronald Kasrils, *Bertrand Russell's America: His Transatlantic Travels and Writing*, vol. i. 1896–1945 (London, 1973).
2. Quoted in Rupert Crawshay-Williams, *Russell Remembered* (Oxford, 1970), p. 151.
3. Crawshay-Williams, p. 122.
4. A photograph of this page from his journal is reproduced in Ronald W. Clark, *Bertrand Russell and his World* (London, 1981), p. 13.
5. Although the draft of *The Principles of Mathematics* was completed on 31 December 1899, the work as a whole was not published till 1930; the first volume of *Principia Mathematica* appeared in 1910, volumes two and three in 1912 and 1913.
6. *The Philosophy of Leibnitz* (London, 1900).

7. Anthony Quinton, 'Bertrand Russell', *Dictionary of National Biography, 1961–70* (Oxford, 1981), p. 905.
8. Norman Malcolm, *Philosophical Review*, January 1950.
9. See G. H. Hardy, *Bertrand Russell and Trinity* (Cambridge, 1970).
10. For the details see Hardy.
11. Feinberg and Kasrils, pp. 60–61.
12. Crawshay-Williams, p. 143.
13. John Dewy and Horace M. Kallen (eds.), *The Bertrand Russell Case* (New York, 1941).
14. Bertrand Russell, *The Autobiography of Bertrand Russell* (3 vols., London 1969), vol. iii, pp. 117–18.
15. Crawshay-Williams, p. 41.
16. *Autobiography*, vol. ii, p. 17.
17. 'Russian Journal', entry for May 19 1920; Russell Archives, McMaster University, Hamilton, Ontario; quoted in Ronald W. Clark, *The Life of Bertrand Russell* (London, 1975), pp. 378 ff.
18. *International Journal of Ethics*, January 1915.
19. *Autobiography*, vol. i, p. 126.
20. *Atlantic Monthly*, March 1915.
21. *Autobiography*, vol. ii, p. 17.
22. Quoted in Feinberg and Kasrils, vol. i, p. 73.
23. Russell's views are presented in detail in Clark, Chapter 19, 'Towards a Short War with Russia?', pp. 517–30.
24. Letter to Gamel Brenan, 1 September 1945, quoted in Clark, p. 520.
25. 5 May 1948, Russell Archives; quoted in Clark, pp. 523–24.
26. *Nineteenth Century and After*, January 1949.
27. *World Horizon*, March 1950.
28. Quoted in Sidney Hook, *Out of Step: An Unquiet Life in the Twentieth Century* (New York, 1987), p. 364.
29. See the *Nation*, 17 and 29 October 1953.
30. Crawshay-Williams, p. 29.
31. The exchange was printed in the *Listener*, 19 March 1959.
32. *Listener*, 28 May 1959.
33. *Autobiography*, vol. iii, pp. 17–18.
34. Reprinted in Edward Hyams (ed.), *New Statesmanship: An Anthology* (London, 1963), pp. 245–49.
35. For the circumstances of the Russell–Krushchev–Dulles correspondence, see Edward Hyams, *The New Statesman: The History of the First Fifty Years, 1913–63* (London, 1963), pp. 288–92.
36. Crawshay-Williams, pp. 106–9.
37. The Collins version is given in L. John Collins, *Faith Under Fire* (London, 1966); the Russell version in Ralph Schoenman (ed.), *Bertrand Russell: Philosopher of the Century* (London, 1967). See also Clark, pp. 574 ff; Christopher Driver, *The Disarmers: A Study in Protest* (London, 1964).
38. Bertrand Russell, 'Voltaire's Influence on Me', *Studies on Voltaire*, vi (Musée Voltaire, Geneva, 1958).
39. Quoted in Clark, pp. 586 ff.
40. Quoted in Crawshay-Williams.
41. Crawshay-Williams, pp. 22–23.
42. *Autobiography*, vol. i, p. 16.

43. Feinberg and Kasrils, vol. i, p. 22.
44. Russell, *The Practice and Theory of Bolshevism* (London, 1920).
45. *Daily Herald*, 16 December 1921; *New Republic*, 15 and 22 March 1922; *Prospects of Industrial Civilization* (London, 1923).
46. Crawshay-Williams, p. 58.
47. Clark, pp. 627–28.
48. *Autobiography*, vol. i, p. 63.
49. *Manchester Guardian*, 31 October 1951.
50. Quoted in Clark, p. 592. Clark thinks this particular assertion was Schoenman's work, Russell having originally written 'Mankind is faced tonight by a grave crisis.' But the expression sounds to me very like Russell in his more extreme mood.
51. Quoted in *Time*, 16 February 1970.
52. Crawshay-Williams, pp. 17; ibid., 23; Feinberg and Kasrils, p. 118; letter to Miss R. G. Brooks, 5 May 1930, *Manners and Morals* (London, 1929).
53. 'Companionate Marriage', lecture in New York City, 3 December 1927, quoted in Feinberg and Kasrils, p. 106.
54. *Autobiography*, vol. i, pp. 203–4.
55. Quoted in Clark, p. 302.
56. Letter of 29 September 1918 (in Russell Archives), quoted in Clark.
57. *Autobiography*, vol. i, p. 206.
58. *Autobiography*, vol. ii, p. 26.
59. Dora to Rachel Brooks, 12 May 1922, Russell Archives, quoted in Clark, p. 397.
60. Dora Russell, *The Tamarisk Tree: My Quest for Liberty and Love* (London, 1975), p. 54.
61. Entry of 16 February 1922 in Margaret Cole (ed.), *Beatrice Webb's Diary 1912–1924* (London, 1952); Dora Russell, p. 53.
62. *New York Times*, 30 September 1927.
63. *Autobiography*, vol. ii, p. 192.
64. Dora Russell, p. 198.
65. Dora Russell, pp. 243–45.
66. Dora Russell, p. 279.
67. Quoted in Clark, p. 446.
68. Dora Russell, p. 286.
69. *Autobiography*, vol. iii, p. 16.
70. Letter of 11 October 1911, quoted in Clark, p. 142.
71. Hook, p. 208.
72. Peter Ackroyd, *T. S. Eliot* (London, 1984), pp. 66–67, 84; Robert H. Bell, 'Bertrand Russell and the Eliots', *The American Scholar*, Summer 1983.
73. Hook, p. 363.
74. Quoted in *Time*, 16 February 1970.
75. Dora Russell, p. 291.
76. *Autobiography*, vol. ii, p. 190.
77. Ralph Schoenman, 'Bertrand Russell and the Peace Movement', in George Nakhnikian (ed.), *Bertrand Russell's Philosophy* (London, 1974).
78. Hook, p. 307.
79. Clark, p. 584.
80. Quoted in Clark, p. 612.

81. The statement, published in the *New Statesman* after Russell's death, is given as an appendix in Clark, pp. 640–51.
82. *Autobiography*, vol. ii, p. 19.
83. Crawshay-Williams, pp. 127–28.
84. Clark, p. 610.
85. Clark, pp. 620–22.
86. *Autobiography*, vol. iii, pp. 159–60.
87. Hardy, p. 47.
88. *Autobiography*, vol. ii, p. 34.
89. Crawshay-Williams, p. 41.

Chapter Nine: Jean-Paul Sartre: 'A Little Ball of Fur and Ink'

1. Annie Cohen-Solal, *Sartre: A Life* (trans., London, 1987), p. 113.
2. Sartre, *Words* (trans., London, 1964), pp. 16–17.
3. *Words*, pp. 21–23.
4. *Words*, p. 73.
5. Quoted in Cohen-Solal, p. 40.
6. Sartre, *War Diaries: Notebook for a Phoney War, November 1939–March 1940* (trans., London, 1984), p. 281.
7. Cohen-Solal, p. 67.
8. Cohen-Solal, pp. 79–80.
9. 1945 article, reprinted in *Situations* (London, 1965).
10. Ernst Jünger, *Premier journal parisien 1941–43* (Paris, 1980).
11. Simone de Beauvoir, *The Prime of Life* (trans., London, 1962), p. 384. The Malraux quote is from Herbert Lottman, *Camus* (London, 1981 edition), p. 705.
12. Cohen-Solal, pp. 166–69. The text has disappeared.
13. Quotations from interviews in Cohen-Solal, pp. 176 ff.
14. De Beauvoir, *The Prime of Life*, p. 419.
15. *Lettres au Castor et à quelques autres* (2 vols., Paris, 1983).
16. *L'Être et le néant* (Paris, 1943); *Being and Nothingness* (trans., London, 1956, 1966).
17. Guillaume Ganotaux, *L'Age d'or de St-Germain-des-Prés* (Paris, 1965).
18. Sartre, *L'Existentialisme est un humanisme* (Paris, 1946); *Existentialism and Humanism* (London, 1973).
19. *Les Temps modernes*, 1 September 1945.
20. See Cohen-Solal, pp. 252–53. For the Picasso episode see Jacques Dumaine, *Quai d'Orsay 1945–51* (trans., London, 1958), p. 13.
21. *Samedi Soir*, 3 November 1945.
22. Christine Cronan, *Petit Catechisme de l'existentialisme pour les profanes* (Paris, 1946).
23. Herbert Lottman, 'Splendours and Miseries of the Literary Café', *Saturday Review*, 13 March 1965; and his 'After Bloomsbury and Greenwich Village, St-Germain-des-Prés', *New York Times Book Review*, 4 June 1967.
24. For a list of them see Cohen-Solal, pp. 279–80.
25. Lottman, *Camus*, p. 369.
26. Claude Francis and Fernande Gontier, *Simone de Beauvoir* (trans., London, 1987), pp. xiv, 6, 25 ff.
27. Ibid., p. 25.

28. Cohen-Solal, pp. 74–75.
29. Translated as *The Second Sex* (London, 1953).
30. Quoted in Cohen-Solal, p. 76.
31. *War Diaries*, pp. 281–82.
32. *War Diaries*, p. 325; Francis and Gontier, pp. 98–100.
33. Francis and Gontier, p. 1, note.
34. *War Diaries*, p. 183.
35. Quoted in Francis and Gontier, pp. 236–37.
36. *Lettres au Castor*, vol. i, pp. 214–15.
37. *L'Invitée* (Paris, 1943); *She Came to Stay* (Cleveland, 1954).
38. De Beauvoir, *The Prime of Life*, pp. 205, 193.
39. Quoted in Cohen-Solal, p. 213.
40. Francis and Gontier, pp. 197–200.
41. John Weightman in the *New York Review of Books*, 13 August 1987.
42. Francis and Gontier, p. xiii.
43. Cohen-Solal, pp. 373 ff.
44. Cohen-Solal, p. 466.
45. Simone de Beauvoir: *La Force des choses* (Paris, 1963); Lottman, *Camus*, p. 404.
46. *Les Temps modernes*, August 1952. For the quarrels see Lottman, *Camus*, Chapter 37, pp. 495 ff. Sartre's attack is reprinted in *Situations*, pp. 72–112.
47. Jean Kanapa: *L'Existentialisme n'est pas un humanisme* (Paris, 1947), p. 61.
48. Quoted in Cohen-Solal, p. 303.
49. *Le Figaro*, 25 April 1949.
50. *Saint Genet, Comedien et Martyr* (Paris, 1952); trans., New York, 1963, 1983.
51. Sartre wrote a little book about the first, *L'Affaire Henri Martin* (Paris, 1953).
52. *Libération*, 16 October 1952.
53. Quoted in Walter Laqueur and G. L. Mosse, *Literature and Politics in the Twentieth Century* (New York, 1967), p. 25.
54. *Les Lettres francaises*, 1–8 January 1953; *Le Monde*, 25 September 1954.
55. *Libération*, 15–20 July 1954.
56. *Situations X* (Paris, 1976), p. 220.
57. Report in *Paris-Jour*, 2 October 1960.
58. 'Madame Gulliver en Amerique' in Mary McCarthy, *On the Contrary* (New York, 1962), pp. 24–31.
59. Interview in *France-Observateur*, 1 February 1962.
60. David Caute, *Sixty-Eight: The Year of the Barricades* (London, 1988), pp. 95–96, 204.
61. Cohen-Solal, pp. 459–60; Francis and Gontier, pp. 327 ff.
62. *Nouvel-Observateur*, 19 and 26 June 1968.
63. Cohen-Solal, p. 463.
64. *L'Aurore*, 22 October 1970.
65. Letter to de Beauvoir, 20 March 1940.
66. Unpublished mss, 1954, now in the Bibliothèque nationale, quoted in Cohen-Solal, pp. 356–57.
67. James Boswell, *Life of Dr Johnson*, Everyman Edition (London, 1906), vol. ii, p. 326.
68. John Huston, *An Open Book* (London, 1981), pp. 295.
69. Cohen-Solal, pp. 388–89.
70. Francis and Gontier, pp. 173–74.

71. *War Diaries*, pp. 297–98.
72. Jean Cau, *Croquis de Memoire* (Paris, 1985).
73. Mary Welsh Hemingway, *How It Was* (New York, 1976), pp. 280–81.
74. Cohen-Solal, p. 377.
75. For example, three issues of *Nouvel-Observateur*, March 1980, on the eve of Sartre's death.

Chapter Ten: Edmund Wilson: A Brand from the Burning

1. See Leon Edel (ed.), *Edmund Wilson: The Twenties* (New York, 1975), Introduction.
2. Ella Winter and Granville Hicks (eds.), *The Letters of Lincoln Steffens* (2 vols., New York, 1938), vol. ii, pp. 829–30.
3. Don Congdon (ed.), *The Thirties: A Time to Remember* (New York, 1962), pp. 24, 28–29.
4. Lionel Trilling, *The Last Decade: Essays and Reviews 1965–75* (New York, 1979), pp. 15–16.
5. Trilling, p. 24.
6. Article reprinted in *The Shores of Light* (New York, 1952), pp. 518–33.
7. Leon Edel (ed.), *Edmund Wilson: The Thirties* (New York, 1980), p. 206.
8. Ibid., pp. 208–13.
9. Ibid., p. 81.
10. Ibid., pp. 678–79.
11. Ibid., pp. 57, 64, 118, 120, 121–22, 135.
12. Ibid., pp. 160–86; letter to Dos Passos, 29 February 1932.
13. Ibid., pp. 378 ff.
14. Mary McCarthy's background and childhood is described in Doris Grumbach, *The Company She Keeps* (London, 1967).
15. Her essay 'The Vassar Girl', reprinted in Mary McCarthy, *On the Contrary* (London, 1962), pp. 193–214, is a brilliant evocation of the Vassar spirit.
16. Reprinted in *Cast a Cold Eye* (New York, 1950).
17. Lionel Abel, 'New York City: A Remembrance', *Dissent*, viii (1961).
18. Printed in *Rebel Poet*, and quoted in Terry A. Cooney, *The Rise of the New York Intellectuals: Partisan Review and Its Circle* (Wisconsin, 1986), p. 41.
19. *Partisan Review*, xii (1934).
20. In *New Masses*, August 1932.
21. For Rahv's various political positions, see A.J.Porter and A.J.Dovosin (eds.), *Philip Rahv: Essays on Literature and Politics, 1932–78* (Boston, 1978).
22. Quoted in Cooney, pp. 99–100.
23. Quoted in Cooney, p. 117.
24. See 'The Death of Gandhi' and 'My Confession', in McCarthy, pp. 20–23, 75–105.
25. Title of article by Harold Rosenberg, *Commentary*, September 1948.
26. See *New York Times Book Review*, 17 February 1974.
27. Norman Podhoretz, *Breaking Ranks: A Political Memoir* (New York, 1979), p. 270.
28. Leon Edel (ed.), *Edmund Wilson: The Fifties; from Notebooks and Diaries of the Period* (New York, 1986), pp. 372 ff (esp. entry of 9 August 1956).
29. *Edmund Wilson: The Twenties*, pp. 64–65.
30. Ibid., pp. 15–16.

31. *Edmund Wilson: The Thirties*, p. 593.
32. Ibid., pp. 6, 241 ff, 250 ff, etc.
33. Ibid., pp. 296–97, 523; Leon Edel (ed.), *Edmund Wilson: The Forties* (New York, 1983), pp. 108–9.
34. *Edmund Wilson: The Fifties*, pp. 582, 397, 140.
35. For example, Chapter 13 of Mary McCarthy, *The Group* (New York, 1963).
36. Quoted in Grumbach, pp. 117–18.
37. *Edmund Wilson: The Forties*, p. 269.
38. Reprinted in Lewis M. Dabney (ed.), *The Portable Edmund Wilson* (London, 1983), pp. 20–45.
39. *Edmund Wilson: The Forties*, pp. 80–157 and passim.
40. *Edmund Wilson: The Fifties*, pp. 101, 135–38, 117.
41. Isaiah Berlin's account of Wilson's 1954 visit, published in the *New York Times*.
42. *The Twenties*, p. 149; *The Thirties*, pp. 301–3; *The Fifties*, pp. 452 ff, 604, etc.; Berlin memoir.
43. Edmund Wilson, *The Cold War and the Income Tax: A Protest* (New York, 1963), p. 7.
44. Ibid., p. 4.
45. *The Portable Edmund Wilson*, p. 72.

Chapter Eleven: The Troubled Conscience of Victor Gollancz

1. Ruth Dudley Edwards, *Victor Gollancz: A Biography* (London, 1987).
2. For the Gollancz brothers see *Dictionary of National Biography, Supplementary Volume 1922–30* (Oxford, 1953), pp. 350–52.
3. Edwards, p. 48.
4. Quoted in Edwards, p. 102.
5. Quoted in Edwards, p. 144.
6. Douglas Jerrold, *Georgian Adventure* (London, 1937).
7. For the firm see Sheila Hodges, *Gollancz: The Story of a Publishing House* (London, 1978).
8. Edwards, pp. 171–72, 175.
9. Edwards, p. 180.
10. Quoted in Edwards, p. 235.
11. Edwards, p. 382.
12. Quoted in Edwards, p. 250.
13. Quoted in Edwards, p. 208.
14. Sidney and Beatrice Webb, *Soviet Communism: A New Civilization* (2 vols., London, 1935).
15. Letter to Stephen Spender, February 1936.
16. Cole's books were published in 1932 and 1934; Strachey's in 1932.
17. Quoted in Edwards, p. 211.
18. November 1932; quoted in Edwards, p. 211.
19. Edwards, pp. 251, 247; Miller's censored book was called *I Found No Peace*.
20. For the LBC see John Lewis, *The Left Book Club* (London, 1970).
21. See Hugh Thomas, *John Strachey* (London, 1973).
22. See Kingsley Martin, *Harold Laski* (London, 1953).
23. *Daily Worker*, 8 May 1937.
24. *Moscow Daily News*, 11 May 1937.

25. Letter to J.B.S.Haldane, May 1938, quoted in Edwards, p. 257.
26. Edwards, p. 251.
27. Edwards, p. 250.
28. George Orwell, *Collected Essays, Journalism and Letters* (4 vols., Harmondsworth, 1970), vol. i *1920–40*, p. 334 note.
29. Kingsley Martin, *Editor: A Volume of Autobiography 1931–45* (London, 1968), p. 217; for Muenzenberg see Arthur Koestler, *The Invisible Writing* (London, 1954).
30. Claud Cockburn, *I Claud: An Autobiography* (Harmondsworth, 1967), pp. 190–95.
31. Martin pp. 215 ff; C.H.Rolph, *Kingsley: The Life, Letters and Diaries of Kingsley Martin*, (London, 1973), pp. 225 ff; Orwell, vol. i, pp. 333–36.
32. Edwards, pp. 246–48.
33. Orwell, vol. i, p. 529.
34. Edwards, p. 313.
35. Edwards, p. 387.
36. Quoted in Edwards, p. 269.
37. Edwards, p. 408.
38. *Dictionary of National Biography, Supplementary Volume, 1961–70* (Oxford, 1981), p. 439.

Chapter Twelve: Lies, Damned Lies and Lillian Hellman

1. William Wright, *Lillian Hellman: The Image, the Woman* (London, 1987), pp. 16–18.
2. Wright, pp. 22–23, 327.
3. The autobiography is in three parts: *An Unfinished Woman* (Boston, 1969); *Pentimento* (Boston, 1973); *Scoundrel Time* (Boston, 1976).
4. Wright, p. 51.
5. There are two biographies of Hammett: Richard Layman, *Shadow Man: The Life of Dashiell Hammett* (New York, 1981), and Diane Johnson, *The Life of Dashiell Hammett* (London, 1984).
6. Johnson, pp. 119 ff.
7. Johnson, pp. 129–30.
8. Johnson, pp. 170–71.
9. Wright, p. 285.
10. Wright, p. 102.
11. See Mark W. Estrin, *Lillian Hellman: Plays, Films, Memoirs* (Boston, 1980); Bernard Dick, *Hellman in Hollywood* (Palo Alto, 1981).
12. Quoted in Wright, p. 326.
13. Quoted in Wright, p. 295.
14. See Harvey Klehr, *The Heyday of American Communism* (New York, 1984).
15. Wright, pp. 129, 251 ff, 361–62.
16. Wright, p. 161.
17. Wright, pp. 219–20.
18. *New York Times*, 2 March 1945.
19. Wright has a full account of all this, pp. 244–56.
20. Johnson, pp. 287–89.
21. Wright, p. 318.

22. *Commentary*, June 1976; *Encounter*, February 1977; *Esquire*, August 1977; *Dissent*, Autumn 1976.
23. Wright, p. 395.
24. See Wright, pp. 295–98, 412–13.

Chapter Thirteen: The Flight of Reason

1. Quoted in David Pryce-Jones, *Cyril Connolly: Diaries and Memoir* (London, 1983), p. 292.
2. Orwell's essay, 'Such, Such Were the Joys' was first published in *Partisan Review*, September–October 1952; reprinted in George Orwell, *Collected Essays, Journalism and Letters* (4 vols., Harmondsworth, 1978 edition), vol. iv, pp. 379–422. Connolly's account is in *Enemies of Promise* (London, 1938).
3. Gow made this charge in a letter to the *Sunday Times* in 1967; quoted in Pryce-Jones.
4. Both republished in Orwell, *Collected Essays*.
5. Orwell, *Collected Essays*, vol. i, p. 106.
6. Orwell, *The Road to Wigan Pier* (London, 1937), p. 149.
7. Orwell, *Homage to Catalonia* (London, 1938), p. 102.
8. Quoted in Pryce-Jones, p. 282.
9. Orwell, *Collected Essays*, vol. i, p. 269.
10. Orwell, *Collected Essays*, vol. iv, p. 503.
11. Mary McCarthy, *The Writing on the Wall and other Literary Essays* (London, 1970), pp. 153–71.
12. Orwell, *Collected Essays*, (1970 edition), vol. iv, pp. 248–49.
13. Michael Davie (ed.), *The Diaries of Evelyn Waugh* (London, 1976), p. 633.
14. Mark Amory (ed.), *The Letters of Evelyn Waugh* (London, 1980), p. 302.
15. Evelyn Waugh, Introduction to T. A. MacInerny, *The Private Man* (New York, 1962).
16. Pre-election symposium, *Spectator*, 2 October 1959.
17. Evelyn Waugh, review of *Enemies of Promise*, *Tablet*, 3 December 1938; reprinted in Donat Gallagher (ed.), *Evelyn Waugh: A Little Order: A Selection from his Journalism* (London, 1977), pp. 125–27.
18. These marginal notes are analysed in Alan Bell's article, 'Waugh Drops the Pilot', *Spectator*, 7 March 1987.
19. *Tablet*, 3 December 1939.
20. 'The Joker in the Pack', *New Statesman*, 13 March 1954.
21. Quoted in Pryce-Jones, p. 29.
22. Quoted in Pryce-Jones, p. 40.
23. Pryce-Jones, pp. 131, 133, 246.
24. Cyril Connolly, 'Some Memories', in Stephen Spender (ed.), *W. H. Auden: A Tribute* (London, 1975), p. 70.
25. 'London Diary', *New Statesman*, 16 January 1937.
26. 'London Diary', *New Statesman*, 6 March 1937.
27. 1943 broadcast as part of Orwell's *Talking to India* series; quoted in Pryce-Jones.
28. 'Comment', *Horizon*, June 1946.
29. *Tablet*, 27 July 1946; reprinted in Gallagher, pp. 127–31.
30. This is the version (there are others) given by John Lehmann in the *Dictionary of National Biography, 1971–80* (Oxford, 1986), pp. 170–71.

31. *New Statesman*, 13 March 1954.
32. Leon Edel (ed.), *Edmund Wilson: The Fifties* (New York, 1986), pp. 372 ff.
33. Barbara Skelton, *Tears Before Bedtime* (London, 1987), pp. 95–96, 114–15.
34. In 1971 interview, quoted in Paul Hollander: *Political Pilgrims: Travels of Western Intellectuals to the Soviet Union, China and Cuba, 1928–78* (Oxford, 1981); see also Maurice Cranston, 'Sartre and Violence', *Encounter*, July 1967.
35. Michael S. Steinberg, *Sabres and Brownshirts: The German Students' Path to National Socialism 1918–35* (Chicago, 1977).
36. Humphrey Carpenter, *W.H.Auden* (London, 1981), pp. 217–19.
37. Edward Hyams, *The New Statesman: The History of the First Fifty Years, 1913-63* (London, 1963), pp. 282–84.
38. For the facts of Mailer's background and career, see Hilary Mills, *Mailer: A Biography* (New York, 1982).
39. *Atlantic Monthly*, July 1971.
40. Mills, pp. 109–10.
41. Norman Podhoretz, *Doings and Undoings* (New York, 1959), p. 157.
42. The whole business of the stabbing is fully described in Mill, Chapter X, pp. 215 ff.
43. Mailer's speech is reprinted in his *Cannibals and Christians* (Collected Pieces, New York, 1966), pp. 84–90.
44. Jack Newfield in the *Village Voice*, 30 May 1968; quoted in Mills.
45. Mills, pp. 418–19.
46. Kathleen Tynan, *The Life of Kenneth Tynan* (London, 1987).
47. Tynan, pp. 46–47.
48. See Ronald Bryden, *London Review of Books*, 10 December 1987.
49. *Declaration* (London, 1957).
50. For Agate's (censored) account of their relationship, see his *Ego 8* (London, 1947), pp. 172 ff.
51. Tynan, p. 32.
52. Quoted in Tynan, p. 76.
53. Tynan, p. 212.
54. Tynan, pp. 327, 333.
55. Tynan, p. 333.
56. Shakespeare, *Sonnets*, 129.
57. For an account of Fassbinder's rise and many other curious details see Robert Katz and Peter Berling, *Love is Colder than Death: The Life and Times of Rainer Werner Fassbinder* (London, 1987).
58. Katz and Berling, Introduction, p. xiv.
59. Katz and Berling, p. 19.
60. Katz and Berling, pp. 33–34, 125.
61. Quoted in Katz and Berling, p. 5.
62. Fern Marja Eckman, *The Furious Passage of James Baldwin* (London, 1968); see also obituaries in *New York Times, Washington Post, Guardian, Daily Telegraph* and Bryant Rollings, *Boston Globe*, 14–21 April 1963.
63. Quoted in Eckman, pp. 63–64.
64. 'The Harlem Ghetto', *Commentary*, February 1948.
65. See, for instance, those in his collection *Notes of a Native Son* (New York, 1963).

66. Norman Podhoret, *Breaking Ranks: A Political Memoir* (New York, 1979), pp. 121 ff.
67. See 'Alas, Poor Richard!' in Baldwin's collection *Nobody Knows My Name* (New York, 1961).
68. See Baldwin's autobiographical novel, *Go Tell It on the Mountain* (London, 1954), 'East River, Downtown' in *Nobody Knows My Name*, and his essay in John Handrik Clark (ed.), *Harlem: A Community in Transition* (New York, 1964).
69. Quoted in Eckman, p. 65.
70. 'Fifth Avenue Uptown: A Letter from Harlem', *Esquire*, June 1960.
71. Eckman, p. 163.
72. 'Letter from a Region of My Mind', *New Yorker*, 17 November 1962.
73. Bertrand Russell, *Human Knowledge: Its Scope and Limits* (London, 1948).
74. See S. P. Stitch (ed.), *Innate Ideas* (California, 1975).
75. See Chomsky's *Cartesian Linguistics* (New York, 1966) and his *Reflections on Language* (London, 1976). For an illuminating analysis of Chomsky's theories of language and knowledge, and the political conclusions he draws from them, see Geoffrey Sampson, *Liberty and Language* (Oxford, 1979).
76. Noam Chomsky, *Problems of Knowledge and Freedom: The Russell Lectures* (London, 1972).
77. Noam Chomsky, *For Reasons of State* (New York, 1973), p. 184.
78. Noam Chomsky, *American Power and the New Mandarins* (New York, 1969), pp. 47–49.
79. Chomsky's contribution to the Pol Pot controversy is scattered in many places, often in obscure magazines. See his collection *Towards a New Cold War* (New York, 1982), pp. 183, 213, 382 note 73, etc. See also Elizabeth Becker, *When the War Was Over* (New York, 1987).

Index

Abildgaard, Theodor, 103
Abraham, Pierre, 196
Adenauer, Konrad, 192, 210
Adorno, Theodor, 188
Agate, James, 327–8
Alexander II, Tsar of Russia, 123
Algren, Nelson, 237, 295
Althusser, Louis, 63
Amis, Kingsley, *Lucky Jim*, 228
Ammers, K. L., 177, 188
ancien régime, 2, 82, 138, 183
Andersen, Hildur, 95, 100
Anderson, Maxwell, 147; *Key Largo*, 293
Anderson, Sherwood, 255
Annenkov, Pavel, 70, 72
anti-Americanism, 210–12, 218, 245, *see also* Vietnam; anti-Britishism, 264–5
Apollinaire, Guillaume, 234
Arbuckle, Roscoe ('Fatty'), 290
Archer, William, 86, 105
Argyll, 2nd Earl of, 76
Arnold, Matthew, 30

Aron, Raymond, 227, 228, 240, 241, 246
Arsenev, Valerya, 118
Asch, Nathan, 148
Asquith, Raymond, 204
Attlee, Clement, 281, 285
Auden, W. H., 179, 188, 265, 315, 319
Audiberti, Jacques, 234
Augustine, Saint, 52
Austen, Jane, 265; *Northanger Abbey*, 32
Avedon, Richard, 334, 336
Aveling, Edward, 79, 80

Baader, Andreas, 332; Baader-Meinhof group, 332
Bacall, Lauren, 166
Bakunin, Michael, 72, 78
Baldwin, James, 324, 333–7; *The Fire Next Time*, 336, 337; *Giovanni's Room*, 335; *Go Tell It On the Mountain*, 335
Balfour, Arthur, 223

Balzac, Honoré de, 147, 234
Banholzer, Paula, 185, 187
Bankhead, Tallulah, 296
Bardach, Emilie, 100–2
Barnes, Djuna, 148
Barney, Natalie, 148
Barnum, P. T., 142, 233
Barruel, Abbé, *Memoirs Illustrating the History of Jacobitism*, 32
Barry, Griffin, 216
Barry, Philip, *The Philadelphia Story*, 293
Barthes, Roland, 183
Bauer, Bruno, 57, 70
Baum, Vicki, *Grand Hotel*, 290
Baumgardt, David, 179
Bayle, Pierre, 4
Bazykin, Aksinya, 116
Bazykin, Timofei, 100–7
BBC, 206
Beatty, Warren, 305
Becher, Johannes, 195
Becket, Saint Thomas à, 27
Beckett, Samuel, 329
Bedwell, John, 47, 48
Beerbohm, Max, 87, 197
Beethoven, Ludwig van, 333
Bell, Bishop, 287
Benda, Julien, 233; *La Trahison des clercs*, 232
Benjamin, Walter, 170, 175
Benn, Sir Ernest, 271–2
Bennett, Arnold, 325
Bentley, Eric, 182, 183, 190, 196, 301
Berenson, Bernard, 155, 165
Bergsoe, Vilhelm, 103, 104
Berkeley, Bishop, 338
Berkeley, Martin, 295
Berlau, Ruth, 186, 187, 195
Berlin, Sir Isaiah, 265
Berners, Lord, 317

Bernstein, Carl, 305
Berzin, Jan Antonovic, General, 156
Birgukov, P. I., 123
Bishop, John Peale, 261–2
Bjornson, Bjornstjerne, 91, 93, 95, 97, 105
Blanc, Louis, 56
Blanqui, Louis Auguste, 56
Bludov, Count Dmitri, 123
Blue Books, 67, 130, 198
Boothby, Brooke, 16
Bornemann, Ernest, 188
Boswell, James, 16, 20, 314
Bourbon dynasty, 173
Bourdet, Edouard, *The Captive*, 292
Braden, Spruille, 158
Brailsford, H. N., 281, 282
Braine, John, 209
Brandes, Georg, 86, 88, 89, 90, 96, 97, 98, 100
Brando, Marlon, 335
Brecht, Bertolt, 173–96, 233, 239, 275, 298, 326, 330, 332; *The Caucasian Chalk Circle*, 179, 182; *Drums in the Night*, 175, 184; *Freiheit und Demokratie*, 188; *The Good Woman of Setzuan*, 178; *Der Jäsager*, 178; *The Life of Galileo*, 178, 179, 184; *Die Massbahme*, 177; *Mother Courage and her Children*, 178, 180, 182, 192; *The Resistable Rise of Arturo Ui*, 179; *The Rise and Fall of the City of Mahagonny*, 177; *Senora Carrara's Rifles*, 188; *The Threepenny Opera*, 177, 185, 195, 330 *The Trial of Lucullus*, 178, 192; *Versuhe*, 176
Brecht, family: parents, 174; Walter (brother), 174; children, 187

Bredel, Willi, 178
Bremer, Karl-Heinz, 229
Bronner, Arnolt, 196
Brooks, Van Wyck, 265
Browder, Earl, 260
Browning, Elizabeth Barrett, 33
Bruun, Christopher, 103
Buddha, 107; Buddhism, 188
Budenz, Louis, 295
Bulgakov, Valentin, 135, 136
Bull, Dr Edvard, 101, 106
Bull, Francis, 92
Bull, Ole, 84
Burke, Edmund, 2, 10
Burns, Emile, 280
Burns, Mary, 75
Butler, Reg, 209
Buttinger, Joe, 303
Byron, George Gordon, Lord, 30,
    31, 35, 41, 43, 44, 46, 49, 50,
    51, 85, 114, 143, 151, 152, 173,
    175, 190

Calvinism, 4, 8
Cambodia, *see* Pol Pot
Camden, Lord, 15
Campaign for Nuclear
    Disarmament (CND), 208,
    209, 219, 286; Committee of
    One Hundred, 209, 220
Campbell, Alan, 295
Campbell, J. R., 282
Camus, Albert, 234, 240–1, 250
capitalism, capitalists, 4, 57, 64,
    68–9, 73, 131, 142, 175, 177,
    181, 192, 210–11, 254, 255,
    256, 259, 276, 289, 309, 339,
    340
Carl XV, King of Sweden, 88, 93
Carlyle, Thomas, 63, 139, 140;
    Jane, 140

Carney, William, 296
Castlereagh, Lord, 32, 49
Castro, Fidel, 245, 319, 327
Catherine the Great, Empress of
    Russia, 108
Catholicism, 5, 8, 144–5, 155, 159,
    201, 242, 259, 260, 316;
    Vatican Index of Prohibited
    Books, 233
Cau, Jean, 240, 241, 250
Cavett, Dick, 324
Cerf, Bennett, 324
Challaye, Félicien, 282
Challoner, W. H., 65
Chaplin, Charlie, 189
Charmette, Comte de, 19
Chateaubriand, Vicomte de,
    Memoires d'outre-tombe, 248
Chayevsky, Paddy, 334
Chazelas, George, 230
Chekhov, Anton, 125
Chertkov, V. G., 132, 135, 136,
    137, 219, 220, 250
Chesterton, G. K., 253
Childers, Erskine, *The Riddle of the
    Sands*, 158
Christianity, 53, 58, 105, 107, 124,
    130, 144, 146, 148, 151, 201,
    223, 270, 277, 285, 317, 323; *see
    also* religion, Calvinism,
    Catholicism, Protestantism,
    Orthodox Church, Quakerism
Choiseul, Duc de, 15
Chomsky, Noam, 337–41;
    Syntactic Structures, 338
Chou En-Lai, 220, 221
Churchill, Sir Winston, 223, 264
Clairmont, Allegra, 43, 190
Clairmont, Charles, 39
Clairmont, Claire, 35, 36, 37, 39,
    41, 43, 50, 190
Clark, Kenneth, 315

Clark, Ronald, 204, 207, 221
Cockburn, Claud, 283, 284, 310
Cohen-Solal, Annie, 240
Cohn-Bendit, Daniel, 246
Cole, G. D. H., 277
Coleridge, Samuel Taylor, 32, 168
Colet, John, 27
Collins, Canon John, 208, 209, 286
Collins, Norman, 274
communism/collectivism,
    communists, 98, 179, 190, 206,
    208, 231, 233, 257, 260, 277–8,
    298, 310
Communist League, 54, 61; Marx
    writes manifesto, 54
Communist Party, 156–7, 173, 176,
    177, 179, 180, 181, 190, 191,
    192, 194, 230, 231, 233, 241,
    242–3, 244, 246, 255, 257, 259,
    260, 276, 278, 279–80, 281,
    282–3, 284, 294–6, 308, 309,
    321
Confucius, 107
Connolly, Cyril, 261, 265, 306, 307,
    308, 312–18, 322, 324, 327,
    330; *Enemies of Promise*, 313,
    315; *Horizon*, 316, 318; *The
    Unquiet Grave*, 316
Connolly wives: Jean (*née*
    Bakewell), 314; Barbara, *see*
    Skelton, Barbara
Conrad, Joseph, 129, 147, 148, 149
Coolidge, Calvin, 197, 215
Corneille, Pierre, 4
Corsica, constitution of, Rousseau
    writes, 15, 25
Cort, David, 290
Counter-Reformation, 173
Cowan, Arthur, 304
Cowley, Malcolm, 166, 295
Crane, Stephen, 147, 197
Crankshaw, Edward, 112

Crawshay-Williams, Rupert, 208,
    210, 211, 219, 222
Crocker, Lester, 9, 18
Crossman, R. H. S., MP, 283
Cullen, Countee, 334

Dana, Charles Anderson, 56
Danby-Smith, Valerie, 167
Dante, *The Divine Comedy*, 265
Darwin, Charles, 52
Davis, Gary, 242
Davy, Georges, 235
Dawson, Geoffrey, 282
Dearmer, Revd Percy, 277
de Beauvoir, Simone, 211, 227,
    230, 234, 235–9, 240, 242, 245,
    248, 249, 250, 251, 262, 263,
    275, 302; *Les Mandarins*, 235,
    237; *The Prime of Life*, 237; *The
    Second Sex*, 235
Debs, Eugene, 156
Dediger, Vladimir, 211
De Falla, Manuel, 159
de Gaudillac, Maurice, 235
de Gaulle, Charles, 152, 210, 244,
    245, 247
de Kock, Paul, 109
Delacroix, Eugène, 21
democracy, 98, 129, 131, 210–11,
    212, 242, 243, 245, 277, 339,
    340
Dempsey, Jack, 176
Demuth, Helen ('Lenchen'),
    79–80, 92, 186
Demuth, Henry Frederick, 80–81,
    187
Descartes, René, 4, 109, 228, 243,
    338
Dessau, Paul, 192
Deutscher, Isaac, 211
de Viane, Elise, 291

Dickens, Charles, 113, 129, 140; *Bleak House*, 46

Diderot, Denis, 6, 14, 17, 26, 95, 234, 240, 302

Dietrich, Marlene, 166

Dilworth, Jim, 168

Diogenes, 201

Dorman-Smith, 'Chink', 155

Dos Passos, John, 144, 146, 155, 156, 157, 161, 228, 255, 257, 302; *Chosen Country*, 161

Dos Passos, Katy (*née* Smith), 144, 161

Dostoyevsky, Fyodor, 147

Douglas, Norman, 147

Dreyfus, Alfred, 232, 233, 235, 282

Duclos, C. P., 11

Dudley, Helen, 214

Dudley Edwards, Ruth, 269–70, 286

Due, Christopher, 94

Dulles, John Foster, 208

Dumas, Alexandre, 109

du Maurier, Daphne, 273

Dundy, Elaine, 328

Duran, Gustavo, General, 155, 159

Duranty, Walter, 295

Dutt, R. Palme, 280, 281

Dyakov, Mitya, 125

Eastman, Max, 160

Eckman, Fern Marja, 333, 336

Eddington, Sir Arthur, 224

Edel, Leon, 253

Einstein, Albert, 195, 200, 201

Eisenhower, Dwight D., 207

Eisler, Hanns, 176

Eldon, Lord, 42

Elias, Julius, 101

Eliot, George, 27

Eliot, T. S., 147, 204, 215, 218, 222, 265; Vivien, 218

Elouard, Paul, 231

Emerson, Ralph Waldo, 139–43

Encyclopédie, 6, 24

Engels, Friedrich, 54, 56, 60, 61, 63, 64–5, 66, 67, 72, 73, 75, 76, 78, 80, 189, 256, 341

Enlightenment, 3, 6, 8, 32

Épinay, Madame d', 10, 11, 14, 16, 19, 26

Epstein, Jason, 322

Epting, Karl, 229

Erasmus, 27, 318

Essex, Earl of, 173

Evans, John, 47

existentialism, 225, 229–30, 232, 233–4, 324, 339

Fadayev, Aleksandr, 233, 243

Fanon, Franz, 245, 246, 337

Farrell, James T., 260

Fassbinder, Rainer Werner, 330–3

Fassbinder, wives: Ingrid (*née* Caven), 331, 333; Juliane (*née* Lorenz), 331, 333

Faulkner, William, 161, 228, 290

Faux, Claude, 250

Feiffer, Jules, 305

Fet, Afanasi, 114, 126, 128

Feuchtwanger, Lion, *Erfolg*, 175, 178

Feuerbach, Ludwig, 107

Field, Frederick Vanderbilt, 294

Fischer, Ruth, 176

Fisher, Geoffrey, 271

Fitinghoff, Rosa, 100

Fitzgerald, F. Scott, 160, 168, 265; Zelda, 160

Flanner, Janet, 324

Flaubert, Gustave, 27, 113, 129, 147, 232, 248; *Madame Bovary*, 226
Fleming, Annie, 317
Foggi, Paolo, 43
Fontenelle, Bernard de, 4
Ford, Ford Madox, 147, 148, 150, 160
Ford, Henry, 256
Forster, E. M., 265
Foster, John, 140
Foster, William Z., 255
France, intellectuals in, 6, 232, 234; American intellectuals in, 147–8
Franco, Generalissimo Francisco, 159, 283
Frank, Waldo, 256–7
Freeman, John, 206, 207
Freisler, Roland, 279
French Resistance, 229–30
French Revolution, 2, 3, 28, 50, 248
Freud, Sigmund, 82, 249
Freyberger, Louise, 80
Fröbel, Julius, 123

Gaitskell, Hugh, 210
Galileo Galilei, 178
Gallico, Paul, 334
Gandhi, Mohandas K., 49, 260
Gardiner, Gerald, 286
Gardiner, Muriel, 303–4
Garibaldi, Giuseppe, 103
Garrick, David, 12, 20
Gaskell, Peter, *The Manufacturing Population of England*, 65
Gauss, Christian, 253
Gautier, Théophile, 234
Gay, John, *The Beggar's Opera*, 177
Gellhorn, Martha, 155, 159, 163–5, 294, 296, 303

Genet, Jean, 242
Georg, Stefan, 182
George IV, King of England, 29
George V, King of England, 204
George VI, King of England, 224
Gibbon, Edward, *Decline and Fall of the Roman Empire*, 265
Gide, André, 228, 230
Gladstone, W. E., 66–7, 89, 341
Glazer, Nathan, 302
Godwin, William, 32, 35, 38, 39, 40, 45, 46, 47
Goethe, Johann Wolfgang von, 100, 333; *Faust*, 55, 179
Golding, Louis, 273
Goldwyn, Samuel, 293
Gollancz, Alexander, 270
Gollancz, Sir Herman, 270
Gollancz, Sir Israel, 270
Gollancz, Livia, 269
Gollancz, Ruth (*née* Lowy), 270, 275, 278
Gollancz, Sir Victor, 269–87, 288, 302, 307, 308
Goncourt, Edmond, 232
Gorer, Geoffrey, 284
Gorky, Maxim, 108, 109; *The Mother*, 177
Gottling, Willi, 193
Gow, A. S. F., 307
Graetz, Heinrich, *History of the Jews*, 270
Graham, Katherine, 305
Grant, Ulysses S., 197
Greco, Juliette, 234
Greeley, Horace, 140
Greene, Graham, 329
Grey, Lord, 204
Grimm, Friedrich Melchior, 6, 14, 26
Grotewohl, Otto, 193
Guinness, Sir Alec, 326

Haire, Norman, 216
ha-Levi, Elieser, Rabbi, 53
Hammett, Dashiell, 290–3, 294,
    295, 296, 297, 298, 300, 304,
    305
Hammett, family: Josephine (*née*
    Dolan), 291; daughters, 304
Harich, Wolfgang, 195
Harisson, Tom, 276
Hart, Moss, and George S.
    Kaufman, *The Man Who Came
    to Dinner*, 293
Hauptmann, Elizabeth, 185, 195
Haydon, Benjamin Robert, 44
Hayward, Nancy, 166
Healey, Dan, 40, 46
Hegel, Hegelianism, 54, 56–7, 58,
    59, 60, 62, 64, 139, 177, 189
Heidegger, Martin, 227, 228, 229,
    231
Heine, Heinrich, 56
Heinzen, Karl, 71
Heller, Gerhardt, 229
Hellman, Lillian, 156, 288–305;
    Another Part of the Forest, 297;
    The Autumn Garden, 297–8;
    The Children's Hour, 292–3,
    299; *Days to Come*, 293, 294;
    Dead End, 293; *The Little Foxes*,
    293, 296, 297, 301; *Pentimento*,
    296, 300, 304; *Scoundrel Time*,
    299, 301, 302; *Toys in the Attic*,
    300; *An Unfinished Woman*,
    297, 300; *Watch on the Rhine*,
    293–4, 303
Hellman, parents: Julia (*née*
    Newhouse), 289; Max, 289,
    297
Hemingway, Ernest, 143–72, 176,
    188, 226, 250, 288, 294, 296,
    302, 321, 324; *Across the River
    and Into the Trees*, 166; *By-line*,

149, 160; *Death in the
    Afternoon*, 144, 149; *A Farewell
    to Arms*, 150; *The Fifth Column*,
    144; *For Whom the Bell Tolls*,
    152, 157, 158, 159; *The Green
    Hills of Africa*, 149; *In Our
    Time*, 150; *Men Without
    Women*, 150; *A Moveable Feast*,
    146, 149, 154, 161; *The Old
    Man and the Sea*, 166, 172; *The
    Sun Also Rises*, 150, 155, 162,
    168; *Three Stories and Ten
    Poems*, 149; *To Have and Have
    Not*, 161, 163
Hemingway, parents: Edmund,
    143, 144, 145, 171; Grace, 143,
    144, 145–6, 168
Hemingway, siblings: Carol, 154;
    Leicester, 169; Marcelline,
    146; Sunny, 144; Ursula, 161
Hemingway, wives: Hadley (*née*
    Richardson), 144, 162, 163,
    164, 169; Pauline (*née* Pfeiffer),
    144, 162, 163, 164, 168;
    Martha, *see* Gellhorn, Martha;
    Mary (*née* Welsh), 165, 166,
    167, 170
Hemingway, children, 168;
    Gregory, 162, 165, 166, 167,
    168; Jack, 162, 168; Patrick,
    162, 168, 169
Henderson, W. O., 65
Henriksen, Hans Jacob, 92–3
Hersey, John, 305
Hildebrand, Bruno, 66
Hirschfeld, Dr Magnus, 216
Hitchener, Elizabeth, 39, 48
Hitler, Adolf, 53, 76, 99, 132, 176,
    179, 180, 205, 209, 227, 240,
    278, 279, 282, 284, 294, 319,
    330
Hoffman, Abbie, 323

Hogg, Thomas Jefferson, 33, 38, 39, 40, 44, 46
Holbach, Baron d', 6
Holbrook, Josiah, 141
Holmes, Oliver Wendell, 140
Holst, Henrikke, 100, 103
Homer, 141
Homolka, Florence, 187
Hook, Sidney, 190, 217, 218, 220, 222, 302
Hoppner, Richard, 43, 44
Horder, Lord, 276
Horkheimer, Max, 188, 233
Hotchner, A. E., 169
Houdetot, Sophie, Comtesse d', 19, 27
House Un-American Activities Committee, 179–80, 295, 298–9
Howe, Irving, 302, 322
Hugo, Victor, 27, 143, 173, 213, 232
Huizinga, J. H., 18
Hulton, Edward, 325
Hume, David, 9, 14, 15, 23, 26, 33, 228, 240, 338
Hume, Thomas, 42
Hunt, Leigh, 32, 44, 46, 47
Husserl, Edmund, 227
Huston, John, 249
Huxley, Aldous, 228
Huysmans, J. K., 234

Ibsen, Henrik, 79, 82–106, 108, 116, 148, 159, 219, 270, 317, 324; early plays, 84; *Brand*, 85, 91, 93; *A Doll's House*, 79, 85, 99, 102; *An Enemy of the People*, 98; *Ghosts*, 85; *Hedda Gabler*, 86, 99, 100; *John Gabriel Borkman*, 86; *The League of Youth*, 96; *The Master Builder*, 86, 102; *Peer Gynt*, 85, 95, 105; *Pillars of Society*, 85; *The Pretender*, 84–5; *Rosmersholm*, 86; *Solhaug*, 87; *The Wild Duck*, 84, 86
Ibsen, family: parents, 83–4, 91, 92; Hedvig (sister), 91; Nicolai Alexander (brother), 91–2; Ole Paus (brother), 92
Ibsen, Suzannah (*née* Thoresen), 94, 95, 102, 105
Ibsen, children, 92–3, 95; Sigurd (son), 93, 94, 95, 97, 103–4
Illuminati, 32
Imlay, Fanny, 39
Industrial Revolution, 4, 69, 130
Ingersol, Ralph, 294
intellectuals:
    and alcohol, 70, 84, 90, 105, 116, 164, 165, 168–72, 234, 239–40, 250–1, 263, 290–1, 322, 332; drugs, 332–3
    and children: legitimate, 23, 38, 42–3, 76–8, 94, 97, 133, 168, 212, 215, 219; illegitimate, 21–2, 43–4, 80–1, 92–3, 187
    and dress, 12, 30, 88–9, 127, 153, 176, 180, 215, 332
    and family, 264; lack of family feeling, 18, 26, 54, 70, 91–2, 125, 146, 174, 227; parental discord, 5, 34, 325, 335; mother troubles, 145–7, 253, 275
    and money, 5, 18, 29, 45–8, 57–8, 73–7, 93, 115, 121, 141, 181, 191, 198, 222–3, 249–50, 258, 266–7, 272–3, 275, 284, 299–300, 304, 321; debts, 36, 40, 41, 46–8, 75, 84, 115, 119, 250, 291; generosity, 222, 250,

intellectuals: and money (*cont.*)
329; miserliness, 26, 92–3,
222; profligacy, 74, 108
and politics, 23–5, 29–30, 32,
49–50, 55–69, 86, 105, 128–31,
155–8, 189–95, 203–12, 230,
241–8, 255–8, 277–80, 285,
294–7, 306–9, 311–12, 316–17,
321–2, 327, 332, 336–7, 339–41
and religion, 33, 53, 58, 105,
107–8, 124, 130, 139, 143–5,
148, 151–2, 174, 198, 242, 248,
270–1, 285, 306, 334; atheism,
33, 198, 201
and sex, 17, 35–6, 44–5, 101,
115–18, 119–20, 133–4, 140–1,
153, 171, 201, 212, 215–16,
236, 262, 294, 295, 327–30;
homosexuality, 17, 26, 136,
327, 331, 335; masturbation,
17, 26, 327, 329; penile
obsession, 9, 101, 160, 216;
policy of 'openness', 35, 119,
133–4, 215–16, 237–9, 261–2
and women, 11, 35, 39–40,
99–102, 111, 115–18, 122, 155,
161–7, 169, 184–7, 212, 218–19,
236, 261–2, 273–5, 320, 328,
331; mistresses/lovers, 6, 17,
19–21, 22, 40, 43–4, 79–80, 92,
116, 185–6, 195, 213–18,
235–40, 250–1, 261–2, 274, 275,
290–1, 294, 329; wives, 36–9,
76, 80, 94–5, 118–22, 125,
133–7, 140–1, 155, 162–7,
212–13, 215–18, 258–60, 262–4,
275, 314–15, 318, 320, 328–9,
331–2
and the 'workers', 49, 60–1, 127,
130, 131, 156, 180, 188, 195,
241, 247–8, 259, 276, 296–7,
307–8

intellectual characteristics:
anger, aggressiveness, violence,
35, 54, 61–2, 69–72, 73, 89–91,
104–5, 109, 123, 164–5, 167,
168, 175, 204, 263, 265–6,
273, 286–7, 291, 301, 323–4,
333, 335–6; espousal of
principle of violence, 204,
205–6, 245–7, 318–20, 321–3,
332, 336–7, 339
canonization of, 27, 131–2,
300–2, 305
cowardice, 94, 102–5, 175, 317,
329; courage, 110, 152–3
cruelty, 20, 26, 31, 126,
166–7
deceitfulness, dishonesty,
17–18, 26, 38, 65–9, 126,
154–5, 157, 160, 181, 188–9,
191–2, 196, 200, 203, 206–7,
213, 220, 226, 235, 240, 243,
244, 245, 258, 269–70, 271,
272, 277–83, 288–90, 292, 296,
297, 299–300, 302–4, 307, 309,
335, 337; passion for truth,
253, 256–7, 307
egocentricity, egotism, 6, 10–11,
19, 20, 26, 36–7, 44, 48, 50,
72–3, 78, 91, 94, 96–7, 107,
109, 110, 119, 125, 133–5, 145,
184, 225, 249, 258, 261, 274,
286–7, 321, 328; lack of, 30
genius for self-publicity, 127,
153, 176, 177–8, 182–3, 189,
233, 291, 298–9, 321, 322–3,
327
hypocrisy, 22, 26, 44, 99, 213,
236, 336
ingratitute, rudeness, 11–13, 19,
26, 33, 34, 40–1, 95, 126, 183
intolerance, misanthropy, 87,
91, 95, 99

intellectual characteristics (*cont.*)
love of power, 72, 113, 179, 280, 330
manipulativeness, exploitativeness, 13, 17, 39–40, 75, 79–80, 95–6, 101–2, 125, 175, 179–84, 185, 186–7, 188, 190, 193–5, 219–22, 236–9, 244, 261, 272, 273–4
quarrelsomeness, 10–11, 14, 18, 70–1, 73, 96, 99, 106, 126, 159–61, 177, 208–9, 240–1, 265, 274, 283, 285, 296, 301, 324
self-deception, gullibility, 37–8, 114, 156–7, 191, 202, 242, 269, 273, 274, 276, 277, 341
selfishness, ruthlessness, 3–4, 19, 31, 96, 119, 126, 128, 184, 186, 187, 190, 207, 221, 274, 314; unselfishness, 30–1, 125, 219, 292
self-pity, 5, 9–10, 278; paranoia, 14–16, 26, 208
self-righteousness, 37, 107–8, 109, 146, 209, 223, 267, 273
shiftlessness, spongeing, 5, 7, 73, 317, 336
snobbery, 76, 88–9, 108, 210, 224, 313–4; intellectual snobbery, 199, 202
vanity, 10, 19, 26, 87–9, 107, 170, 208, 285–6, 321, 326, 328
Irving, Washington, 139
Isaiah, 107
Isherwood, Christopher, 188
Ivancich, Adriana, 166

Jaegar, Henrik, 88
James, Henry, 50, 140, 147, 252, 253, 254, 264
Janson, Kristofer, 98

Jaspers, Karl, 62
Jeanson, Francis, 241
Jensdatter, Elsie Sophie, 92
Jerrold, Douglas, 272, 274
Jessel, Lord, 314
Jesus Christ, 107, 109, 130, 144, 271
Jews, Judaism, 32, 53, 57–8, 80, 128, 209, 230, 243, 248, 270–1, 271–2, 289; anti-Semitism, 57–8, 62, 73, 123, 133, 258
John, Augustus, 209
Johnson, Revd Hewlett, 281
Johnson, Lyndon B., 211, 323, 324
Johnson, Samuel, 16, 24, 96, 249
Johnsrud, Harold, 259
Jollivet, Simone, 227
Jones, James, 324
Joyce, James, 147, 149, 160, 228, 253, 261

Kamenev, Lev, 190
Kanapa, Jean, 241
Kant, Immanuel, 27, 139
Katz, Otto, 282, 283
Kay, J. P., *Physical and Moral Conditions of the Working Classes . . .*, 66
Kazin Alfred, 302
Keats, John, 30, 41, 51
Kennedy, John Fitzgerald, 205, 210, 211, 212
Khrushchev, Nikita, 191, 207, 208, 210, 220, 300
Kieler, Laura, 102
Kielland, Kitty, 100
Kierkegaard, Sören, 94
Kilian, Isot, 195
King, Martin Luther, 336
Kipling, Rudyard, 147, 149, 151, 161, 188, 253, 311

Knudtzon, Frederick, 90, 104
Kober, Arthur, 290
Koestler, Arthur, 219, 237, 240,
    241, 286; *Darkness at Noon*, 25
Kohlberg, Alfred, 207
Kolman, Arnost, 206
Kosakiewicz, Olga, 238
Kosakiewicz, Wanda, 238, 239, 251
Kovalevsky, Maxim, 71
Kriege, Hermann, 61, 62
Kugelmann, Dr Ludwig, 73

Lafargue, Paul, 78, 79
La Guardia, Fiorello, 201
Lanchester, Elsa, 185
Lanham, Charles, General, 146,
    165
Laski, Harold, 279–80, 281, 284
Lassalle, Ferdinand, 62
Lassithiokatis, Hélène, 239, 251
Laughton, Charles, 179
Lawless, John, 46
Lawrence, D. H., 147, 265
Lawrence, J. H., 44
Left Book Club, 279–82, 283, 284,
    285, 309, 311
le Gallienne, Richard, 87
Leibnitz, Gottfried, 4, 200
Leigh, Augusta, 35
Lemaître, Jules, 6
Lenin, V. I., 24, 52, 71, 99, 130,
    176, 179, 187, 191, 202, 205,
    255, 295, 309; Marxism-
    Leninism, 177
Lessing, Doris, 209
Levasseur, Thérèse, 5, 15, 19–21,
    22, 235
Lévi-Strauss, Claude, 27
Levy, Benn, 276
Lévy, Raoul, 230
Lewis, John, 280, 284

Lewis, Mildred, 291
Lewis, Sinclair, 160
Lewis, Wyndham, 147, 160
Liebknecht, Wilhelm, 78
Lind, Dr James, 32
Lippmann, Walter, 255
Liverpool, Lord, 32
Locke, John, 4, 33, 213, 338
Loeb, Harold, 155, 160, 162
Lofthuus, Christian, 92
Longuet, Charles, 78
Lorange, August, 90
Lorre, Peter, 179
Louis XVI, King of France, 2
Lowell, Robert, 265
Lubbock, Lys, 316
Lund, Dr Robert, 186
Lynd, Sheila, 274, 280, 284

Macaulay, Thomas Babington, 46
McCarthy, Desmond, 317
McCarthy, Senator Joseph, 211
McCarthy, Mary, 245, 259, 260–1,
    263, 302–3, 304, 310;
    A Charmed Life, 263
McCartney, Paul, 175
McCracken, Samuel, 304
MacDonald, Joan, 7
Macdonald, Ramsay, 277
Mackenzie, Compton, 209, 253,
    325
MacLeish, Archibald, 156, 160, 168
Macmillan, Harold, 209, 210
McTaggart, J. E., 224
Maeterlinck, Maurice, 270
Mahan, John, 277
Mailer, Norman, 305, 320–4, 328,
    329, 330; *The Naked and the
    Dead*, 321; *The White Negro*,
    322, 323, 327, 336, 337, 338
Mailer, Fanny, (mother), 320

Mailer, wives: Beatrice (*née*
    Silverman), 320; Lady Jean
    (*née* Campbell), 320; Beverley
    (Bentley), 320; Norris (*née*
    Church), 320; Adele (*née*
    Morales), 323
Makarios, Archbishop, 220
Malcolm X, 336
Malesherbes, Chrétien, 15
Malleson, Lady Constance, 214,
    215, 216, 218, 317
Malleson, Miles, 198
Malraux, André, 152, 229, 230,
    231, 244
Mancy, Joseph and Anne-Marie,
    226
Mann, Thomas, 188, 228
Mao Tse-Tung, 52, 72, 220, 245,
    247, 319, 340, 341; Madame
    Mao, 178
Marat, Jean-Paul, 56
Marcuse, Herbert, 188
Marie-Louise, Princess, 87
Marryat, Frederick, 147
Marseille, Walter, 205, 207
Marsh, Mae, 154
Martin, Henri, 242–3
Martin, Kingsley, 204, 282, 283,
    284, 285, 309, 319
Marx, Jenny (*née* von
    Westphalen), 54, 70, 71, 74,
    75, 76–7, 78, 79, 80, 186
Marx, Karl, 4, 29, 52–81, 83, 85, 92,
    97, 116, 129, 130, 133, 140,
    142, 146, 173, 178, 179, 186,
    187, 188, 189, 219, 255, 258,
    265, 267, 269, 287, 295, 305,
    340, 341; *Capital*, 55, 56, 63–5,
    67, 68–9, 73, 85, 130, 176, 265,
    276; *Communist Manifesto*, 54,
    55, 56; *The German Ideology*,
    55, 72; *Oulanen*, 54

Marx, parents: Heinrich and
    Henrietta (*née* Pressborck), 53,
    74
Marx, children: 76–7, 78, 79, 80–1;
    Eleanor, 67, 78, 79, 80; Jenny,
    76, 78, 79; Laura, 54, 76, 78,
    79
Marxism, 52–3, 69, 130, 131, 177,
    178, 188, 195, 205, 227, 230,
    241, 255, 257, 258, 259, 267,
    280, 282, 294, 341; Marxism-
    Leninism, 177
Mason, Jane, 163, 169
Masons, 32
Matthews, Herbert, 152, 155
Maude, Aylmer, 115, 134
Maugham, W. Somerset, 317
Maupassant, Guy de, 147
Mauriac, François, 242
Melby, John, 294, 297
Merleau-Ponty, Maurice, 230, 240
Mill, John Stuart, 27, 198; *The
    Subjection of Women*, 117
Millay, Edna St Vincent, 261–2,
    268
Miller, Arthur, 294
Miller, Webb, 278
Mills, C. Wright, 324
Minz, Jehuda, 53
Monks, Noel, 165
Monnet, Jean, 242
Montaigne, Michel, 4
Montaigu, Comte de, 5, 6
Montmorency-Luxembourg, Duc
    and Duchesse de, 13, 20
Moore, G. E., 199
Moore, George, 147
Morel, E. D., 223
Morison, Stanley, 272, 274
Morrell, Lady Ottoline, 213, 214,
    217, 218
Mortimer, Raymond, 283

Moses, 107
Muenzenberg, Willi, 282, 283
Muggeridge, Malcolm, 211
Mumford, Lewis, 255
Muntyanov, S. I., 130
Murray, Gilbert, 270
Mussolini, Benito, 25, 319

Nabokov, Vladimir, 329; *Lolita*, 286
Napoleon Bonaparte, 2, 28, 132
Nasser, President of Egypt, 208, 220, 245, 319
Nazism, Nazi Party, 176, 177, 200, 210, 212, 229, 230, 245, 282, 290, 294, 296, 309, 319, 330, 340
Neher, Carola, 185, 190
Nekrasov, Nikolai, 108
Nellhaus, Gerhard, 182
Newton, Sir Isaac, 201
Nicolson, Sir Arthur, 200
Nin, Andreas, 156
Nixon, Richard Milhous, 197
Nizan, Paul, 227, 228
Nobel Prize, 222
Nugent, Catherine, 46

Ohlendorf, Otto, 319
Olaf IV, King of Norway, 83
Orthodox Church, 130
Orwell, George, 240, 283, 284, 285, 306–10, 311, 319; *Animal Farm*, 285, 309, 310; *Down and Out in Paris and London*, 308; *Homage to Catalonia*, 157, 283, 284; *Nineteen Eighty-Four*, 25, 285, 309, 310; *The Road to Wigan Pier*, 283
Orwell, Sonia (*née* Brownell), 316

Osborne, John, 209; *Look Back in Anger*, 326
Owen, Robert, 56

Pabst, G. W., 177
Pachter, Professor Henry, 190
pacifism, 131, 192, 203–4, 209, 214, 223, 271, 337
Parker, Dorothy, 150, 160, 265, 295
Parks, Rosa, 289
Pascal, Blaise, 4, 107
Pasternak, Boris, 191
Paulhan, Jean, 228, 235
Pauli, Georg, 89, 90
Paulsen, John, 87, 96
Paus, Christian, 91
Peacock, Sir Peter, 325, 329
Peacock, Thomas Love, 39; *Nightmare Abbey*, 32
Peel, Sir Robert, 49
Petersen, Clemens, 95
Petersen, Laura, 100
Petrarch, Francesco, 4
Phelps, Revd William Lyon, 293
Phenomenology, 227
Philips, Lion, 60, 74
Picasso, Pablo, 181, 233
Pieck, Wilhelm, 180, 191
Pinay, Antoine, 243
Pinter, Harold, 197
Plato, Platonism, 23, 139, 204, 270, 338
Plimpton, George, 323
Podhoretz, Norman, 322, 335
Poland, constitution for, Rousseau writes, 15
Pollitt, Harry, 276, 280, 281, 282, 284, 308, 315
Polonsky, Yakov, 115
Pol Pot, 25, 246, 341
Porter, Cole, *Leave It to Me*, 293

Porter, Herman, 334

Postan, M. M., 219

Pound, Ezra, 147, 149, 150, 160, 309, 310

Priestley, J. B., 208, 275, 286

Pritchett, V. S., 320

Prometheus, 2

Protestantism, 53; *see also* Calvinism, Orthodox Church, Quakerism

Proudhon, Pierre-Joseph, 61, 62, 78, 230

Proust, Marcel, 197

Pushkin, Aleksander, 123, 258

Quakerism, 203

Queensberry, Marquess of, 96

Quintanilla, Pepe, 157

Rabelais, François, 4

Raff, Helene, 100, 101

Rahv, Philip, 259–60, 261

Ramsay, Allan, 12

Ransom, Arthur, 147

Rauh, Joseph, 298–9

Raymond, John, 313

Read, Herbert, 209

Reform Bill, 198

Reid, Betty, 280, 284

religion, decline of, 1–2, 28; in the United States, 138; *see also* Christianity, Jews

Renan, Ernest, 253

Resistance, *see* French Resistance

Reston, James, 305

revolution, revolutionary politics, 23, 26, 29, 54, 59, 70, 71, 98, 129, 138, 255, 324

Rey, Evelyne, 239

Rhodes, Cecil, 272

Richardson, Samuel, *Clarissa*, 7

Riddevold, Revd H., 98

Ridgeway, Matthew, General, NATO commander, 243

Roberts, Dr William, 47

Robespierre, Maximilien, 2

Robles, José, 156, 157, 161

Rolland, Romain, 244

Romantic movement and philosophy, 3, 8, 12, 30, 65, 139, 144, 153

Roosevelt, Franklin D., 294, 295

Rosicrucianism, 32

Ross, Christian, 90

Ross, Lillian, 169

Roughhead, William, *Bad Companions*, 292

Rousseau, Jean-Jacques, 2–27, 28, 30, 31, 38, 44, 46, 50, 53, 58, 73, 80, 82, 83, 91, 92, 95, 109, 114, 132, 153, 154, 159, 173, 175, 226, 234, 235, 240, 251, 276, 287, 305, 317, 336, 340; *Confessions*, 3, 5, 6, 9, 16, 17, 18, 19, 22, 154, 302; *Dialogues avec moi-même*, 16; *Discours sur l'inégalité*, 4, 7, 21; *Émile*, 3, 4, 5, 8, 18, 21, 23; *Narcisse*, 4; *La Nouvelle Heloïse*, 7, 12, 21; *Rêveries du promeneur solitaire*, 16, 22; *Social Contract*, 7, 21, 23, 24, 236

Rousseau, parents: Isaac, 4, 5, 18; Suzanne (*née* Bernard), 4

Rousseau, children, 21–2

Rousset, David, 242

Rubin, Jerry, 323

Rubinstein, Hilary, 274

Rulicke, Käthe, 195

Ruskin, John, 63

Russell, Bertrand, 3rd Earl, 178, 197–224, 225, 237, 243, 244,

Russell, Bertrand, 3rd Earl (*cont.*)
245, 246, 247, 248, 251, 302,
317, 318, 320, 321, 338, 339,
341; Bertrand Russell Peace
Foundation, 220, 221, 223;
Analysis of Mind, 223; *The
Conquest of Happiness*, 200;
Essays in Analysis, 197; *German
Social Democracy*, 197; *A
History of Western Philosophy*,
200; *Human Knowledge*, 200;
*Introduction to Mathematical
Philosophy*, 223; *Justice in
Wartime*, 211; *The Practice and
Theory of Bolshevism*, 205;
Principia Mathematica, 199
Russell, 1st Earl (Lord John
Russell), 198
Russell, 2nd Earl (BR's brother),
212–13
Russell, wives: Alys (*née* Whitall),
203, 204, 212, 213, 224; Dora
(*née* Black), 215–16, 217, 218,
219; Peter (*née* Margery
Spence), 202, 216, 217, 218,
224; Edith (*née* Finch), 218,
221
Russell, children, 215, 217
Rustin, Bayard, 336

Sacco and Vanzetti, 210
Sanchez, Thorwald, 169
Sand, George, 27, 234
Santayana, George, 215
Saroyan, William, *The Time of Your
Life*, 293
Sartre, Jean-Paul, 211, 225–51, 261,
263, 274, 275, 298, 300, 301,
302, 305, 314, 318, 320, 321,
324, 329, 338, 341; *Les Chemins
de la liberté*, 229; *Critique de la
raison dialectique*, 239; *L'Être et
le néant*, 230, 231; *Huis-clos*,
231; *Les Mots*, 226, 248; *Les
Mouches*, 231; *La Nausée*, 228,
236; *Récherches philosophiques*,
228, *Les Séquestrés d'Altona*,
239; *War Diary*, 225
Sartre, parents: 225–6; *see also*
Mancy, Joseph and Anne-
Marie
Sartre, Arlette, 239, 250–1
Sauvy, Alfred, 245
Schandorph, Professor Sophus, 90
Schapper, Karl, 56
Schiller, Johan Christophe
Friedrich von, 27; *Maid of
Orleans*, 188
Schneekloth, Martin, 94, 97
Schoenman, Ralph, 209, 211, 219–
23, 224, 250
Schramm, Konrad, 71
Schreiner, Olive, 78
Scribner, Charles, 153
Segovia, Andres, 159
Selassie, Haile, 220
Semionov, Vladimir, 194
Shakespeare, William, 330; *Richard
II*, 173; *Troilus and Cressida*, 70
Shaw, George Bernard, 79, 86,
243, 270, 293; *St Joan*, 188
Shaw, Irwin, 295
Shelley, Percy Bysshe, 27, 28–51,
85, 151, 185, 190, 212, 218,
238; *An Address to the Irish
People*, 49; *Alastor*, 48; *The
Cenci*, 29; *Defence of Poetry*, 28;
Epipsychidion, 42; *A Hymn to
Intellectual Beauty*, 29; *Julian
and Maddalo*, 49, 151; 'Lines
from the Eugenean Hills', 29;
The Mask of Anarchy, 29, 49,
188; *The Necessity of Atheism*,

Shelley, Percy Bysshe (*cont.*)
33; 'Ode to the West Wind',
29; *Original Poetry by Victor
and Cazire*, 33; 'Ozymandias',
29; *A Philosophical View of
Reform*, 44; *Prometheus
Unbound*, 29; *Queen Mab*, 36,
41, 42, 48; *The Revolt of Islam*,
29, 35; *St Irvyne*, 33; *Swellfoot
the Tyrant*, 29; 'To a Skylark',
30; *The Triumph of Life*, 50; *The
Witch of Atlas*, 51; *Zastrozzi*, 33
Shelley, family: Sir Timothy, 31,
33–4, 35, 47; Lady Shelley, 34;
Helen (sister), 33, 35;
Elizabeth (sister), 34–5; Mary
(sister), 35
Shelley, wives: Harriet (*née*
Westbrook), 34, 36, 37, 38, 39,
40, 41, 42, 44, 45–6, 48, 50,
212; Mary (*née* Godwin), 30,
32, 35, 36, 37, 38, 39, 41, 42–3,
44, 46, 48, 50–1
Shelley, children, 36, 38, 42–3, 51,
217
Shelley, Sir Bysshe, 31
Shevlin, Durie, 166
Sitwell family, 266
Skelton, Barbara, 261, 318
Slater, Humphrey, 219
Slatter (Oxford bookseller), 33, 47
Smith, Adam, 67
Smith, Bill, 144, 161
Smith, Logan Pearsall, 203, 314
Smith, Randall, 296
socialism, socialists, 59, 60, 98,
130, 156, 210, 223, 230, 245,
255, 271, 276, 277, 279, 280,
281, 285, 307, 309, 310, 327,
339
Socrates, 107, 178, 201
Somervell, D. C., 271

Sontum, Hildur, 100
Sorokine, Nathalie, 238
Soviet Union, 52, 81, 137, 156, 178,
180, 190, 192–5, 204–6, 207,
210, 239, 243–4, 254, 255, 256,
257–8, 259, 274, 277, 279, 280,
282, 284–5, 294, 295, 296, 297,
300
Spanish Civil War, 145, 152, 153,
155, 156, 157, 159, 178, 227,
296, 302, 308, 315, 319
Spender, Sir Stephen, 303
Spinoza, Baruch, 107
Staël, Germaine de (Madame de),
4
Stalin, Joseph, 52, 53, 71, 81, 131,
156, 177, 190–1, 193, 205, 219,
233, 240, 243, 255, 258, 260,
265, 277, 279, 281, 282, 283,
284, 288, 294, 295, 297, 300,
302, 305, 309, 310, 319, 341;
Stalin Peace Prize, 174, 191
Stanislavsky, Konstantin, 193
Steffens, Lincoln, 148, 149, 254
Steffin, Margarete, 186
Stein, Gertrude, 147, 148, 160
Steinbeck, John, 254
Steiner, Dr Herbert, 304
Stendhal, 147
Stern, Ada, 168
Stewart, Donald Ogden, 295
Stoppard, Tom, 197
Stowe, Harriet Beecher, 165; *Uncle
Tom's Cabin*, 334, 336
Strachey, John, 208, 219, 277, 279–
80, 281, 285, 308
Strachey, Lytton, 253; *Eminent
Victorians*, 223
Strauss, Richard, 177
Stravinsky, Igor, 192
Styron, William, 261, 305, 324
Sue, Eugène, 109

Suhrkamp, Peter, 181
Sukarno, President of Indonesia, 220
Suter, Gody, 194
Synge, J. M., *Riders to the Sea*, 188
Syutayev, V. K., 125

Tasso, Torquato, 4
Techow, Gustav, 71, 72
Thalheimer, August, 281
theatre, 85, 173, 183–4, 326–7
Thomas, J. Parnell, 179
Thomson, Virgil, 169
Thoresen, Magdalene, 87, 94, 95, 96
Thrane, Marcus, 103
Tintoretto, 248
Tito, President of Yugoslavia, 245
Tocqueville, Alexis de, 138
Tolstoy, Leo, 27, 86, 87, 89, 98,
    107–37, 143, 147, 153, 165–6,
    167, 208, 211, 212, 215, 219,
    220, 237, 240, 247, 287, 302,
    318, 321, 324, 341; *Anna
    Karenina*, 112–13, 114, 117,
    120, 121, 124, 125n; *Childhood*,
    111; *Confessions*, 121; diaries,
    115, 116, 118, 119–20, 126,
    127, 134–5; doctrinal tracts,
    124; 'The Kreutzer Sonata',
    133, 134; *A Letter to the
    Chinese*, 131; *The Significance of
    the Russian Revolution*, 131;
    War and Peace, 112, 113, 114,
    124, 125n, 129, 130, 150, 265;
    Youth, 109, 111
Tolstoy, family: father and
    grandfather, 108, 116;
    Countess Alexandra, 123;
    Dimitri (brother), 125; Nikolai
    (brother), 110, 115, 125; Sergei
    (brother), 110, 114, 115, 125

Tolstoy, Sonya (*née* Behrs), 118–20,
    121, 124, 125, 126, 127, 128,
    131, 133, 134–6, 137, 165, 167,
    250
Tolstoy, children, 116–17, 120,
    121, 125, 133, 137; Alexandra,
    133, 136; Masha, 133; Tanya,
    121, 133
totalitarianism, 23, 25, 131, 231,
    309, 332, 339, 340
Transcendentalism, 139, 140
Trilling, Diana, 301
Trilling, Lionel, 254
Tronchin, Dr, 9, 10, 14
Trotsky, Leon, Trotskyism, 255,
    260, 282, 310
Tunney, Gene, 176
Tupanov, Colonel, 180
Turgenev, Ivan, 108, 115, 116, 126,
    129, 147, 240
Twain, Mark, 147
Twysden, Dorothy (Lady), 162, 166
Tynan, Kenneth, 169, 183, 184,
    324–30, 333; *Oh! Calcutta!*,
    327, 328
Tynan, parents: Rose, 325, 329; *see
    also* Peacock, Sir Peter
Tynan, wives: Elaine, *see* Dundy,
    Elaine; Kathleen (*née* Gates),
    325, 327–8

Ulbricht, Walter, 193, 195
Undset, Ingvald, 105
United States, disagreement with
    policies of, *see* anti-
    Americanism, Vietnam
United States, intellectuals in,
    138–40, 142–3, 148, 254;
    American intellectuals in
    Paris, 147–8
Unwin, Sir Stanley, 222

Vercellis, Comtesse de, 15

Vian, Boris, 234, 239; Michelle, 239

Victor, Pierre (Benny Levy), 250

Victoria I, Queen of England, 59, 197, 198

Vidal, Gore, 291, 324

Viertel, Virginia, 166

Vietnam, 211, 245, 246, 323, 339, 340, 341; Vietnam War Crimes Tribunal, 178, 211, 212, 220, 221, 245

Ville, Marie, 238

Villon, Francois, 177, 188, 244

Vishinsky, Andrei, 279

Viviani, Emilia, 44

Volkonsky family, 108

Voltaire, 2, 6, 9, 14, 21, 26, 53, 143, 173, 209, 233, 234, 240, 244

Wagner, Richard, 181, 184

Wahl, Jean, 235, 249

Wallace, Mike, 323

Walpole, Hugh, 147

Warens, Françoise-Marie de, 5, 17, 19, 22

Watson, Peter, 315

Waugh, Evelyn, 171, 265, 306, 310, 311–13, 314, 315, 316, 317; *Brideshead Revisited*, 144, 326; *Robbery Under Law*, 311; *Sword of Honour*, 316

Webb, Beatrice, 215, 277

Webster, John, *The Duchess of Malfi*, 179

Weideman, Jerome, 334

Weigel, Helene, 181, 185, 186–7, 188, 189, 195, 196, 239, 275

Weightman, John, 239, 242

Weill, Kurt, 175, 177–8

Weishaupt, Adam, 32

Weisstein, Gottfried, 88

Weitling, William, 61, 70

Wells, H. G., 270

Wesker, Arnold, 209

West, Nathanael, 289

West, Rebecca, 253

Westbrook, Eliza, 38, 40

Westmorland, Earl of, 77

Wharton, Edith, 253

Whitehead, Alfred North, 199, 224

Whitman, Walt, 270

Whitton, William, 35

Wilde, Oscar, 96, 325

Wilhelm I, Emperor of Prussia, 71

Williams, Edward and Jane, 50, 51

Williams, John and Owen, 47, 48–9

Williams, William Carlos, 148

Williams-Ellis, Clough and Amabel, 219, 223

Willich, August von, 71

Willingham, Calder, 324

Wilson, Edmund, 150, 153, 252–68, 269, 295, 300, 307, 318; *The American Jitters*, 256; *Apologies to the Iroquois*, 267; *Axel's Castle*, 253, 256, 315; *I Thought of Daisy*, 253; *Memoirs of Hecate County*, 266; *Patriotic Gore*, 267; *The Scrolls of the Dead Sea*, 266, 267, 268; *To the Finland Station*, 257, 258, 266

Wilson, parents, 253

Wilson, wives: Mary (*née* Blair), 253, 258; Margaret (*née* Canby), 254, 258, 262; Mary, *see* McCarthy, Mary; Elena (*née* Mumm), 262, 264

Wilson, Harold, 221, 285

Wilson, Woodrow, 204

Wittgenstein, Ludwig, 200, 286

Wodehouse, P. G., 310

Wolf, Wilhelm, 74

Wollstonecraft, Mary, 35
Wood, John S., 299
Woolf, Leonard, 277, 282, 319
Woolf, Virginia, 228
World Congress for Peace
    (Helsinki, 1965), 221
World War I, 153, 154–5, 174, 200,
    203, 211, 253, 271, 290
World War II, 153, 169, 179, 203,
    211, 228–30, 244, 283, 285,
    325
Wright, Richard, 295, 334, 335

Wright, William, 290, 303
Wuolojocki, Hella, 188

Yeats, W. B., 254, 321

Zaphiro, Denis, 167, 170
Zilboorg, Gregory, 297
Zinoviev, Grigori, 190
Zoff, Marianne, 184, 185
Zola, Emile, 147, 233, 234, 253
Zucker, E. A., 101
Zuckmayer, Carl, 176

About the author

2 Meet Paul Johnson

3 In His Own Words

About the book

6 The High Priests of Knowledge:
A Conversation with Paul Johnson

9 A Critical Eye on *Intellectuals*

Read on

11 An Excerpt from *Heroes*, the Sequel
to *Intellectuals* and *Creators*

14 Have You Read?
More by Paul Johnson

Insights,
Interviews
& More ...

Meet Paul Johnson

© Mark Gerson

PAUL JOHNSON is a historian whose work ranges over the millennia and the whole gamut of human activities. His *History of Christianity* and *History of the Jews* describe the religious dimension, his *Modern Times* encapsulates the twentieth century, and his *Art: A New History* is the story of visual culture in all its forms, from the cave painters to today. He contributes a weekly essay to the *Spectator*, a monthly column to *Forbes*, practices the gentle art of watercolor painting (at a rate of one an hour), and lives in London and Somerset. He has four children and eight grandchildren. ∿

In His Own Words

On his origins

"I'm from Manchester. My family's origins are in North Lancashire, one of those Catholic pockets that resisted the pernicious effects of the 16th century. There is not a drop of Protestant blood in my veins" (*Trushare*).

On his Jesuit education

"At my school [Stonyhurst] the reading of verse, and of prose for that matter, was taken seriously.

"The Jesuits were a Counter-Reformation order, and the central aspect of that powerful cultural and theological movement was presentation: the displaying of the Catholic faith in the most colourful, flamboyant and dramatic way possible. Hence the cultivation of painters like Tintoretto, Veronese and, above all, Caravaggio and his followers. Every square inch of the interiors of Jesuit churches was covered in scenes of sanctity and martyrdom, rendered with intense realism, high masses were celebrated with gorgeous ritual, and sung by massed choirs thundering out the sumptuous sounds of the new baroque music, and the drama too was added to press home the message.

"All this was reflected in my time at Stonyhurst. The very names of the classes, in ascending order, were a transfiguration of the old Latin syllabus: Rudiments, Figures, Grammar, Syntax, Poetry and Rhetoric" (*The Spectator*).

On his hair

"People have always assumed I've got a temper, simply because I used to have very red hair. . . . It's a form of racism I've always suffered from. James Baldwin was once moaning on to me about how he'd been discriminated against, but I said, 'Look here, Baldwin. If, like me, you've been born left-handed, red-haired, and an English Catholic, there's nothing you don't know about prejudice' " (*New York Sun*).

On his height

"For most of my life, being six foot one, I have loomed over the majority of men and almost all women. Now, at the local Sainsbury's, where queues are constant as they are too mean to employ enough staff, I find I am often out-topped by young fellow-queuers, sometimes even by girls. . . .

"When I was a young man living in Paris, one of my girlfriends was a ▶

six-footer, an American called Euphemia, whom the goggling French thought a *gratte-ciel*. But that was most unusual. My French girls tended to be around five foot two or three. Quite enough, as tall French females, in my experience, tend to be exceptionally tiresome. So, paradoxically, do English girls of five foot or less" (*The Spectator*).

On his art collection

"I have a rule that I will never buy paintings that I could have painted better myself. That rules out virtually all modern art. The last great painter in my view was John Singer Sargent, who died in 1925. After that, the 20th century was a dismal century in the history of art. When future generations look back on it, they'll think we were all mad" (*The Sunday Telegraph*).

On his modesty

"I have had my portrait painted. It was not my idea. One fault I do not possess is vanity. Indeed I am extremely vain about not being vain" (*The Spectator*).

On his writing

"In the art of writing, one of the central problems is what to put in and what to leave out. In the past, I have always been one for putting in. I felt myself full of good things I did not want the reader to miss. So my books got longer and longer. This gigantism spent itself, and from the gross satisfaction of putting everything in I turned to the more delicate pleasure of deciding what to leave out. I discovered I could write down everything a reasonable person needed to know about the Renaissance in 40,000 words, and I have since done Napoleon and Washington at the same length. It has proved to be great fun.

"I don't give a damn for grammar, or syntax either. Having learned to 'parse' as a small boy, and done ten years of Latin and eight of Greek, I take it all for granted. But I love semantic and grammatical niggles and rejoice in the way some people get red in the face with rage at the lapses of others" (*The Spectator*).

On his marriage

"I think our division of labor is quite good. I mean, I let her [his wife, Marigold] do everything. She decides what clothes I wear and buy, and when I need a haircut, and all things like that. The only thing I have the upper hand on is which pictures to buy and where to hang them. There I am the absolute master" (*New York Sun*).

On history

"The study of history is a powerful antidote to contemporary arrogance. It is humbling to discover how many of our glib assumptions, which seem to us novel and plausible, have been tested before, not once but many times and in innumerable guises; and discovered to be, at great human cost, wholly false" (*The Recovery of Freedom*).

On intellectuals

"Nothing appeals to intellectuals more than the feeling that they represent 'the people.' Nothing, as a rule, is further from the truth" (*National Review*).

On making enemies

"When I had my seventieth birthday, Marigold said to me, 'Don't you have enough enemies? Do you really want any more? Why not stop making them?' I took her advice" (*The Daily Telegraph*).

The High Priests of Knowledge
A Conversation with Paul Johnson

This interview originally appeared in U.S. News & World Report, *March 27, 1989. Reprinted with permission.*

PAUL JOHNSON: Intellectuals have the arrogance to believe that they can use their brains to tell humanity how to conduct its affairs. In so doing, they turn their backs on natural law, inherited wisdom and the religious background that have traditionally defined the aims of society. Their approach, beginning with Jean-Jacques Rousseau and on through Jean-Paul Sartre, Ernest Hemingway, and James Baldwin, is moralistic but not in a religious sense. In fact, those who have been most influential have often challenged religion. They find it hard to admit that there is a higher authority than their own judgment; they have a deep-rooted and tremendously powerful arrogance.

This separates them from other men of letters throughout the last two hundred years, men such as Evelyn Waugh, a great writer with a most powerful intellect who could humble himself in the presence of the Deity. I regard Waugh, Edmund Burke, Samuel Johnson, Rudyard Kipling, and others like them almost as anti-intellectuals. They view the established churches and the practices and customs of society as an important part of human wisdom. If I were writing my book over again, I would write it as a dialogue between

> " I regard Waugh, Edmund Burke, Samuel Johnson, Rudyard Kipling, and others like them almost as anti-intellectuals. "

these thinkers and those I call intellectuals, rather than focusing on intellectuals alone.

ROBIN KNIGHT: But doesn't the rise of intellectuals simply reflect the rise of modern science and scholarship?

P.J.: Secular intellectuals really emerged in the eighteenth century, particularly in France, where Rousseau, although a bohemian figure, received enormous hospitality from the ruling class. The aristocracy felt guilty about their privileges and thought that having him in their châteaux was a talisman against disaster. The 1930s was another period when intellectuals were important, and, more recent, there were the 1960s, which coincided with a huge expansion of higher education, so there were jobs galore at the universities. It was a period when intellectual gurus such as Sartre and Bertrand Russell appeared to have a worldwide audience.

Old-style intellectuals tended to be politically oriented, while in more recent decades, intellectuals have tended toward hedonism, which can rapidly get out of control and develop characteristics that are terrifying, including violence. James Baldwin and Norman Mailer, to name two, attempted to legitimize violence in certain forms. It is a curious fact that intellectuals, though generally nonviolent, nevertheless have a certain attraction to violence. They will defend the most violent courses of action taken by foreign governments of which they approve. In Cambodia, in fact, there was a purge carried out by a group of intellectuals whom I call Sartre's children because they had all been educated in Paris in the 1950s and influenced by Sartre's ideas as well as by Marxism. Sartre was one of the first philosophical figures to produce arguments in favor of terrorism and is very much to blame for what has become the ambivalence of some intellectuals toward terrorists.

R.K.: You're saying, then, that intellectuals are out of touch with the real world of actions and consequences?

P.J.: Intellectuals are always talking about the workers, the masses, or humanity—they love the word humanity—but they don't come into contact with ordinary people very much. Most come from pretty secure middle- or upper-middle-class backgrounds and see ordinary people more as individuals who do things for them rather than as acquaintances, let alone intimates. They dismiss the middle class as bourgeois, ▶

mercenary, and materialistic, while seeing themselves as rebels against society. But once they're with one another, they are very conformist. Someone has referred to intellectuals as a "herd of independent minds"; they are easily stampeded. At the same time, individual beasts sometimes do get ejected from the herd, and then they gore one another. For the most part, however, they move together and plug one another's books. As a result, they are liable to create an intellectual consensus that can easily become a general consensus because they are very influential, powerful people who have a gift for words and access to the media. That's why I think they are so dangerous.

R.K.: *What exactly do they threaten?*

P.J.: In the rise of intellectuals, truth has become a prime casualty. They think that there is only Truth with a capital *T*, which they feel that they have found and must deliver to others. In that respect, Karl Marx, who thought that he had a direct line to metaphysical truth, is the archetypal intellectual. Intellectuals are simply not inclined to take the scientific approach, to look for evidence that conflicts with their hypothesis just as carefully as they look for evidence that confirms it.

But perhaps that matters less than it used to because the literate public is increasingly unlikely to listen to these gurus, in part because there is a pervasive feeling throughout the Western World that utopia is not attainable. One should listen to and read intellectuals but not necessarily take great notice of what they say, particularly when they gang up and produce manifestos. Winston Churchill used to say, "Experts should be on tap but never on top." That's very good advice. ∾

A Critical Eye on *Intellectuals*

Intellectuals *was first published in 1989. Its biographical essays formed what Kingsley Amis called "a valuable and entertaining Rogues' Gallery of Adventures of the Mind." A bestseller in many languages, the book nonetheless caused a severe commotion for its habit of describing clever people "so as to bring out their bad behavior" (Bernard Williams,* New York Review of Books*). Below are excerpts from some of the negative reviews accorded the book upon its publication.*

"On every page there is something low, sniggering, mean, and eavesdropped from third-hand. How right that it should have drawn an enthusiastic endorsement from Norman Podhoretz, another moral and intellectual hooligan who wishes he had the balls to be a real-life rat fink."
—Christopher Hitchens, *The Nation*

"Johnson writes with zest, and he provides excellent gossip. . . . Heaven knows this book is not boring. But it is monotonous, in that Johnson's attitude of censure rarely lets up. . . . He is Will Durant with a beadle's whip."
—Joseph Sobran, *National Review*

"Great fun to read—a sort of scandal sheet for readers more interested in dead celebrities than live ones. . . . What does it matter if, say, Lillian Hellman's life was a pack of lies? Her work, such as it was, is her legacy."
—*Chicago Tribune*

> " Johnson's attitude of censure rarely lets up. . . . He is Will Durant with a beadle's whip. "

A Critical Eye on *Intellectuals* (continued)

"The reader, in fact, suspects that most 'intellectuals' in this volume were chosen on the arbitrary basis of having difficult personalities and a taste for radical ideas that Mr. Johnson, a former editor of *The New Statesman* turned conservative, apparently finds distasteful. Why else include Marx, but neither Darwin nor Freud? Why focus on Hemingway, Tolstoy, and Shelley, who are respected as writers, not as thinkers nor theoreticians? (And if one is going to feature poets and novelists, why not also include the likes of T. S. Eliot and Saul Bellow?) Why look at Sartre instead of Camus, Lillian Hellman instead of Mary McCarthy?"

— Michiko Kakutani, *New York Times*

An Excerpt from *Heroes*, the Sequel to *Intellectuals* and *Creators*

Below is an excerpt from the introduction to Heroes, *a galaxy of heroes from Alexander the Great to Ronald Reagan.* Heroes *will be published hardcover by HarperCollins in Fall 2007.*

IN THE WESTERN DESERT OF EGYPT, sometimes twenty miles from the nearest road, you come across solitary tombs of stone, much weathered by the wind. Some are buried in the sand. They may be 3,000 or even 4,000 years old and testify to the veneration once felt for men of outstanding virtue or generosity or heroism by their younger contemporaries, who built the tomb to mark their respect and perpetuate the memory of the dead. The names have long since been obliterated by time and weather, but a certain sanctity hovers around the spot still. So too, in the Alps and the Tyrol, in the Pyrenees and the Carpathians, little shrines by the wayside, and rustic ornamental fountain-heads over springs, commemorate the lives of local men or women who once struck their neighbors as remarkable. Occasionally the name survives. Usually time has imposed anonymity. But the spirit of virtue hovers over the illustrious dead. The names do not matter. It is the principle of honoring the good or the brave which strikes us. Samuel Johnson records in his *Journey to the Western Islands* his visit to

An Excerpt from *Heroes* (continued)

Iona, and in particular what he believed to be the cemetery of the ancient Scottish kings and other famous men. The place filled him with awe and he commented: "By whom the subterraneous vaults are peopled is now unknown. The graves are very numerous and some of them undoubtedly contain the remains of men who did not expect to be so soon forgotten."

I have visited Iona, and meditated like Dr. Johnson on these sepulchers of unknown notabilities. Such places stir the imagination and make me think more kindly of the human race. I have seen such shrines in Greece, some going back to before the Greeks came. Homer called those thus honored Ῠρω-ες, heroes, defined as 'a name given to men of superhuman strength, courage, or ability, favored by the gods; at a later time regarded as intermediate between gods and men, and immortal.' Graveyards of distant kings are always impressive, and I relish seeing them: the great Fourth Dynasty pharaohs at Giza, for example, or the tombs of the Holy Roman Emperors in Palermo Cathedral, especially the vast but simple block of black marble that covers the last resting-place of the most formidable of them all, Henry VI; or the tombs of the Angevins in the abbey church of Fontevrault, between Chinon and Poitiers. There rest our great law-giving King Henry II, his wife Eleanor of Aquitaine, his son Richard the Lionheart, and Isabella, second wife of Henry's bad son, King John. The remains of France's kings were once carefully preserved in the royal abbey of St. Denis, their hearts being kept separately in reliquaries. But all was desecrated by the sansculottes of the Revolution, precious items being sold off for cash: thus the shriveled heart of Louis XIV, the Sun King, ended up at Stanton Harcourt, where it was sacrilegiously eaten by a Cambridge professor. The sanctuary of the English kings, Westminster Abbey, has fared better, and still houses intact the remains of its founder, Edward the Confessor, and many of his successors. . . .

We should not take it for granted that the original heroes were all men, anymore than we should assume that the primitive races worshipped only gods. Goddesses make their appearance in the archaeological evidence from the very earliest times, and we can be sure that heroines followed swiftly in the steps of heroes. Humanity invented gods as the originators of natural events they could not understand, and feared, and gods were therefore terrible personages. What ordinary mortals needed to identify with were creatures, recognizably human but of great capacity and accomplishment, who stood halfway between the deities and the rest. These demigods were heroes, and they had to include heroines, like Pallas and Medea, for pure goddesses were too frightening to be domesticated and reduced to human scale. Once heroines come into existence, even if, like Medea, they sometimes take the form of witches, the concept of the hero ceases to be the exclusive preserve of military men or those who rejoice in superhuman physical strength. For the purpose of this collection of biographical essays, I have taken the concept of the heroic individual in its widest possible sense, even if I have included a number famed for military exploits, such as Alexander and Caesar among the males, and Boudicca and Joan of Arc among the females. The fact is, anyone is a hero who has been widely, persistently over long periods, and enthusiastically regarded as heroic by a reasonable person, or even an unreasonable one. I have put into this collection one or two heroes and heroines of my own, believing that an element of idiosyncrasy is a legitimate part of hero worship. Indeed it is only by asking ourselves how we, personally, judge heroism that we begin to get to the essence of the matter. . . . ❧

Have You Read?
More by Paul Johnson

CREATORS

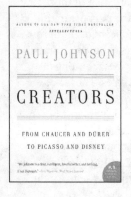

In this volume of essays on outstanding and prolific spirits, Paul Johnson looks at writers from Chaucer and Shakespeare to Mark Twain and T. S. Eliot, artists like Dürer, and architects such as Pugin and Viollet-le-Duc. He explains the different ways in which Jane Austen, Madame de Staël, and George Eliot struggled to make their voices heard in the masculine hubbub. Victor Hugo allows him to ask, "Can imaginative genius coexist with low intelligence?" Johann Sebastian Bach gives him the opportunity to focus on the role of genetics in creativity and to explore the strange world of the organ loft. Louis Comfort Tiffany takes him into the technology of glass-making and the tragic vagaries of aesthetic fashion. Some essays make illuminating comparisons: of Turner with his contemporary, the Japanese master Hokusai, and of the two great dress designers, Balenciaga and Dior. The final essay examines those two inventive geniuses, Picasso and Disney, and asks which had the greater influence on the visual arts of the twentieth century—and beyond. Paul Johnson believes that creation is a mysterious business that cannot be satisfactorily analyzed. But it can be illustrated in such a way as to bring out its salient characteristics. That is the purpose of this instructive and witty book.

GEORGE WASHINGTON: THE FOUNDING FATHER (Eminent Lives)

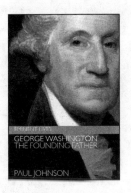

George Washington is by far the most important figure in the history of the United States. Against all military odds, he liberated the thirteen colonies from the superior forces of the British Empire and presided over the process to produce and ratify a Constitution that (suitably amended) has lasted for more than two hundred years. In two terms as president, he set that Constitution to work with such success that, by the time he finally retired, America was well on its way to becoming the richest and most powerful nation on earth.

Despite his importance, Washington remains today a distant figure to many Americans. Previous books about him are immensely long, multivolume, and complicated. Paul Johnson has now produced a brief life that presents a vivid portrait of the great man as young warrior, masterly commander in chief, patient Constitution maker, and exceptionally wise president. He also shows Washington as a farmer of unusual skill and an entrepreneur of foresight, patriarch of an extended family, and proprietor of one of the most beautiful homes in America, which he largely built and adorned.

Trenchant and original as ever, Johnson has given us a brilliant, sharply etched portrait of this iconic figure—both as a hero and as a man.

"Johnson, a noted British historian, submits a beautifully cogent, enthrallingly perceptive, and, given the vast accumulation of published material on his subject, startlingly fresh take on the ultimate American icon."
—*Publishers Weekly* (starred review)

ART: A NEW HISTORY

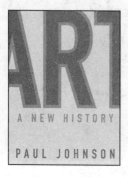

In *Art: A New History*, Paul Johnson turns his great gifts as a world historian to a subject that has enthralled him all his life: the history of art. This narrative account, from the earliest cave paintings up to the present day, has new things to say about almost every period of art. Taking account of changing scholarship and shifting opinions, he draws our attention to a number of neglected artists and styles, especially in Scandinavia, Germany, Russia, and the Americas.

"A gorgeously illustrated and provocative interpretation of the evolution of Western art. . . . Thrilling in its scope, fluency, and zest."
—*Booklist* (starred review)

MODERN TIMES: THE WORLD FROM THE TWENTIES TO THE NINETIES (Revised Edition)

Modern times, says Paul Johnson, began on May 29, 1919, when photographs of a solar eclipse confirmed the truth of a new theory of the universe—Einstein's Theory of Relativity.

Churchill, Roosevelt, Hitler, Stalin, Mao, Hirohito, Mussolini, and Gandhi are the titans of this period. There are wartime tactics, strategy, and diplomacy; the development of nuclear power and its use at Hiroshima and Nagasaki; the end of WWII and the harsh realities of the uneasy peace that followed. The rise of the superpowers—Russia and the United States; the emergence of the Third World and the Cold War. We see the economic resurgence of Europe and Japan; existentialism; Suez; Algeria; Israel; the New Africa of Kenyatta, Idi Amin, and apartheid;

the radicalizing of Latin America; the Kennedy years, Vietnam, Watergate, and the Reagan years; Gorbachev and perestroika; and Saddam Hussein and the Gulf War. And then the Space Age, the expansion of scientific knowledge, the population explosion, religion in our times, genetic engineering, and sociobiology. Incisive, stimulating, and frequently controversial, *Modern Times* combines fact, anecdote, incident, and portrait in a major full scale analysis of how the modern age came into being and where it is headed.

"Truly a distinguished work of history . . . *Modern Times* unites historical and critical consciousness. It is far from being a simple chronicle, though a vast wealth of events and personages and historical changes fill it. . . . We can take a great deal of intellectual pleasure in this book."

—Robert A. Nisbet,
New York Times Book Review

THE QUEST FOR GOD: PERSONAL PILGRIMAGE

In this probing, challenging, and personal account of his feelings about God and religion, Paul Johnson shares with others the strength and comfort of his own faith. Informed by his great knowledge of history, *The Quest for God* is written with force, lucidity, and eloquence by the author of *Intellectuals, Modern Times, A History of the Jews*, and other works.

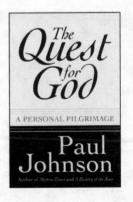

"Nuanced and always informative, Johnson is guaranteed to stimulate even when he does not convince." —*Kirkus Reviews*

A HISTORY OF THE AMERICAN PEOPLE

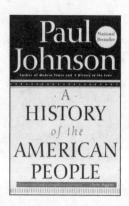

This is an in-depth portrait of a great people, from their fragile origins through their struggles for independence and nationhood, their heroic efforts and sacrifices to deal with the "organic sin" of slavery and the preservation of the Union to its explosive economic growth and emergence as a world power and its sole superpower. Johnson discusses such contemporary topics as the politics of racism, education, Vietnam, the power of the press, political correctness, the growth of litigation, and the rising influence of women.

This challenging narrative and interpretation of American history by the author of many distinguished historical works is sometimes controversial and always provocative. Johnson's views of individuals, events, themes, and issues are original, critical, and admiring, for he is, above all, a strong believer in the history and the destiny of the American people.

"A fresh, readable, and provocative survey. He is full of opinions . . . and Johnson can be very wise."
                                        —*Los Angeles Times*

THE BIRTH OF THE MODERN:
WORLD SOCIETY 1815–1830

The extraordinary, bestselling chronicle of the period that laid the foundations of the modern world.

"Fascinating. . . . A savory social history, spiced with lively gossip. . . . It is never dull. In many ways a tour de force."
—Eugen Weber, *New York Times Book Review*

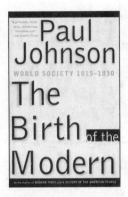

THE CIVILIZATION OF ANCIENT EGYPT

Paul Johnson explores the growth and decline of a culture that survived for 3,000 years and maintained a purity of style that rivals all others. His study, accompanied by 150 full-color illustrations, looks in detail at the state, religion, culture, and geographical setting, and how they combined in this unusually enduring civilization. From the beginning of Egyptian culture to the rediscovery of the pharaohs, the book covers the totalitarian theocracy, the empire of the Nile, the structure of dynastic Egypt, the dynastic way of death, hieroglyphs, the anatomy of pre-perspective art and, finally, the decline and fall of the pharaohs. Johnson seeks, through an exciting combination of images and analysis, to discover the causes behind the collapse of this great civilization while celebrating the extraordinary legacy it has left behind.

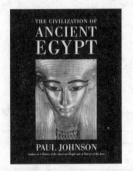

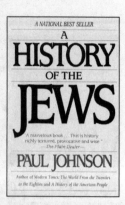

Have You Read? *(continued)*

A HISTORY OF THE JEWS

A national bestseller, this brilliant 4000-year survey covers not only Jewish history but the impact of Jewish genius and imagination on the world.

"An extraordinary amount of useful information."   —*New York Review of Books*

Don't miss the next book by your favorite author. Sign up now for AuthorTracker by visiting www.AuthorTracker.com.